The Political Economy of Sugar Production in Colonial Kenya

The Political Economy of Sugar Production in Colonial Kenya

The Asian Initiative in Central Nyanza

Godriver Wanga-Odhiambo

LEXINGTON BOOKS
Lanham • Boulder • New York • London

Published by Lexington Books
An imprint of The Rowman & Littlefield Publishing Group, Inc.
4501 Forbes Boulevard, Suite 200, Lanham, Maryland 20706
www.rowman.com

Unit A, Whitacre Mews, 26-34 Stannary Street, London SE11 4AB

British Library Cataloguing in Publication Information Available

Library of Congress Cataloging-in-Publication Data

Library of Congress Cataloging-in-Publication Data Available.

ISBN 978-1-4985-1163-6 (cloth : alkaline paper)
ISBN 978-1-4985-1164-3 (electronic)

∞™ The paper used in this publication meets the minimum requirements of American National Standard for Information Sciences Permanence of Paper for Printed Library Materials, ANSI/NISO Z39.48-1992.

Printed in the United States of America

For my mentor Professor Robert Maxon of West Virginia University.
And in memory of my father.

Contents

List of Figures

List of Tables

List of Abbreviations

AAO	Assistant Agricultural Officer
AO	Agricultural Officer
CAADP	Comprehensive Africa Agricultural Development Program
CCDB	Central Commodity Distribution Board
CEC	Committee of Economic Co-operation
CK	Central Kavirondo
CKDAR	Central Kavirondo District Annual Report
CN	Central Nyanza
CNDAR	Central Nyanza District Annual Report
CO	Colonial Office
DAO	District Agricultural Officer
DC	District Commisioner
DO	District Officer
FAO	Food and Agriculture Organization
DAAR	Department of Agriculture Annual Report
EAC	East African Community
EAINC	East Africa Indian National Congress
GMR	Guaranteed Minimum Return
GDP	Gross Domestic Product
IBEA Co.	Imperial British East Africa Company
K-L	Kisumu-Londiani
K-LDAR	Kisumu-Londiani District Annual Report
KDQR	Kisumu District Quarterly Report
KNA	Kenya National Archives
KSM	Kisumu
KUSPW	Kenya Union of Sugar Plantation Workers
Lab	Labour Department Files

LegCo	Legislative Council
LIR	Labour Inspection Report
LNC	Local Native Council
LRNo.	Land Registration Number
NADAR	Native Affairs Department Annual Report
NFCS	Nyanza Farmers' Cooperative Society
NK	North Kavirondo
NIFA	Nyanza Indian Farmers Association
NZA	Nyanza
PAO	Provincial Agricultural Officer
PADO	Provincial Agrarian Development Officer
PC	Provincial Commissioner
PS	Permanent Secretary
SAO	Senior Agricultural Officer
SK	South Kavirondo
TE Pass	Temporary Entry Pass

Acknowledgments

The success of this book was a result of the support of many individuals. I wish to acknowledge Professor Robert Maxon of West Virginia University who not only chaired my Ph.D. dissertation but also encouraged me to pursue this field of economic history and eventually publish this book. I am greatly indebted to Professor Maxon, for his unfailing support and reading my drafts. I sincerely wish to thank Mrs. Araba Dawson-Andoh of Ohio University (formerly of West Virginia University) for her support with sources during the research and the writing up of this book. My appreciation also goes to Dr. Emanuel Uwalaka, of St. Louis University for reading my drafts and offering valuable suggestions, Douglas Edgerton of Le Moyne College for sharing his honest views of my work and encouragement.

Special thanks to the workers of Kenya National Archives in Nairobi and the many interviewees in rural Nyanza who welcomed me into their homes and shared their stories, and to the current workers of Kibos Sugar Company. I am also indebted to my students, Leilani Leechalad, Suzy McRae and Ronald Logan VanDyke (Le Moyne College) for their help with tables.

Special thanks to my friends Ms. Zoa William and Ms. Chrissie Rizzo for their support and encouragement to finish the work. Finally, I am indebted to my family; Eng. C.L. Odhiambo, my beloved children, Yophi, Bianca-Joy and Noel who persevered during my long absence in pursuit of academia and the writing of the book "Thank you for your understanding."

Preface

During 2015, Kenya's sugar industry was at the center of much controversy as a result of a trade agreement the country's leadership negotiated with Uganda. The agreement allowed Kenyan entrepreneurs to export beef and milk to Uganda while allowing Uganda to dump 145,000 metric tons (MT) of sugar into Kenya. This drew strong protest from sugar producing regions in western Kenya as the agreement was likely to depress prices of Kenyan produced sugar and potentially harm the Kenyan sugar industry already struggling to survive so as to favor certain economic interest groups.[1] What many Kenyans, including sugar growers and processors, did not know was that a similar situation happened in the 1950s. Then the colonial state sought to secure exports of wheat grown by European farmers to Uganda in exchange for dumping cheap Uganda sugar into Kenya. These incidents illustrate an unfortunate commonality that has characterized Kenya's sugar industry in the colonial and independent periods: a lack of political support and a failure to put the interests of sugarcane growers and processors first while favoring special interest groups with political clout.[2]

This example not only illustrates a past parallel with a current critical economic issue, but also makes clear the important need to closely examine the history of sugar production in Kenya. This history is the subject of the book, specifically the experience of Asian sugar producers in the Kibos-Muhoroni area of what was then Nyanza Province. While sugarcane planting and sugar production expanded markedly after Kenya's independence in 1963, many of the issues and problems surrounding this industry, such as processing capacity, marketing, yields of sugar, lack of government support, and discrimination against growers mark both periods. What was and is an important commodity one that has seen steady expanding demand since the early part of the twentieth century, has yet to fulfill its potential.

Kenya's sugar industry today, just as during the colonial period when it was dominated by Asian farmers and factory owners in western Kenya, primarily produces for the domestic market rather than for export. Sugar forms an important part of the Kenyan diet in almost all households, but demand continues to outpace the supply of locally produced sugar. This insufficiency of an important food supply is not unusual in the context of contemporary Africa, but for Kenya it presents a challenge for farmers and policymakers alike, just as it did in the colonial period. Like cotton, (unlike coffee and tea) local sugar production in western Kenya, has never come close to meeting demand. Today, Kenya has eleven sugar mills, which have an annual production capacity of 600,000 MT of refined sugar while the annual domestic need is approximately 800,000 MT.

Just as in the colonial period, therefore, there exists an opportunity for growth and the potential to contribute much more substantially to the national economy. The sugar industry already contributes in significant ways. It currently supports approximately six million Kenyans, which is about 16 percent of the nation's population. Most are small-scale farmers; others are employed by the sugar mills in transport and processing activities. Kenya sugar also contributes about 7.5 percent of the nation's GDP, and the industry accounts for some 15 percent of agricultural GDP. Additionally, the sugar industry provides several by-products that can contribute in other ways to national economic development. Industrial waste from sugar processing may be used for the manufacture of molasses and filter-mud can be used as fertilizer. The sugar industry also has the capacity to generate electricity that could be used in the sugar belt region to help provide rural electrification. Unfortunately, the potential has so far not been fully tapped with only the Mumias Sugar Company having managed to invest in this energy supply. If all the sugar factories could generate excess electricity and export a portion to the national grid, the cost of electricity importation would be reduced and the cost of rural electrification in the sugar belt region would be affordable.[3]

Nevertheless, this potentially valuable industry faces many difficult problems and obstacles. There are certainly opportunities to expand planting of sugarcane, and the average yield per acre in Kenya compares favorably to the global average. Expanded planting and improved delivery in recent years has resulted in an increase in cane supplied by farmers for processing. Yet Kenya's supplies of processed sugar continue to lag behind demand. Sugar imports from Uganda and other member countries of the Common Market for Eastern and Southern Africa (COMESA) not only serve to address the consumption gap but provide stiff competition for the Kenya growers. In addition to the lack of consistent government support, pervasive corruption associated with the processing and distribution of sugar presents another huge obstacle for the contemporary Kenyan sugar industry.

To infuse new life in the sugar industry and make the industry competitive, capable of fulfilling its potential, represents a challenge for the growers, processors, and the government. One way to begin to do this is through studies of the history of the industry and seeing what lessons can be drawn from the past. The many lessons from the colonial sugar industry are crucial at this challenging time for Kenya's sugar industry. Through an analysis of the Asian sugar industry in Kibos-Muhoroni during the colonial period, this book intends to offer just such a perspective, examined in the subsequent chapters. The experience of the past will shed light on such questions as marketing and distribution, transportation, and whether the viability of peasant or estate production of sugar is more sustainable.

NOTES

1. Neville Otuki, "Kenya Uganda Sugar Deal," *Daily Nation*, August 11, 2015.
2. Peter Anyang' Nyongo', "Modernize Sugar Production Methods to Compete with Fellow Cane Growers," *The Standard*, September 13, 2015.
3. Today, as in the past, Kenya's sugar belt lies in the western portion of the country. The belt occupies substantial portions of the counties of Kisumu, Homa Bay, Kakamega, and Bungoma or what was until 2013 Nyanza and Western Provinces. The Kibos-Muhoroni area falls within present-day Kisumu County.

ONE

Historical Background to Asian Farming in Kenya

The migration of Asians to Kenya and their eventual involvement in the sugar industry is a complex story. Many Asians first came to Africa as soldiers, hired by the colonial powers to help in the pacification of their newly conquered territories. Although these soldiers were promised settlement on agricultural land, the proposed settlement was never implemented as soon as anticipated. Nonetheless, the sepoys (Indian soldiers) and other Indians looking for economic opportunity came to the east coast of Africa ready to work and establish themselves in the colonial protectorate that is now Kenya.

This chapter lays the foundation of the Asian participation in sugarcane farming in Kenya. It discusses how Asians came to Kenya in several waves, impelled by the "push" factors that drove them from their homelands and the "pull" factors that drew them to Kenya specifically. Drawn by employment opportunities in the colonial government, including construction work, rail construction, carpentry, and soldiering, they were soon involved in economic ventures such as trading and farming as well. The chapter also lays out the demographic characteristics of these immigrants.

Once in Kenya, their settlement patterns were determined by the social, economic, and political conditions within the colony. Pushed by land policy that favored Europeans toward the Kibos-Muhoroni area of Nyanza Province, they faced challenges like any immigrants, and the chapter addresses how many overcame these odds, adapting to their new country.

The chapter also shows the contradictory nature of the colonial state throughout the years of Asian settlement and farming. Africa remains an underdeveloped continent many years after its nation-states gained their independence, an economic stagnation that is due to several factors, including colonial policies, population growth, land scarcity, bad governance, and poor ecology.[1] Of particular importance are the colonial agricultural policies that emphasized cash crop production, growing agricultural products for export at the expense of growing food for domestic consumption. This led to inadequate food production in Kenya, which continues to suffer from insufficiency of some foodstuffs, leading to continued poverty and underdevelopment.[2] Sugar production fell within this category and was greatly impacted by land policies in colonial Kenya.

AGRARIAN COLONIAL POLICY—AN OVERVIEW

To understand Kenya's food insufficiency, in particular the scarcity of sugar, it is imperative to look at the colonial policies that pushed Kenya to dependency status in food production and the sugar industry. The colonial state enacted land policies that impacted agriculture since the economic policies favored European settlers whereby the settlers got more land than the Africans and Asians, while at the same time many Africans lost arable land to the settlers, pushing them to unfertile reserves. Yet availability of fertile land remains a crucial component in any production. Moreover, the colonial state did not believe the African economy to be viable enough to make the colony self-sustaining; the colonial state believed that settler farming could sustain the state; hence, no efforts were made to improve Asian farming, partly due to racial prejudice and also since they were viewed as possible competitors to the European settlers in terms of land and labor. Due to such factors the colonial state entrenched measures that discouraged Asian farming especially in the fertile highlands, while European settlers were granted tens of thousands of acres there.

Despite the fact that many of the colonial administrators advocated only for the settler agriculture, at least one official did believe in Asian agriculture. Fredrick D. Lugard on several occasions pointed out that Asians could play a crucial economic role in East Africa. He noted in his diary the unique opportunity the Indian agricultural settlement along the trade route would offer. Lugard explained how Asians could contribute to agricultural innovations in East Africa. Expounding on this point Lugard wrote in his diary:

> The African too is extremely imitative; the presence in his midst of fully clothed people would be to him an example of decency which he will speedily imitate. His wants would become identical with theirs and thus, while his status was improved and a new encouragement given to trade, he would be compelled to exert himself and labor. . . . Moreover the methods of agriculture, the simple implements of the Indian

> ryot and the use of bullock, the sinking of wells, the system of irrigation and manuring the soil, would soon be imitated by Africans and produce of his land will be multiplied. . . . It is not as imported coolie labor that I advocate the introduction of Indian but as a colonist and a settler.[3]

Lugard believed that Indians would be easily given land in Kenya through the system of "Jagirs" where tracts of land would have been given to veteran soldiers in East Africa as it was done in Burma. Unfortunately in East Africa—and Kenya in particular—settler politics interfered with the Jagir policy. Furthermore, the soldier-settlement scheme that was implemented later in Kenya after the war only gave land to the European settlers and ignored the Indian sepoys who also fought in World War I especially in the East African campaign. Moreover, colonial policy changed after World War II. Before the war the colonial state was unresponsive and negligent; after the war, the state became more supportive and responded adequately as the colony embarked on postwar development plans. The contradictions emerge as the colonial state started to support the Asian sugar manufacturing industry that was solidly under Asian capital by the late 1940s.

However, for over a century the Asian community played, and has continued to play, a significant economic role in many countries such as Uganda and Kenya. In these countries the Asians were involved in sugarcane farming. In Kenya, the Asian cane production came into play in an initially predominantly settler-dominated agriculture and later in peasant farming. The Asian agency in sugarcane production was due to the problem of sugar insufficiency in Kenya, yet there was a viable need for such an industry due to local demand. The lack of self-sufficiency in sugar production in particular has continued to impact the economy of Kenya. This trend is worrying yet there are areas in Kenya where successful sugarcane farming could be practiced satisfactorily. The colonial Nyanza Province was such a place as the region had varied natural endowments that contribute to varied economic potential.[4] The Asian sugar industry of Kibos-Muhoroni added to the diversification of Nyanza's economy. The intensification of agricultural production in parts of central Kenya and the Nyanza Province during the colonial period led to diversification and participation of small-scale farmers, especially of Africans, within the colonial economic system who engaged in farming, operating mills and shops and buying lorries for transportation. Such initiatives were also common among Asians.[5]

In Kenya farming was an important part of the colonial economy. Ironically the colonial state pursued only the settler and peasant farming under the dual policy that the state first implemented under Governor Sir Robert Corydon in the 1920s. This dual policy propounded support for both European and African agriculture but in practice favored European settler production.[6] The European coffee and wheat farmers were supported to maintain quality control of their produce for export purposes. Unfortunately, under this policy the Asians were left out, due to the belief that they could not be good farmers but rather could excel in entrepreneurship. Despite that, the economic role of Asians in Kenya was not limited to trade and middle-level employment in the colonial administration. Many served in the colonial government. For instance, almost the entire clerical staff on the Kenya-Uganda railway was Asian. Initially very few were engaged in agriculture. Later the pioneer Asians showed farming prowess and many would later own large tracts of sugarcane land near Kisumu; this land was effectively managed by Sikhs of the Jat caste who had been farmers back in India. These Sikh cane growers around Kibos and Eldoret in the Kenya highlands demonstrated the resilience and profit-orientedness of the Asian farmers.[7]

Asian resilience in agriculture was partly curtailed by settler politics; however, some leaders recognized that Asians could contribute to the growth of the economy. Apart from Lugard and Governor Corydon who had supported them initially, Asian farming was also supported by Sir Charles Eliot who on taking the governorship in 1900 asserted that "The Indians seem likely to become intermediaries between Europeans and natives, and if they were allowed by the Indian government to settle, would probably do much to develop the natural resources of the country by cultivating rice and practicing other forms of agriculture."[8] This was made possible as the majority of the Asian immigrants came from rural agricultural backgrounds in India.

It may seem that the Asians' cultural institutions back at home such as the caste system could have hindered progress among Asians and acted as an impediment where family ties and other practices come into play; this was not the case. Although it is believed that cultural determinism determined the scale and success of economic ventures, this did not apply to Asians in Kenya. The need to break with culture seemed important to the Asians in East Africa as a prerequisite to economic progress. Possibly what was needed was a change in totality of social, cultural, and habits, and more so psychological attitude and way of life to attain success.[9]

Among the Asian immigrants, some had to discard their cultural beliefs, especially the restriction of their work ability based on the caste system. This enabled them to venture into various economic activities in Kenya. However, for Asians to succeed in colonial Kenya, they needed the support of the administration. It is evident that any culture or society is capable of achieving economic advancement as long as they receive support and recognition by the policy makers. Policy makers must be able to recognize within a particular culture those elements that contribute to rising productivity and income.[10] However, given the fact that Kenya had a three-tier society, issues pertaining to cultural diversity and rivalry emerged among the farmers of the various races within the colony.

Although Asian economic endeavors have been broad-based with greater emphasis on their role in entrepreneurship, farming was a major economic activity as well, and struggles for equality and the contradiction of the land policies in Kenya resulted to huge debates such as the Asian question. This was important because colonial politics, especially settler politics, and the ecology of the area where the Asian cane farmers eventually settled impeded the Asian sugar production in the Kibos-Muhoroni area. Comparatively, Asian cane growers in Uganda seemed to have thrived, possibly because the colonial state in Uganda did not inhibit the Asian sugarcane farmers. Given the fact that colonial Kenya was a three-tier society where Europeans, Asians, and Africans were all involved in economic and political struggles and land issues, which in essence affected production, the competition was stiff. However, due to the economic prospects many immigrants moved to Africa during the colonial period.

MIGRATIONS AND SETTLEMENT

East Africa has seen many immigrants for a very long time. In Kenya, many foreigners migrated and settled to trade during the colonial period. Moreover, many countries in Africa experienced internal movements by different communities for economic reasons. For example, immigrants' communities from the northern part of Ghana moved into southern Ghana and were able to bring more land under cocoa farming; and in some cases they were able to take over the farms whose owners had become bankrupt due to either poor markets or crop diseases. These internal migrants played a very significant role in cocoa farming. Their success was not only due to remarkable economic stimuli given by the government but due to their determination to succeed and their willingness to migrate, invest, and reinvest in land.[11] Likewise in East Africa and Kenya in particular, many immigrants came from Asia for mainly economic opportunities.

The history of the interactions between Asia and the East African coast predates European imperialism in East Africa. This interaction started as far back as the first millennium CE when immigrants from present-day India and Pakistan settled in Zanzibar and the East African coast and engaged in the extensive Indian Ocean trade.[12] This interaction brought many Asians to the coast including those from China who traded in porcelain ware. Asian migration and settlement can be traced from the second century, when there was an extensive trade network in the Indian Ocean that attracted merchants and seamen from India, Persia, and Arabia.[13] These groups established a long history of commercial links with East Africa and played an active role in the financial and commercial life of the area. It was this trade connection that led Asians to settle along the coast of East Africa, especially in Zanzibar and Mombasa. The presence of Asian merchants and seafarers was made possible by geographical proximity and the knowledge and availability of the monsoon trade winds. Evidence of this earliest contact was found in the Greek sailors' guide *Periplus of the Erythraean Sea* in the first century CE, which indicated the path of the trade network found in the Red Sea and Indian Ocean.[14] The guide showed how the east coast of Africa was connected with the outside world through trade and the role that the Asian merchants played along the coast by supplying cotton cloth and beads and exporting ivory, gum copal, and incense. The guide also indicated where certain goods could be found along the east coast of Africa, goods that had been imported from the interior. Indian merchants used to finance many of the Arab traders and slave caravans into the interior of East Africa long before the advent of Europeans in East Africa.[15] Much of the ocean shipping business was also owned by Indians. Evidence of early Asian contact with the east coast was established by the excavations of Asian-made glass beads along the coast especially at Husuni Kubwa near Kilwa. The Asians were able to use small sailing ships through the Persian Gulf and the Red Sea into Egypt, Axum, and Kush. They also employed a direct ocean route to the coast of East Africa, at that time referred to as Rhapta.[16]

Before 1500, Arabs, Indians, and other people from Asia such as the Chinese dominated the lucrative Indian Ocean trade. However, many of the Asian traders along the east coast of Africa were Hindus and Muslims. This changed with the coming of the Portuguese around 1500. The Portuguese wanted to monopolize trade along the east coast and plunder its resources without competition from these traditional traders. Later, with the establishment of Omani Arab rule at the coast, many more Asian immigrants moved to East Africa; these included the Indians, Persians, and Omani Arabs.

The migration of Asians into East Africa could be divided into several periods. The movements were in four phases starting from the second millennium CE. The migrations and interest in Africa started very early.

By 1498 when Portuguese exploration started and Vasco Da Gama explored the coast, Asians were already in Zanzibar, Pemba, Malindi, and Mombasa. By 1833, Asian traders were already established at the coast as customs collectors. These early Asian immigrants were attracted by the Indian Ocean trade and eventually settled at the coastal towns. This first group of Asians was basically trading with the coast and their settlement increased with the transfer of Seyyid Sayyid's capital from Oman to Zanzibar in 1840. Many Asian residents of Oman followed him to Zanzibar to engage in the lucrative international trade. Other Asians, especially the Baluchi troops from India, were employed by Seyyid Sayyid when he was fighting to establish his rule in Mombasa and Pate. The sultan preferred the Indian troops to the Arabs. The sultan would eventually leave 150 Baluchi troops to safeguard Fort Jesus.[17]

Similarly, British interests at the coast indirectly encouraged Asian merchants. This was because Britain during the nineteenth century aimed at controlling the sea route to India and abolishing the slave trade that existed through the Indian Ocean (the oriental slave trade) and replacing it with legitimate trade. Moreover, the British presence during the short-lived "Owen Protectorate" over Mombasa in 1824 to 1826 initiated a policy that supported Asian merchants trading under British flag in legitimate commerce, in an effort to combat slave trade.[18] This encouraged more traders to come to the coast. Furthermore, the opening of a British consulate in Zanzibar in 1841 also encouraged Asian traders. This was because the government of India helped in the growth of British diplomatic influence over Seyyid Sayyid in Zanzibar, which in turn helped the Asian traders carry on commerce peacefully.

Migration to East Africa continued even with the change of political control over the East African coast from one power to another, and the Asian merchants continued to play a key role in commerce and the general economic life of the region. For instance, during Sayyid's rule between 1840 and 1856, Asian trade and migration to Zanzibar and Mombasa increased. The group that came during this period became bankers, shopkeepers, merchants, and financial advisers to the sultan and also some were moneylenders known as "Banyans."[19] Specifically, the Indians followed the Arabs to Zanzibar and settled there. This trend continued up to the 1890s. For example, in Zanzibar alone the Asian population by 1887 was estimated to be 3,086. This was because Seyyid Sayyid's interest in commerce attracted the Asians, whom the sultan needed to facilitate in the marketing of cloves and coconut in Zanzibar and Pemba.

Asian traders eventually followed the Arab caravans, which moved into the interior in search of goods such as ivory and slaves from the source. A few Asians set up trading centers along these caravan routes. For example, Musa Muzuri, a Khoja from Surat, traded in ivory in Unyamwezi and eventually set up his base at Koroqwe.[20] Other centers included Bagamayo, Tambora, Dar es Salaam, Tanga, Pangani, and Ujiji. Mombasa also experienced an increase in the population of Asians; by 1873, this was estimated to be about 533.[21] Therefore, even before the coming of the Europeans and the advent of the railway, a few Asian traders had already penetrated into the mainland of East Africa. For example, Alidina Virsram, a leading pioneer merchant, had established trading posts in Uganda twenty years before the railway reached Lake Victoria. The Asians monopolized trade along the coast together with other traders such as the Arabs. Local people encouraged them since they were able to provide commodities that were much needed. This encouragement was seen, for example, during the Abushiri rebellion in Tanganyika, when the Asians were assured of their security as the Europeans (Germans) were attacked. However, many of these early immigrants did not bring their families (wives) along. Such a situation led to the rise of new races at the coast such as Afro-Shirazi and the Jotawa; this situation was to change later.

Moreover, the penetration of Asians into the interior of Kenya, as with the rest of East Africa, was facilitated by the declaration of imperial interests over the region by both Britain and Germany. After the conclusion of the Berlin Conference in February 1885, Britain, as was the case with all European powers, quickly consolidated its imperial gains on the ground by the establishment of administrative networks in areas they had perceived as their "spheres of influence." In Kenya, since the British did not have immediate plans to establish a fully operational administrative network, this task was handed over to the Imperial British East Africa Company (the IBEA Company), which was granted a royal charter in September 1888 that gave it a free hand to open up and develop the interior through the establishment of administrative, transport, and commercial infrastructure.[22] In 1889, the president of the company, Sir William MacKinnon, was granted permission by the Foreign Office to recruit Indian sepoys for service in East Africa. The government of India allowed the company to recruit sepoys from Delhi with a three-year contract.[23] The sepoys were used in conquest and pacification of the area. They were recruited to join the East African Rifles; later this group would be merged with the King's African Rifles in 1902. Many sepoys were encouraged to come to East Africa and join the King's African Rifles, since it was suggested that they would be given plots of land for settlement in areas that were not suitable for Europeans.[24] Unfortunately, the scheme to allocate plots to the sepoys did not materialize.

It was generally held that Asians were good entrepreneurs and farmers who could help spur the economic productivity of the region. This was because Britain, like all the colonial powers, believed that the colonies had to be self-sustaining. As a result, more Asians moved into East Africa following the takeover of East Africa by Germany and Britain, which was followed by the signing of the Anglo-German treaty of 1890. Therefore, the IBEA Company, which had a strong background in India where the British had already established a colony, recognized the desirability of incorporating manpower from India as it established itself in Kenya. The company therefore supported Asian migration and even toyed with the idea of establishing permanent Asian settlements in the region, and this, it was also hoped, could help alleviate the problems of overpopulation back in India. During this period, the IBEA Company assumed control of the interior and later accorded Asians a prominent part in the development of the East Africa Protectorate. The founders of the IBEA Company, led by Sir William Mackinnon, all hoped to provide an outlet to India's surplus population by encouraging Asian traders and the agricultural class to come to East Africa.[25] Moreover, as the company established itself in the hinterland it relied on indentured labor from India to serve in the nascent administrative network and also in the military and the police force.

After the successful signing of the 1890 Anglo-German treaty that formally ended the reign of the Omani Arab aristocracy over much of the coastal region, save for the narrow coastal strip, plans were put in place to construct a railway within the British sphere of influence. Railways were to the British empire what roads were to the Romans, the arteries of power.[26] That being the case, in 1891, Captain J. R. L. Macdonald, who had worked for seven years in India on railway construction and defense, was contracted to survey the route for the railway and, in his report of March 1893 he strongly recommended the employment of Asian labor. This, he argued, could help solve the problem of obtaining African labor, which was viewed as difficult, given the general hostility of the Africans to company administration. Macdonald's sentiments were shared by other company administrators like Fredrick Lugard who believed that Asian settlement was most likely to follow the construction of the railway. This, it was believed, would help to develop the economic viability of the region.[27] It became clear from the onset that the company considered Asians very favorably, and this encouraged their migration to Kenya in large numbers in the succeeding decades. Besides, Britain needed access to the protectorate of Uganda, which was landlocked, and so the Kenya-Uganda railway was seen as an important artery.

When the Foreign Office in London eventually assumed direct control of the territory between Uganda and the coast on June 15, 1895, and renamed it the East Africa Protectorate, plans to construct the railway were put in place. In October 1985, a Uganda Railway committee was appointed and was chaired by Sir. Percy Anderson. The following year, a bill authoring the construction was introduced in parliament in London and eventually the construction plan was implemented. As construction of the railway from the port of Mombasa to Lake Victoria began, Macdonald, as chief engineer and surveyor of the route, reiterated that Asians from Madras and South India would be best suited to work on the construction site at the coast while those from Punjabi, Panthans, and northern India would work at higher altitudes such as the highlands.[28] Due to these recommendations, the British colonial administration employed the use of Indian indentured labor to build the railway. The Indian colonial government then stipulated that each worker must be free to remain in East Africa at the end of the contract. Asian labor was preferred over Africans because it was argued that their labor would be continuous while that of Africans was temporary because when the rains started, most Africans left and went back to work on their farms.

Because of these factors, the second wave of Asian migration was much larger than the first one. It occurred between 1890 and 1914, after the start of colonial rule. The second wave, made up of mostly of the rail workers and traders, also penetrated to the interior of East Africa. As the Asians built the Uganda railway from Mombasa to Kisumu, many established shops along the route in several towns. The first group of construction workers arrived in Mombasa and numbered 350 and laid the first rail in 1896. This first group included blacksmiths, carpenters, and masons. This number increased and by 1901 there were 32,000 Indian workers involved in the construction of the railway. Five years later in December 1901 the railway had reached Kisumu covering 582 miles.[29] Between 1895 and 1914, the British brought 37,747 Asians, mainly from Punjab, on three-year contracts.[30] Out of these, 16,312 were repatriated at the expiration of the contract or were dismissed from work, 6,454 were injured during the construction of the railway line and became invalids, and 2,493 died.[31] With the construction of the railway, the administrative boundaries in colonial Kenya were changed in 1902 for easier administration of the railway line; this brought the western part of the highlands into Kenya.

Asians did not stop coming to East Africa with the completion of the railway, they continued to be imported under three-year contracts until a new Indian emigration bill was passed in 1922. The railway greatly facilitated Asian immigration since it eased the transportation into the interior of East Africa. Even those who had settled at the coast for a long period could now easily move into the interior, establishing shops along the railway line. Those who came after the railway was completed worked on road constructions such as the Mazeras-Mackinnon road, built in 1906. The Germans in Tanganyika also employed Asian labor, though only a small number (approximately 200) because the Germans relied heavily on African labor force. Of the Asians that came to East Africa Protectorate, only 7,000 opted to remain in East Africa upon expiration of their three-year contract.[32] This number was relatively small, and European settlers claimed that the Asians were "crowding" the East Africa Protectorate. Some of these Asians who did not go back practiced farming while others engaged in various economic activities. But even among those that returned to their home as per their contracts many later returned bringing along friends and relatives; those who opted to remain back home continued to spread the information about opportunities in East Africa.

The availability of economic opportunity working in rail transportation and the supply of goods to the European settlers and Africans attracted another 10,000–20,000 Asian migrants between 1914 and 1921. In 1914 there were 34,000 Asians in East Africa, and by 1921 the Asian population had risen to 54,000, with the East Africa Protectorate (Kenya) having most of the immigrants.[33] The British employed many to work as subordinate staff; others worked as clerks or customs officials in the colonial administration in Kenya and Uganda.[34] These Asian recruits initially were mainly Parsees and Goans. Also for the IBEA Company to establish its control in East Africa, it had to recruit Asian troops. Apart from active military duty, the colonial government also relied heavily on indentured labor from India to serve in the police force and help in restoring law and order, since the Africans resisted.

The third wave, which took place during the interwar period, was facilitated by a free immigration policy. This movement led to an increase in population. The majority came because of the 1930s economic depression that impacted British India, leading to the relocation of Asians to other countries, such as Burma, Ceylon, and the East African countries including Kenya. For example, there was a tremendous increase in population between 1921 and 1931: by 1931 the number of Asians in East Africa had increased to 98,000; by 1939 the Asian population in East Africa had risen to 105,000.[35] The increase was also partly attributed to natural reproduction of those who had settled earlier. The majority of the Asians settled in the interior of Kenya.

The number of Asians was increased by the fourth wave of migration, which started at the end of World War II. After the war the Asian population in East Africa was 260,000, out of which Kenya alone had a population of 130,000 Asians, and the increase in numbers continued until after Kenya attained independence in 1963, which marked the final extensive migration into Kenya. During the early migrations, Asian workers left their families behind in Asia to work the land. But during the colonial period these workers began to bring their wives and families to Africa with them. Most of these early Asian women settled along the coastal towns such as Mombasa and Zanzibar and were mainly Muslims.

A majority of these Asian migrants lived in towns and trading centers. These communities consisted of various religious sects such as Hindus, Christians, Sikhs, and Muslims. Each of these sects retained their traditions and maintained a strong cohesive social group.[36] However, this intense communal identification and interaction also impacted them negatively in that it tended to isolate the Asians from other races in Kenya. On the other hand, Asian adaptability and flexibility led to other success in that although they came from a very stratified background; once in Kenya, their caste system was not much adhered to. For example, although the Shahs, Patels, Ismailis, and Sikhs were traditionally farmers back in India, almost all became businessmen once in Kenya, with the exception of the Sikhs of Jat who did not abandon their desire to engage in farming. Nonetheless, by the twentieth century most of the immigrants who came to East Africa were from the urban areas and included shopkeepers, doctors, lawyers, accountants, and teachers.

ASIANS IN KENYA

The Asians who came to Kenya were engaged in various activities from petty traders to prominent pioneer Asian businessmen such as Aldina Visram. The petty merchants commonly known as *dukawallas* (from the Hindi words *dukaan* meaning store and *walla* meaning owner)—played a prominent role in the development of the Kenyan economy. These petty merchants and poor Asians engaged in almost all economic activities in Nyanza. For instance, they monopolized the fishing trade, operating dhows from Wadh Konyimbo, Kisumu, Kendu, Nanga, and Kamito.[37] The same could be said of small farmers, who included Asian cane farmers. These small farmers had land under food crops of all kinds and by the 1950s had increased their cultivated areas with important market products and produced some of Kenya's export crops.[38]

To understand the Asian agricultural initiative in Kenya, we need to look at their social, cultural, and economic setup and their quest for prosperity in Kenya and to show who they were and particularly which group was involved in farming. Many of these Asians came to Kenya through the port of Bombay on the west coast of India. They originated from one of the four areas: the coastal belt around Cutch; Rojaputani and the Central Provinces; Punjab; and Goa. The groups spoke one of the four indigenous languages represented by the regions such as Gujarati, Hindi, Punjabi, and Konkani among Goans.[39] The Asians in Kenya could best be classified according to their religious communities: Muslims, Hindus, Sikh, Christians, and the Parsee.

MUSLIMS

One-third of Kenya's Asian population is Muslim and are divided into three sects of Shias: the Ismaili Khojas, the Bohra, and the Ithna'ashris. There are also Muslims from the Sunni sect called Memons from the coastal belt of western India who speak Gujarati.

Ismaili Khoja

The Khojas were originally Hindu traders living in the upper Sind area of northwest India before converting to Islam. Among the three sects, the Khoja are the majority in East Africa. Many came to Kenya during the colonial period. By 1960 Khojas totaled 17,000 in Kenya. At the same time there were 6,000 Khojas in Uganda and 27,000 in Tanganyika, for a total Khoja population of about 50,000 in East Africa.[40] The word *Khoja* means "honorable discipline." This sect does not make a pilgrimage to Mecca and are led by their famous religious leader, H. H. the Aga Khan. The Khojas are very progressive; both men and women are well educated and in most cases are professionals. During colonial times they served as bankers, clerks, doctors, and restaurant owners, so they did not engage in farming in Kenya.

Bohras

Bohra is a Gujarati word derived from the word *Vohoru* meaning "to trade."[41] Unlike other Muslims, the Bohras offer prayers (*Salat*) three times daily instead of the orthodox five. The Bohras came to East Africa early and by 1871 about 600 of them had settled at Zanzibar and Mombasa. Originally they are from Cutch and Karachi. The Bohras are believed to be conservatives who opposed rapid modernization and were somewhat resistant to higher education.[42] Many Bohras in Kenya during the colonial period were traders, iron mongers, tinsmiths, and watchmakers. A well-known Bohra family that was very enterprising in East Africa generally, and Kenya in particular, was the Jeevanjee family led by pioneer Sir Tayabali H. A. Jeevanjee.

The Ithna'ashris

The third major Muslim group was that of the followers of the "eleventh Imam" (hidden Imam), the Ithna'ashris, who came from Punjab and also spoke Gujarati. They represented a very diverse background and were also engaged in diverse economic activities in East Africa. This group split from the Ismailis and had a bitter conflict back home that acted as their push factor to migrate. The group believes that the Imam is still concealed and so they do not recognize the Aga Khan as their spiritual leader.[43] They were very successful businessmen and became very wealthy in East Africa, especially those who settled in Kenya.

The Memons

This group moved to East Africa due to religious persecution back home. The persecution of this group started with the persecution of Nasserpurai Memon Muslims in the Sind area in the nineteenth century. This sect was largely composed of weavers and cloth dealers. When the persecution started, many fled to Zanzibar, Mombasa, and other coastal towns along the east coast of Africa, where they found sanctuary and continued with their weaving occupation until they were dislodged from their trade due to the importation of American textile during the colonial era.[44]

HINDUS

The Hindus also came early; by 1841 they were already residing at Mombasa. Many came from Cutch and belonged to the Bhattia sect and were strict vegetarians. The Hindus are divided into sects such as Patels who came from Gujarat and speak Gujarati, Lohanas from Cutch who speak Cutchi (both Gujarat and Cutch are Indo-Aryan languages), and the Jains from Gujarat. Jainism was founded by Verdhaman Maharira who many believe was a contemporary of Buddha. Jains are a branch of Hindus, but they deny the authority of the Hindu scriptures although they follow the same rituals and are strict vegetarians who do not eat root vegetables. The Jains are divided into the Vanik and Visa Oshwal classes. There is a very close relation with traditional Hinduism. For instance, Jain women worship the Hindu goddesses along with Tirthankaras, who is the mythical founder of Jainism. Although Jainism initially forbade agriculture as a profession since it entailed destroying microbic life, the Oshwals of Porbandar were farmers.[45] Most Oshwals in Nawanagar were landowners back in India, while some of them were sharecroppers or leased their land.

In Kenya the Hindus became professional traders and engaged in commerce, with such successful businessmen as Dr. Manu Chandaria, A. B. Patel, and S. G. Amin. Hindus excelled in business and were described by western travelers as "Jews of East Africa."[46] *Oshwals,* popularly known as "Shahs" were also educated, but most of those who came to Kenya were poor and illiterate. *Oshwal* refers to a number of castes found in northern and western India. Sometimes they are identified by their geographical locality through the use of a prefix—for example, *Halari Oshwal* indicates an Oshwal from Halari, an area west of Kathiawad that included Nawanagar and Jamnagar. On the other hand, the prefix *Visa* for Visa Oshwal refers to a numerical prefix that means twenty and is used among the Gujarati trading castes.

The majority of Oshwals who came to East Africa were Visa Oshwals who came from the fifty-two villages around Jamnagar, an area that was prone to drought and famine. Once in Kenya, many settled in Nairobi. This group focused on communal development and was rarely involved in politics of the colony. Among the Hindu were other smaller sects such as Bania and Patidar who settled in Kenya. However, the majority of Asians, both Hindus and Muslims, who came to Kenya took to trade and shopkeeping. In East Africa, including Kenya, these traders; both Hindu and Muslims were known as "Banyans" and were seen as a money-making class.

SIKHS

The Sikhs, popularly known in East Africa as "Singh," came from Punjab and were very close to Punjabi Hindus. Many of them were farmers from Punjab, and they were from one of two groups; the Jats and Ramgahia. The Jats were an indigenous agricultural caste back in India. Ninety percent of Jats belonged to the Ramgarhia, which was a lower caste in Punjab. In Kenya they remained farmers growing crops at Kibos and Eldoret and were also used in military services. In Kenya the Ramgarhia Sikhs were basically skilled laborers (*Fundis*). By 1960, there were about 20,000 Sikhs in Kenya. They often worked as mechanics, carpenters, and rail men.[47] The Sikhs were believed to be the most adaptable Asians, and this could possibly be true due to their ideology; for instance, their religious doctrines are unsophisticated and thus tend to be flexible.

In India, all Sikhs were farmers and so the majority of the Jat sect of Sikhs who came to Kenya were farmers.[48] Back at home in India they were known for their resilience and were believed to have a background of heroism and soldiering that still persists in the self-image of the Jat farmer as a farmer-soldier.[49] For them buying land was deemed to be prestigious. In Kenya, the Jats were among the pioneer farmers at Kibos settlement. Later a number of Jats bought considerable tracts of rich land from departing white South Africans and the white settler farmers in the former white highlands of Kenya between Eldoret and Nakuru on the dawn of independence.[50] This indicated their resilience and agency in agriculture. The white settlers in the highlands did not know that many Asians were predominately farmers back home, and the Asian settlers did not attempt to convey this image. On the contrary, the Jats who settled at Kibos and Kisumu succeeded in breaking the stereotype of the Asians as being only interested in business. Some Asians started farming, specifically the Jat Sikhs, who were later joined by few other Asian castes. The Sikhs settled well in Kenya and formed many socioeconomic societies, the most important being the Siri Guru Savha.

CHRISTIANS (CATHOLICS)

This group mainly consisted of Goans. They formed less than 10 percent of the Asian community in Kenya by 1962. They came from a tiny Portuguese colony of Goa, south of Bombay, and disliked being classified as Indians; they ignored their Indian background despite the fact that they spoke an Indian language. They are of mixed blood having intermarried with the Portuguese and adopted Portuguese names such as D'Souza (or de Souza) to indicate their ancestry.[51] They were well educated and modeled their social life on the Western lifestyle.

During colonial period this group held clerical jobs in banks and the civil service, served as teachers in mission schools, and worked as lawyers. They spoke English, Portuguese, and their Asian vernacular known as Konkani, which is very close to the Marathi language. Many became involved with politics and supported the freedom struggle. For instance, Pio Gama Pinto was a trade unionist, journalist and a champion of African nationalism; Fitzval R. S. de Souza was a lawyer, journalist, and founder member of the Kenya Freedom Party; and John Maximian Nazareth was a legislative councilor and the president of East African Indian National Congress.[52] These three contributed so much on issues of equality and the political struggle in Kenya. Some critics of Goans, however, assert that Goans believe that India is a Hindu country and due to their religious background they would rather identify with Portugal than the so-called heathen India.[53] However, as an Asian community, they were reputed for honesty and reliability.

PARSEE

This group was the smallest in number, less than 1 percent of the Kenyan Asian population. They were originally refugees from Persia who ran away from religious persecution from Muslims, and they practiced Zoroastrianism during the sixteenth century.[54] They moved and settled in Bombay and Gujarat before their migration to East Africa. The Parsee were well educated, affluent bankers and lawyers, and very progressive. Due to their progressive nature, they were more socially accepted by Europeans than other Asians, but most did not make Kenya their home. They came on temporary assignments, for instance, to manage Asian banks and then went back home.[55] By the end of colonial rule in Kenya, there were only about 500 Parsees.[56]

These religious beliefs divided the Asians. For example, many believed that it was common for Shahs not to give their daughters to the Patels for marriage. Be that as it may, by 1948, there were many Asian immigrants in East Africa; in Kenya alone their populations were as follows: Hindus 45,238; Muslims 27,583; Sikhs 10,621; Goans 7,145; and Jains 6,146.[57]

THE PUSH AND PULL FACTORS

People normally migrate from places of limited prospects to areas that seem to have greater economic opportunities. The motivations for leaving a place and those for going to another place are known, respectively, as the "push" and "pull" factors.[58] "Push" and "pull" could be viewed as factors for expulsion and attraction. These factors can be categorized as economic, social, and political. In the case of Asians, it seems that the economic motivation acted as a great push and pull factor. Many immigrants were drawn to the coast of East Africa by hopes of eradicating poverty; many came to seek out any opportunity that East Africa could offer financially. It appears that most of early traders were attracted to the coastal towns such as Zanzibar and Mombasa due to commerce; many were lured to trade.[59] The extensive trade opportunities were an important impetus to Asian migration.[60] With the establishment of colonial rule, not only did the trade increase but also the British found it necessary to encourage the immigration of the Asian merchant class so as to open up the protectorate. The colonial state realized that Asian merchants would be helpful in exploiting resources that were much needed for attaining the goal of self-sustainability. Commenting on this stance, a British administrator in 1896 asserted that "Such immigration is good and for the benefit of the country, and we should encourage it by every means in our power. This class of immigrants is possessed of some means and (what often goes with money) some intelligence and enterprise."[61] Therefore, initially immigrants were encouraged to come to the colony.

Most of the Asian early traders became middlemen who then facilitated the exchange of goods in the interior markets. Some of these traders started from humble beginnings and became prominent and great financiers, for example Aliboy M. Jeevanjee and Allidina Visram. However, the majority remained retail traders known as *dukawallas*. The majorities of these businessmen came from urban areas and were motivated to move to Africa due to stiff competition back home; others moved due to the potential of financial improvement.

Needless to say, Asian resilience in trade cannot be underestimated. As the colonial economy developed, they contributed to it immensely. For instance, the African peasants in the rural areas depended on the Asians as middlemen. They sold their produce to the Asians who then resold it to the urban areas and also traded basic items to the Africans from their petty shops (*duka*). Trade acted as a pull factor, and they succeeded in remote areas. Their expertise and mercantile skills earned them recognition even from the settlers who disliked them. For example, commenting on their resilience, Sidney Kelson, a settler, wrote in his diary in 1923:

> Food supplies were readily available . . . I was able to do most of my shopping at the Indian store. Indian traders had long penetrated deep in East Africa and established themselves in townships and isolated trading posts as far afield as the Belgian Congo. They were dotted all over Kenya, often in the most unexpected places . . . While the white man played tennis or golf, the Indian shopkeeper in his little round hat and shirt-tails hanging out, toiled on then bedded down on his shop counter when his long day's work was done.[62]

This seemed to have been a common feature of many of the early Asian migrants' lives. The *dukawallas* indeed opened up Kenya as they opened shops in remote parts of the country and also along the railway line. Most of these traders belonged to the Visa Oshwal Community, who mostly belonged to the "Vaishya" or the trading caste. These immigrants played a significant role in the Kenyan economy, introducing currency into the market and creating a demand that stimulated economic growth and greatly impacted the economic history of Kenya.[63]

However, economic conditions also acted as push factors, given the poor economic conditions back at home. Most migrants came from poor villages and farms, and many came to Kenya in hopes of rising above their poverty due to the economic opportunities offered by the colonial administration. Many young people were also encouraged by family members to go out and seek greener pastures. To the Asians the most important source of information that led to their migration was word of mouth; others wrote letters back home and encouraged others to come.[64] Although the push and pull factors varied with individuals, their impact was felt among various communities. For instance, the hope of better economic opportunity drove various groups such as the Hindus, Muslims, Sikhs, and Christians to come to Kenya.[65]

Other economic pull factors included opportunities for employment in government services, crafts, and especially construction, carpentry, and masonry where the Sikhs thrived.[66] The employment in government and European firms acted specifically as an attraction to educated Asians. The opportunity assured a steady pay and a pension upon retirement. Also the difference in salaries attracted many. During the colonial period teachers in India could earn approximately 4–15 rupees a month compared with those who worked as clerical staff in the Kenya colony who earned 80 rupees a month.[67] Another factor included the establishment of British rule in India, which led to the decline of the cottage industries, especially handicrafts, which was affected by flooding of the Asian market with cheap ready-made goods from British industries. Hence, many who had been self-employed were dislodged from their occupations and so chose to move to Kenya where their skills were much sought after.[68] This was compounded by the demand for skilled workers in Kenya to work in the construction of the railway, as clerical worker services in the colony, and other indentured workers such as artisans and masons.

These jobs offered prospects of earning a better salary, especially for people like the Goans who were highly educated, not traders, but found it difficult to find employment in the saturated job market of Goa.[69] Most professionals were also attracted by job prospects and higher salaries and came mainly from Bombay, Poona, and Goa with a few from Calcutta. As such, many Goans moved to Kenya where they offered services such as clerks, typists, accountants, doctors, lawyers, priests, and teachers. By 1904, for example, most clerical posts in the provinces and districts in Kenya were filled by Goans.[70] Although they were paid less than Europeans, still the wages which they received were better than those offered in Goa. There was variation in wage earnings among the civil service employees in colonial Kenya; and because the colony had the three-tier arrangement, the Asians were paid less than the Europeans while the Africans were paid less than the Asians.

There was deep economic discrimination in the Kenyan job market. For instance, by 1950, approximately 63 percent of Asians were engaged in civil service work earning between £180 and 359 per annum while for the same jobs the Europeans earned between £600 and 2000 per annum.[71] Seeking economic opportunities with family or caste members was a huge attraction to East Africa, which many considered to offer better opportunities and a higher standard of living than India.[72] The steamship companies took advantage of the migration boom, using various incentives to lure passengers to their ships. For instance, the fare from Bombay to Mombasa was lowered to between 10 and 15 rupees; for the slightly higher price of 35 rupees a passenger could have meals included with their fare.

Asians also migrated due to environmental factors that had greatly impacted their home areas. Possibly, the dry arid Kutch area made farming difficult and so most people resorted to commerce as a source of livelihood.[73] The Sind, which was also an important migration spot, was also more of a desert, which made any livelihood difficult to earn. These environmental factors contributed to recurrent famines. For example, the famine of 1899–1901 was so severe that it resulted in many deaths and forced many Asians, especially the Patels, to move to Kenya. Furthermore, between 1899 and 1900 Gujarat also experienced severe famine leading to acute food shortages; hence many left their homes. Many immigrants also came from Kathiawar (Saurashtra), Kutch, and Maharashtra. These areas had fertile soil but were dependent on the monsoon winds, which occasionally failed, leading to drought and famine. These droughts and famine were a common feature, recurring every two to three years.[74] Therefore, the environment of these areas was among the push factors that made many Asians seek a livelihood elsewhere.

Other push factors included other environmental disasters that affected India in the nineteenth century. For example, the plagues of the 1890s killed many people and forced many to move to Kenya. The plagues were recurrent and had great ramifications. For instance, the 1896 plague in Bombay destroyed businesses, disorganized trade, and greatly affected the mill industry there so much that Bombay suffered a mass exodus of people;[75] about 13,343 residents of Bombay left the city that year.[76] Other disasters included cyclones, locust invasions, and floods. Many of the affected people moved to the east coast of Africa and many settled at Mombasa.

Climatic conditions were another possible push factor. India is an area of high humidity; it appears as though some immigrants might have found the climate in East Africa, and Kenya in particular, better than India. This could be said of most of the Goans who had migrated after recurrent outbreaks of malaria epidemics in Goa.[77] Yet even in Kenya the immigrants faced challenges posed by malaria. Similarly, the Oshwals moved because there were a number of calamities that befell Nawanagar; there was irregular rainfall that led to the several instances of famine between 1880 and 1940, and there were other natural disasters such as cyclones, locust infestation, and floods.[78] One Nawanagar ruler, Jam Ranjitisinhji, was concerned by his people's migration and wrote about it in 1910: "the subjects of this state are noted for a spirit of commercial enterprise which makes them seek fortunes elsewhere as they find only a limited scope within their own territory."[79] Such sentiments show how many migrants moved because of poverty and the lack of economic alternatives.

There were also social factors that pushed the Asians out of their homes. These included population pressure, marriage obligations, religious intolerance, and the desire for adventure. This could explain why although some immigrants came from fertile areas that were well watered and where farming was possible: although the land was good for farming the areas were prone to overpopulation. For example, in Gujarat, an area where many immigrants came from, the population density by 1911 was 532 people per square mile while in 1941 it rose to 640 people per square mile.[80]

The increase in India's population was partly caused by the introduction of modern medicine with the establishment of the Raj; this resulted in a fall in the death rate and helped to spur the immigration of many people.[81] India's population growth was alarming and needed an outlet. For example, in 1847 India's population was 133 million, and by 1902 the population had increased to 236 million.[82] This surplus population was composed mainly of the young and productive people, who ventured out of their country to other areas such as Fiji, Burma, Ceylon, and East Africa.

The population in India continued to increase and by 1961 was 439 million.[83] This was part of the reason why early colonial officials supported Asian immigration. The increase in population impacted greatly on agricultural production and influenced the decision to move, as the landholdings increasingly became smaller and the soils exhausted.

Similarly, marriage obligation was a factor for migration; the lack of resources in terms of money to pay for marriages acted as a push factor. This was because some people migrated to escape pressures from oppressive customs that demanded that they pay dowries for the marriages of their daughters.[84] This became increasingly difficult particularly in areas where agriculture could not provide resources, let alone support large families, especially the poor. Frustrated, many families opted to go elsewhere to earn money that would eventually be remitted home for their daughters' dowries. On the other hand, marriage also acted as a pull factor since most young men who had come to Kenya earlier would go back to India or send for wives back home after they felt financially stable. This was a common trend among the early migrants. It was not difficult to get a wife even for men in East Africa because marriages were normally arranged between families and young men could depend on their families back home to get wives for them. Then the wives would be "called" to come to Kenya. This phenomenon came to be known as "chain migration," which also included family members like uncles and younger brothers who were called to Kenya once the original migrants had established themselves.[85] This process was bound to repeat itself, leading to more migrations.

Other groups moved because of what was perceived as religious persecution. For instance, the Parsees, who had embraced Zoroastrianism and who initially had fled Persia due to the persecution of the sixteenth century, had settled first in Bombay and later moved to Kenya. Since India was dominated by Hindus and Muslims, fear of another persecution from the well-established religions in India might have encouraged some of them to move to Kenya.[86] Moreover, other events during the nineteenth century in India indicated how religious persecution of the Nasserpuria Memon Muslims in the Sind region occurred, sending many of the sect members fleeing to the East African coastal towns such as Mombasa and Zanzibar. The Memon Muslims were mainly weavers and cloth dealers by profession, and they carried this practice to East Africa and Kenya in particular, where there was no stiff competition. As weavers, many had been dislodged from their occupation by British manufactured goods. As such, the Memons were not only afraid due to their religious affiliation but also felt economically squeezed by the Raj.

Other Asians could be said to have migrated because of their desire for adventure. This was especially found among the young who mostly moved to new locality to escape boring village life. This group was motivated by the spirit of adventure and excitement, having heard success stories from early immigrants. The desire to cross the ocean and visit new lands was overwhelming.[87] Stories of successful people, for example, renowned businessman Nanji Kalidas Mehta, who arrived in East Africa at the tender age of fourteen or Mepabhai Vershi Shah who boarded a German steamer and came to East Africa when he was only thirteen years of age captivated many youngsters. Some Asians also moved due to insecurity that was posed by marauding gangs of robbers commonly known in India as *thagi*. This group affected people who were already stressed by different types of economic challenges.

Asian migrants, especially those from India, had been used as soldiers in Kenya for a long time. Their coming dates back during the reign of Seyyid Sayyid (the sultan of Zanzibar) who employed Baluchi mercenaries recruited from the Makran Coast on the shores of the Arabian Sea. This group was used in his campaigns against the towns of Mombasa and Pate. After the defeat of these areas, Sayyid left behind 150 Baluchi troops as a garrison at Fort Jesus in Mombasa. Furthermore, during the partition of East Africa, more trained Asians were brought to serve in the military by the IBEA Company. The first group to arrive was the 17th Madras Sappers. These soldiers were known as the sepoys. In 1889 Sir William Mackinnon, the president of the IBEA Company, sought permission from the government of India to recruit the sepoys for service in East Africa.[88] The Indian government allowed the recruitment of 200 men from around Delhi to serve in the military on a three-year contract. This group consisted mainly of Sikhs. Although initially the Indian government had opposed sporadic recruitment, especially from Punjab, more were still recruited.

Later, in July 1895, after the protectorate was transferred to the British crown, 300 more Punjabis were added to the force to serve alongside 300 Swahilis and 100 Sudanese soldiers. The Indian sepoys also joined the military force known as the East African Rifles that was established to man the area. This East African Rifles existed for six and a half years.[89] They were used in the conquest and pacification of the region. For instance, they took part in expeditions against Mbaruk of Takaungu along the coast and the Ogaden Somali in Jubaland between 1898 and 1901.[90] Another group, consisting of the Fourth Bombay Rifles, also was brought. The East African Rifles was later merged into the King's African Rifles in 1902.[91] It is important to note that after conquest, the Colonial Office advocated for a scheme to settle the Asian soldiers as indicated earlier. The sepoys were to be offered land for settlement. Unfortunately, this scheme was not implemented.

By 1913, when it was believed that the African troops had sufficient training, it was decided to discontinue the use of the Asian troops on health grounds, since many were falling sick. A year later, with the outbreak of World War I, more troops were brought in an attempt to capture Tanga and blockade Dar es Salaam, the German's main East African port. In the battle of Tanga, many Asian soldiers were used, and approximately 820 Indian soldiers died in the battle. As recognition toward the war effort, the settlement scheme issue was revisited, when General W. H. Manning, the inspector general of the King's African Rifles, prepared a memorandum to the Foreign Office in which he proposed that Indian soldiers should be offered land in areas that were not suitable for European settlement. Once again the plan was abandoned. The dismissal of these agricultural grants was one reason why initially Asian agriculture did not feature prominently in colonial Kenya.[92] Right from the onset of Asian settlement, it was clear that the colonial administration did not support Asian colonization of agricultural land in Kenya.

Lastly, another reason for migration was the security and encouragement that the immigrants received. Some immigrants were driven by the fact that they felt they were British citizens and as such believed they could be offered better security and freedom to excel under the British flag. Such a decision could have been influenced by the fact that during the early twentieth century, the British encouraged the migration and participation of the Asians in the colonial economic development of Kenya. Other prominent Asians also encouraged the migrations. A case in point was the advice and encouragement of the Ismaili leader H. H. the

Aga Khan, who encouraged his followers to take the advantage of the opportunities offered by the British empire to migrate and make Kenya their new home.[93]

SETTLING IN: THE CHALLENGES OF IMMIGRANTS

The journey to Kenya took three weeks or more depending on the weather. Coming from a background of poverty, many had borrowed money to undertake the voyage. Others spent their entire savings on the fare alone, which by 1900 was about 10 rupees by dhow and 35 rupees by steamers. Many used a dhow, a traditional Arab sailing vessel with one or more lateen sails, because it was cheaper. Therefore, many arrived at the coast with nothing or only few rupees at hand, and with little information of the prevailing conditions in the colony.[94] Moreover, the journey was not easy; they endured hardships such as sleeping on the decks, where they were clustered in groups according to their religion. They had to bring and cook their own food, even with the uncertainty of the journey's length. The voyage could take as short as twenty-six days if the weather was good or months if it was bad. Sometimes the journey took as long as nine months.[95] Lack of medical services during the journey made it more difficult and many fell sick.

Once they arrived at the coast, in most cases, they were quarantined for a week.[96] This was necessary since some came from the plague zone areas in India. These early immigrants often went through several weeks or months of uncertainty and deprivation while trying to settle and gain some employment. The initial years were difficult as many experienced poverty and loneliness.[97] Once they arrived in Kenya many faced discrimination as they were quickly separated from the Africans and Europeans by the colonial administration. This type of segregation in townships put the Asians in airtight compartments and hindered their assimilation into the foreign country.

However, many were received by friends and relatives who had preceded them. Those who had none were assisted by other Asians or charitable organizations that provided them with food and lodging until they found some means of survival. By the 1920s, for example, there was the existence of informal eating places (messes) run by charities such as the Meghji Ladha and Co., which had branches in Mombasa and Nairobi. It offered accommodations to many Asians, providing free food and lodging until employment was secured.[98] This helped the immigrants to settle. Similarly there were other organizations, such as the Mombasa Indian Association founded in 1901 and the East African Indian National Congress (EAINC), founded in 1914, that assisted the Asians with migration issues, obtaining trading licenses and lending money to the immigrants.[99]

Most of the immigrants were assisted by the patron-client system that was widely entrenched among the Asians. This system enabled the new immigrants (mostly young men) to settle in by serving as an apprentice to someone who had come earlier and who in most cases was older and had already established themselves financially.[100] The young men who came as apprentices also had their fare paid by the patron. A client was dependent on the patron, not only for food and lodging but also his job. In return, the young man not only learned skills while working for the patron but also did errands and even took care of his patron's children. The new apprentices were exploited and worked for long hours, between fourteen and eighteen hours a day.[101] Once the apprentice had accumulated enough skills, he would either be put in charge of a new branch of the business or the patron would help to set him up with his own business.[102] In cases of retail shops, the client would borrow the first stock from his patron's shop. Therefore, the patron-client system helped many immigrants to adopt, adjust, and settle in Kenya. Once in Kenya, many adapted to the new circumstances. For instance, the Shahs, Patels, and Ismails left their traditional occupation of farming and started enterprises. Many struggled to make ends meet. It was not uncommon to find professionals like doctors and lawyers also engaged in some form of business besides their jobs because opportunities were available.

Regardless of any assistance accorded to the new arrivals, the immigrants still faced many challenges. Their problems were as diverse as their reasons for coming. For instance, among the coolies who built the Kenya-Uganda railway, the assistance discussed above was rare. Having been contracted back in their country, many had no time to adjust. Though they were employed, they experienced atrocious living conditions in Kenya. Most of their dwellings were temporary and overcrowded. Moreover, they, like other Asians, faced challenges of tropical diseases, such as malaria, that killed many. For instance, in 1897 during the month of January alone, about 50 percent of the coolies were sick.[103] Some suffered from dysentery, scurvy, ulcers, and jiggers, while others died from the recurrent plagues that broke out in the Asian bazaars of Nairobi in 1902.

Even those who had started small businesses suffered initially. Some were killed and had their shops pillaged. Boycotts by Africans were common. For instance, in 1912, the Duruma people who live at Kilifi (near Mombasa) in the then East Africa Protectorate undertook a successful boycott of the forty-four Asian shopkeepers for allegedly increasing the price of maize (corn). Moreover, over a period of time, the Africans began to resent the Asian presence. This was because Africans believed that most of the Asian traders deliberately took advantage of the African farmers by using false weights and measures. It seems that competition existed because both the African and Asian traders competed in the sale of commodities. Unfortunately, this conflict between African and Asians traders was further fueled by the implementation of postwar British policies in East Africa. This was because the colonial administration in Tanganyika, Kenya, and Uganda extended training programs to Africans and also assisted individual traders in starting business, while it excluded the Asians. This was facilitated by the Joint Boards that were started in Kenya by 1950 with grants from United States that provided loans and training to Africans.[104]

Among the Asians who came to Kenya, the coolies can be said to have suffered most. Not only had they to deal with the diseases but also the man-eating lions of Tsavo that killed quite a number. Food also posed a formidable problem, especially among the Hindus and specifically the Jains who were strict vegetarians and were not accustomed to eating food prepared by anyone outside their caste. This problem was compounded by the fact that early immigrants had insufficient money to buy food such as vegetables and spices. Needless to say, the Asians also faced discrimination, which was enforced by the colonial government. They were separated from the Africans and Europeans by discriminatory legislation in the form of land alienation and segregation policy in townships, both in places of residence and business areas.

Another problem experienced by the immigrants was language. Apart from the Goans and the Patels who were well educated and spoke English fairly well and so could easily get a job with the government, the rest had a difficult time. For example, many Punjabi who lacked knowledge of English and skills in business had to accept menial employment with someone from their own community. They had to learn English and sometimes Kiswahili before venturing into the civil service or setting up their own businesses. In fact, few Asians of any group could speak English, Arabic, or Kiswahili. Hence, they had to learn new languages apart from making other social adjustments; the immigrants basically experienced several weeks and months of uncertainty and led a precarious life trying to determine how to get employment and settle.

After a year or two, most had settled in well and were able to eke out a living on their own, engaging in various economic activities even in the most remote parts of the colony. Their success could be attributed to their stoic discipline, perseverance, and resilience; many worked for long hours.[105] Luxuries were shunned and any profits accrued were reinvested in their economic ventures. This was the secret behind their success, besides their resilience and hard work. It was perseverance and industry that allowed the Asians to excel in East Africa.

SETTLEMENT PATTERNS

The earliest Asian immigrants, who were basically traders prior to the establishment of colonial rule, settled in the coastal region where they pursued economic activities. In contrast, the immigrants who came after 1895 exhibited different patterns of settlement due to various restrictions put in place directly or indirectly, and many settled in areas where opportunities seemed available. The colonial state also controlled where the Asians settled because they regulated where trading centers could be established and prohibited Asians from trading elsewhere. However, the Asians mostly followed the railway line into the interior, setting up shops along the route. These settlements, though very modest, were important since they supplied the laborers and the railway staff with food and other basic necessities. This in turn led to flourishing markets and towns such as Voi, Kibwezi, Machakos, Nairobi, and Kisumu. Most Asians settled in the urban centers, including some smaller towns but most concentrated in Nairobi and Mombasa. These towns offered the Asians the most attractive economic and social opportunities such as education and other amenities. Nairobi, for instance, offered health facilities and was located near European settlements, especially near European farms. These areas were not only secure but they also offered opportunities to supply both the white farmers and the African workers with basic necessities such as soap and sugar.

Most traders were also concentrated in towns because the state controlled and restricted their operations in specified areas. This was enforced through a series of government ordinances that prohibited trading in areas not prescribed by government regulations.[106] These regulations were meant to constrain the spatial distribution of the Asian population in Kenya.[107] Consequently, some African districts were closed to Asians for trading during the first quarter of the twentieth century.[108] Moreover, due to settler politics, Asians found it difficult to own land. Ordinances were enacted to bar them from the white highlands, which frustrated their quest for agricultural land. Since the Asians could not own land anywhere they might have wished, their choices were limited to living in towns and major market centers. Moreover, the towns offered desirable educational, social, and cultural amenities.

The Indian social system (caste) also had a significant role in determining where the Asians settled; many settled where they had ties either of family or community. Accordingly, Asians congregated in specific areas due to the process of chain migration. Most people tended to settle in areas that already had a good number of their caste members. Also, new arrivals tended to settle where communal aid could be easily accessed.[109] The family and social groupings played a significant role in social networking, which was important not only for social support but also links to jobs through family connections. Therefore, the settlement patterns in most cases were not made by choice but were determined by sociopolitical and economic conditions in Kenya.

The Asians tended to settle in urban areas, though a few had stores in remote areas of the interior. This concentration could be explained by the fact that few Asians were engaged in agriculture. Moreover, the government was not interested in supporting their agriculture as a measure to stop their quest for the fertile land in the white highlands. Policies were put in place that discouraged Asian farming. For example, by 1902 it became an official government policy to deny Asians access to agriculturally high-potential areas such as the white highlands. Accordingly, Sir Charles Eliot, then commissioner of the protectorate, declared in 1902 that "the protectorate is a white man's country." This position was contradictory, or a change of view. Only two years earlier, when he assumed office, he had supported the Asian enterprise. Having changed his position in 1902, Eliot set in place a policy that excluded Asians from getting any land allocation in the highlands. This meant that the Asians were relegated to earn a living in the urban centers in business activities, with the majority engaged in retail trade.

Furthermore, the government policy that compelled the Asians to conduct their businesses in towns and in specific areas favored a policy of concentration in matters of town development schemes. This led to segregations in towns. The justification for segregation was based on health and sanitation grounds. This was counterproductive since it led to overcrowding in Asian settlements that led to sporadic outbreaks of plague.

The discrimination that characterized the protectorate's land policies would be repeated throughout the colonial period. Not only did the Asians deal with discrimination but also the constant contradictions that bedeviled the colonial state. Winston Churchill saw the complex situation in Kenya and referred to it as the "Sphinx riddle" in its newest form. This was because Churchill acknowledged the crucial role that the Asians played in Kenya, particularly of opening up trade in remote areas, building the railway, and financing various projects within the colony. Their contribution was something the colonial administrators took long to realize.

SETTLEMENT AT KIBOS-MUHORONI

Following the completion of the Kenya-Uganda railway in 1901, the colonial state had to look for a means to make the railway pay for its construction costs and also help in making Kenya a self-reliant, self-sustaining colony. Since Kenya was not endowed with minerals, agriculture was seen as a possible means toward this sustainability. Asian settlement was one avenue pursued, as this chapter makes clear. The initial reaction to Asian immigrants was friendly, and the colonial state welcomed the idea of Asian settlement in the colony. However, this stance gradually changed. Eliot's change of mind may have been caused by the fact that by 1902 there were about thirty European settlers in the country who strongly opposed the granting of any agricultural land to Asians, especially in what became the white highlands. In an effort to keep the Asians out of the white highlands, Eliot recommended that Asian settlement should be confined to the lowlands, which were considered unsuitable for European settlement. Therefore, in April 1903, he inaugurated a special Asian settlement at Kibos, a railway stop in the lowlands of the Nyanza basin. This area where the Asians settled and started farming was known as Kibos-Muhoroni.

Kibos-Muhoroni area is found in Nyanza Province of Kenya. The area is about six miles from Kisumu and includes Kibos, Chemelil, Kibigori, and Muhoroni.[110] It is an area that does not drain well with unreliable rainfall. The soil is rich especially at the foothills of Nandi hills down up to about a mile; there is also sandy loam soil. Far from the Nandi foothills, the area turns swampy with badly drained black cotton soil that varies in fertility. The fertility decreases as it moves toward the Kenya-Uganda rail line.

The area has an elevation of between 3,800 and 4,200 feet. The rainfall average is 50 inches per annum, which is suitable for sugarcane cultivation. There are three main seasons in the year: heavy rains (late March-June), short rains (July-November), and dry season with light rains (December-early March). The sugar-growing area was situated along the Kenya-Uganda railway line from Kibos to the east of Muhoroni. This area was acquired and occupied by Asians starting in 1903. By 1910, a number of Asian cane farmers owned several farms.[111] Some of indentured Asian laborers who had opted to remain in Kenya after the expiration of their contracts together with ex-railway workers formed the group of the first thirty pioneer Asian settlers. Out of these, the majority were Sikhs from Punjab who were traditionally an agricultural community. They were mostly poor. The pioneer settlers were allotted land of which about 60 percent of the plots were freehold while the rest were on 99-year leaseholds.[112]

Between 1903 and 1915, despite experiencing problems, Asian cane farmers obtained land ranging from three acres to one hundred acres. Later, between the 1940s and 1950s, the farms acquired were bigger as more farmers settled in the area. The largest Asian farm was that of Imhazalhi and sons; this was a London-based Asian company that acquired about 4,500 acres in Kibos, under a 99-year lease. This estate was registered as the Kisumu Rubber Estate.[113] These leases expired in 2002. Consequently, the growth of sugarcane farming in Kenya can be attributed to the Asian agency in the Kenyan economy in cane-growing initiatives. The Asian sugarcane farmers adapted to and put the land to maximum use to produce sugarcane and successfully built the sugar industry in colonial Kenya.

CONCLUSION

The migration of Asians to Kenya was a complex process, accomplished in several waves and motivated by diverse push and pull factors. As with many migrants, the overriding motivation was economic betterment. The coming of British colonial rule accelerated the movement into Kenya since, as British subjects, the Asians took advantage of the security it provided and the economic opportunities offered under British rule.

The Asians who migrated to Kenya came from all types of religious background and even differing castes. Once there, their settlement patterns were dictated by the social, economic, and political conditions within the colony, most importantly the land policies that favored Europeans and pushed Asians and others to the less fertile lowlands in Nyanza Province. The Asian settlers faced all sorts of challenges, from tropical diseases and the man eaters of Tsavo to the more mundane challenges of establishing themselves in a new community. The community by helping newcomers settle in through the establishment of charitable organizations that offered food and accommodation to the new arrivals.

Despite being locked out of the fertile white highlands due to racial bias and despite the colonial state's neglect of the sugar industry, the Asian community made the best of their environment, used their creativity and resources to overcome many odds, and searched for ways to provide for themselves. While they tested out various types of business ventures, many eventually settled on the production of sugarcane at Kibos and embarked on a life of farming that would present incessant challenges to success. The chapters that follow will detail both the challenges and the strategies that the Asian community undertook to overcome them.

NOTES

1. Martin Shanguhyia, "The State, Ecology, and Society in Western Kenya: Soil Erosion and Conservation in Vihiga," Ph.D. diss. West Virginia University, 2007, 2.
2. See N. Chazan and T. M. Shaw, "The Political Economy of Food in Africa," in N. Chazan and T. M. Shaw, eds., *Coping with Africa's Food Crisis* (Boulder: Lynne Reinner, 1988), 2.
3. Robert Gregory, *Indian and East Africa: A History of Race Relations within the British Empire, 1890–1939* (Oxford: Clarendon Press, 1977), 50–51.
4. Hugh Fearn, *An African Economy: A Study of the Economic Development of Nyanza Province of Kenya* 1903-1953 (London: Oxford University Press, 1961), 26.
5. For studies on Asian entrepreneurship and transport industry in Kenya, see J. S. Mangat, *A History of the Asians in East Africa 1886–1945* (Oxford: Clarendon Press, 1969); Robert Gregory, *South Asians in East Africa: An Economic Social History 1890–1980* (Boulder, CO: Westview Press, 1993).
6. Robert Maxon, *East Africa: An Introductory History* (Morgantown: West Virginia University Press, 1994), 204.
7. Agehananda Bharati, *Asians in East Africa: Jayhind and Uhuru* (Chicago: Nelson-Hill Company Publishers, 1972), 103.

8. Sir Charles Eliot, "The East Africa Protectorate," as quoted in Mehta P. Bhatt, "A History of Asians in Kenya, 1900–1970," Ph.D. diss., Howard University, 1976, 54.

9. J. L. Sadie, "Social Anthropology of Economic Underdevelopment," in D. E. Novak and R. Lekachman, eds., *Development and Society: The Dynamics of Social Change* (New York: St. Martin's Press, 1964), as quoted by Gerishon Ikiara, "Entrepreneurship, Industrialization and the Nation Bourgeoisie in Africa," in Ulf Himmelstrand, Kabiru Kinyanjui and Edward Mburugu, eds., *African Perspectives on Development: Controversies, Dilemma and Openings* (Nairobi: East African Educational Publishers, 1994), 124.

10. B. Higgins, *Economic Development: Principles, Problems and Policies* (New York: W.W. Norton and Co., 1968), as cited in Ikiara "Entrepreneurship, Industrialization," 125.

11. Polly Hill, *The Migrant Cocoa Farmers of South Ghana: A Study in Rural Capitalism* (Oxford: International African Institute, 1997), 3.

12. Gregory, *South Asians in East Africa*, 1.

13. Mangat, *A History of the Asians in East Africa*, 1.

14. Maxon, *East Africa*, 37.

15. George Delf, *Asians in East Africa* (London: Oxford University Press, 1963), 3.

16. Gregory, *South Asians in East Africa*, 10.

17. L. W. Hollingsworth, *The Asians of East Africa*, (London: Macmillan and Company, Ltd., 1960), 183.

18. Mangat, *A History of the Asians in East Africa*, 3.

19. Gregory, *India and East Africa*, 19.

20. Gregory, *India and East Africa*. 35.

21. Gregory, *India and East Africa*, 37.

22. Haraprasad Chattopadyhyaya, *Indians in Africa: A Social Economic-Study* (Calcutta: Bookland Private, Ltd., 1970), 47–48.

23. H. Moyse Bartlett, *The King's African Rifles* (Aldershot: Gale and Polden, 1956), 96.

24. Chapttopadhyaya, *Indians in Africa*, 354.

25. Gregory, *Indians and East Africa*, 47.

26. Delf, *Asians in East Africa*, 11.

27. Delf, *Asians in East Africa*, 49.

28. Gregory, *India and East Africa*, 49.

29. Gregory, *India and East Africa*, 52.

30. Gregory, *South Asians in East Africa*, 10.

31. Gregory, *India and East Africa*, 52.

32. Mangat, *A History of the Asians in East Africa*, 39.

33. Gregory, *South Asians in East Africa*, 12.

34. Mangat, *A History of the Asians in East Africa*, 40.

35. Gregory, *South Asians in East Africa*, 13.

36. Hollingsworth, *The Asians of East Africa*. 3.

37. E. S. Atieno Odhiambo, "Economic Mobilization and Political Leadership: Oginga Odinga and the Luo Thrift Trading Corporation," in Aloo Ojuka and William Ochieng', eds., *Politics and Leadership in Africa* (Nairobi: East African Literature Bureau, 1974), 143.

38. L. D. Smith, "An Overview of Agricultural Development Policy," in Judith Heyer et al., eds., *Agricultural Development in Kenya: An Economic Assessment* (Nairobi: Oxford University Press, 1976), 124; see also Judith Heyer and J. K. Waweru, "The Development of the Small Farm Areas," in Judith Heyer et al., 191–194.

39. Richard T. McCormack, *Asians in Kenya: Conflict and Politics* (Brooklyn: Theo. Gaus' Sons, Inc., 1971), 13.

40. Delf, *Asians in East Africa*, 6.

41. Chattopadhyaya, *Indians in Africa*, 350.

42. Robert Gregory, *Quest For Equality: Asian Politics in East Africa 1900–1967* (New Delhi: Orient Longman, Ltd., 1993), 16.

43. Gregory, *Quest For Equality*, 16.

44. McCormack, *Asians in Kenya*, 2.
45. Dharam P. Ghai, ed., *Portrait of a Minority: Asians in East Africa* (London: Oxford University Press, 1970), 40.
46. Chattopadhyaya, *Indians in Africa*, 351.
47. Chattopadhyaya, *Indians in Africa*, 351.
48. Chattopadhyaya, *Indians in Africa*, 69.
49. Chattopadhyaya, *Indians in Africa*, 70.
50. Chattopadhyaya, *Indians in Africa*, 71.
51. Delf, *Asians in East Africa*, 9.
52. Gregory, *Quest for Equality*, 18.
53. Bharati, *Asians in East Africa: Jayhind and Uhuru*, 19.
54. Dana April Seidenberg, "The Asians and Uhuru: The Role of a Minority Community in Kenya Politics 1939–1969," Ph.D. diss. Syracuse University, 1979, 7.
55. Bharati, *Asians in East Africa: Jayhind and Uhuru*, 22.
56. Gregory, *Quest for Equality*, 19.
57. Mangat, *A History of the Asians in East Africa*, 142.
58. For more explanation on these concepts, see William Petersen, "A General Typology at Migration," *American Sociological Review* 23 (1958), 256–266.
59. Seidenberg, "The Asians and Uhuru," 1. Gregory discusses the economic motivation that brought Asians to East Africa; he contends that the Shahs, Patels and Ismailis left their farming occupation to do business in East Africa. See Gregory, *South Asians in East Africa*, 22.
60. Bhatt, "A History of Asians in Kenya," 44.
61. KNA: PUB.1/1/1/6, Correspondence of Department of Revenue and Agriculture, Government of India, 4th of August, 1896, "Emigration Records and Correspondence 1856–1900." See also Seidenberg, "The Asians and Uhuru," 6–7.
62. Sidney T. Kelson as quoted in Seidenberg, "The Asians and Uhuru," 21.
63. John I. Zarwan, "Asian Businessmen in Kenya during the Twentieth Century: A Case Study," Ph.D. diss., Yale University, 1977, viii. The author discusses in depth about Asian traders in Kenya.
64. Zarwan, "Indian Businessmen," 49.
65. Zarwan, "Indian Businessmen," 7.
66. Ghai, ed., *Portrait of a Minority*, 18. Seidenberg also discusses Asian artisans who engaged in skilled and semi-skilled occupations. See Seidenburg, "The Asians and Uhuru," 8.
67. Zarwan, "Indian Businessmen in Kenya," 47.
68. Bhatt, "A History of Asians in Kenya," 44.
69. Gregory, *Quest for Equality*, 6.
70. Mangat, *A History of the Asians in East Africa*, 74.
71. Seidenberg, "The Asians and Uhuru," viii.
72. Zarwan, "Indian Businessmen", 58.
73. Mangat, *A History of the Asians in East Africa*, 14. For more environmental reasons see McCormack, *Asians in Kenya: Conflict and Politics*, 1. Hollingsworth, *The Asians of East Africa*, 6; and Ghai, ed., *Portrait of a Minority*, 6.
74. Gregory, *Quest for Equality*, 6.
75. Zarwan, "Asian Businessmen," 57.
76. McCormack, *Asians in Kenya: Conflicts and Politics*, 19.
77. Hollingsworth, *The Asians of East Africa*, 6. For further discussion on climatic effects see also McCormack, *Asians in Kenya: Conflict and Politics*, 1.
78. Seidenberg, "The Asian and Uhuru," 29.
79. Seidenberg, "The Asians and Uhuru," 30.
80. Bhatt, "A History of Asians in Kenya," 43. On population pressure, see also Zarwan, "Asian Businessmen," 30. Zarwan indicated that population growth led to land pressure which brought about changing economic patterns due to inefficient production.
81. Gregory, *Quest for Equality*, 5.

82. Gregory, *South Asians in East Africa,* 17.
83. Bhatt, "A History of Asians in Kenya," 6.
84. Bhatt, "A History of Asians in Kenya," 44.
85. Bhatt, "A History of Asians in Kenya," 51.
86. Gregory, *Quest for Equality,* 19. See also McCormack, *Asians in Kenya: Conflict and Politics,* 2–3.
87. Bhatt, "A History of Asians in Kenya," 47.
88. Delf, *Asians in East Africa,* 39.
89. Delf, *Asians in East Africa,* 39.
90. Hollingsworth, *Asians of East Africa,* 39.
91. Chattopadhyaya, *Indians in Africa,* 354.
92. Chottopadhyaya, *Indians in Africa,* 354.
93. Bhatt, "A History of Asians in Kenya," 46.
94. Gregory, *Quest for Equality,* 8.
95. Gregory, *Quest for Equality,* 7.
96. Gregory, *Quest for Equality,* 7.
97. Gregory, *Quest for Equality,* 8.
98. Zarwan, "Indian Businessmen,"52.
99. Zarwan, "Indian Businessmen," 75.
100. Bhatt, "A History of Asians in Kenya," 46–51.
101. Gregory, *Quest for Equality,* 8.
102. Gregory, *Quest for Equality,* 52.
103. McCormack, *Asians in Kenya: Conflict and Politics,* 19.
104. Gregory, *South East Asia,* 78.
105. Gregory, *Quest for Equality,* 8.
106. Bhatt, "A History of Asians in Kenya," 71.
107. Charles Bennett, "Persistence Amid Adversity: The Growth of Spatial Distribution of Asian Population in Kenya 1902–1963," PhD. diss., Syracuse University, 1976," 93.
108. Bennet, "Persistence Amid Adversity, 93.
109. Zarwan, "Indian Businessmen," 68.
110. KNA: AK/11/50, Sugar Growing Areas.
111. KNA: PC/NZA 3/22/1/2, Kibos Indian Settlement 1910–1912.
112. Joshia O. Osamba, "A Forgotten Minority: The Case of Asian Sugar Plantation Owners in Nyanza Province, Kenya," paper presented to the Historical Association of Kenya Annual Conference, Moi University Eldoret, Kenya, July 1998, 3.
113. KNA: AGR/6/4308, Miwani Farms, May 7, 1910.

TWO

Asian Penetration, Land Politics, and the Genesis of Asian Sugarcane Agriculture

This chapter illustrates how the Asian community in Kenya initiated sugarcane farming, responded to the challenges that farming in the colonial state entailed, and expanded their entrepreneurial reach to other types of economic activity as well.

First, the chapter reviews the widespread economic endeavors that Asian immigrants participated in, including raising livestock such as cattle and pigs; experimenting with crops such as maize and rubber; and engaging in trade. Many were shopkeepers, running both retail and wholesale businesses, while others were engaged in transport entrepreneurship. In transport they were competent and managed to compete effectively with the Europeans in the same business.

Second, the chapter introduces the key issues of land and labor that so deeply influenced the Asian sugarcane growing experiences, showing how European settler politics and the struggle for land between the various races in Kenya affected the onset of Asian farming. It uses the colonial state's land policies as a background of the struggle for land not only between the Asians and the European settlers but also between Asians and the Africans (Luos).

Finally, the chapter discusses the Asian penetration of the Nyanza basin and the establishment of Asian sugarcane farming, specifically in the nine areas, despite the discrimination that they faced from the colonial state.

ECONOMIC ACTIVITIES OF ASIAN IMMIGRANTS

Asians who settled in colonial Kenya experimented with various economic activities. Many worked for the colonial administration, most of them worked as surveyors, clerks, masons, and draughts men. Some of the laborers recruited from India also served as administrators in the protectorate, in the public works department, road construction, post office, and customs office. Those Asians who worked in the civil service, like the colonized Kenyans who worked there, did not find it easy. Kenya used the three-tier system during the colonial period and remained essentially racial in its appointments, with Europeans occupying the highest-paid jobs, claiming that Europeans had a higher cost of living in East Africa than Asians. Asians were found generally in the middle-level positions, a situation where they were later challenged by emerging elite Africans. Furthermore, Asians were discriminated in the civil service and excluded from higher posts, which were reserved for the British on the pretext that it was the British who could transmit the tenets of Western civilization to Africans and that Asian officers would not be accepted by Africans.[1] Despite such discrimination, the Asians accepted public service employment in large numbers throughout the colonial period. For instance, in the early 1900s while the East Africa Protectorate (Kenya) had excess clerical workers across the border in German East Africa (Tanganyika), there was an acute shortage.

However, the vast majority of the Asians who settled in Kenya became involved in business starting with small *dukas* (shops) along the railway line and major roads. In towns they opened their stores within settlements which were known as "bazaars" in urban centers like Mombasa, Machakos, Kisumu, Yala, Kisii, Mumias, and Nairobi among others. Their ability to live on a small portion of their profits and their tendency to shun leisure until they were well established in their business led to success. Asians were resilient and were able to expand their businesses because they were able to penetrate the very remote areas, unlike Europeans. However, they experienced great difficulties such as poor transport due to the bad conditions of roads especially during the rainy season. They also faced other challenges such as hostility from local communities, disease, and theft.

Nonetheless, their success was due to industry and business acumen. For example, those who were self-employed worked for seven days a week from dawn to dusk and were content with small turnovers. Among the successful ones were A. M. Jeevanjee, Allidina Visram, Gulamali Meghji, T. K. Meghji, Nanji Kalidas Mehta, Hasham Jamal, and Muljibhai Madhvani among others. While Asians were involved in different business activities, some of these early traders were supported by the colonial government. For example, Adamjee Alibhoy was invited to Machakos by the provincial commissioner for Ukamba Province, John Ainsworth, to set up a business in 1898 to help open up the province. Others set up businesses on their own since the colonial administration supported their business activities. For example, Gulam Hussein Abdulla Datto established an auctioning firm in Nairobi as well as a successful retail shop, while Deroda Shamji Harji and A. M. Jeevanjee established a successful construction company, building most government offices, post offices, and railway stations along the line from Mombasa to Kisumu.[2]

The success of Asian entrepreneurship attracted many others to migrate to the colony as their success stories reached back home to India. The colonial administration also facilitated this by making the Indian rupee the official currency for the colony in 1898. Moreover, the opening of the National Bank of India as the official bank of the protectorate was a boost to their economic endeavors.[3] Hence, when Sir Edward C. Buck, secretary of the Department of Revenue and Agriculture in the Indian government, visited the protectorate in 1905, he acknowledged the fact that the colony was suitable for tropical products and that the Asian experience could be put to use.[4]

Whereas many were involved in retail and wholesale firms, a few specialized in one type of commodity, mainly hardware or provision of textiles; the majority had a variety of stock such as cattle feed, blankets, tea, *merikani* cloth, beads, hoes, *pangas* (machetes), knives, sweets, and so on, all in one shop. As trade expanded, others became involved in the import and export trade. For example, Karimjee Jeevanjee became involved in the supply of petroleum to East Africa.[5] Many Asians were involved in export of produce and trade despite the fact that they were also perceived negatively as being dishonest. They boosted trade by bringing to the local people new consumer goods, such as bicycles and soap, which were exchanged for the locally grown produce, such as millet. The Asian immigrants were resilient and grabbed any opportunity to excel; whenever there was a need they filled it up. For example, in 1902, a Goan named Augustino De Figueiredo, having seen the demand for clothing, recruited some tailors from Goa and began a tailoring enterprise.

Asians also engaged in processing activities, especially the processing of agricultural products. Some engaged in activities like oil milling, ginning of raw cotton, and extracting tannin from wattle bark. For example, by 1907 Allidina Visram had established a factory in Mombasa that made soap and oil from coconut; this was followed in 1910 by his establishment of four cotton gins and six mills for manufacturing sesame at Mombasa. Others were involved in sawing of timber, which was in high demand as towns were built. Others quarried stone, which was in high demand in the construction of many government buildings and residential quarters. Some were involved in refining of soda ash and making of soft drinks. Others milled maize flour or processed settler produce like wheat flour or were involved in the transportation of the coffee beans and drying of sisal.

Asians were involved in nearly all types of business transactions. They were involved in money-lending, sold sand and charcoal, and ran barbershops. They dominated the manufacturing industry. For example, Shapurji Sidhwa was an engineer who started his own company making mineral water as early as 1907.[6] However, their efforts were later curtailed by big multinational companies like Brooke Bond, which was established by the help of the colonial state and was involved in processing and supplying of tea in Kenya. Most Asian industries did not get sufficient help from the state, so they remained small until the 1950s. However, their economic contributions were immense.

Asians were also involved in the transport business. By 1920, Asians had moved into road transport, a field initially dominated by Europeans. In transport system the Asians started modestly. For example, they used reconditioned cars, lorries, and motorcycles in the conveyance of goods. These were mainly military vehicles that had been converted into civilian use. They bought the vehicles from the government and reconditioned them to suit their businesses. Apart from transporting goods and ferrying produce from the farms of African farmers, they also provided transportation for people. The Asians provided cheaper and more efficient means of transportation than the railway, especially in rural areas where the railway line could not reach. The Asians carried Africans to and from the labor areas and even reached the remotest part of the colony. This way they opened the remote areas to world trade and indirectly stimulated production of cash crops since the farmers felt that their produce could be collected and marketed.

Asians were innovative and posed a great challenge to the Europeans in the transport business. When at one point they were barred by the legislation from operating on specific roads, they devised alternate routes and put their lorries in areas not reached by many Europeans in the transport industry. Their business acumen led to the rapid development in road transport in Kenya by Asian businessmen such as Gulamali Meghji, Nanji Kalidas Mehta, and Noor Mohamed Walji, who dominated the Kakamega-Kisumu road and other routes. They not only revolutionized the transport system in Kenya but also challenged and rivaled the government and European transport industry.[7] Beginning with practically no capital, the Asians' diligence, perseverance, and incredible hard work enabled them to effectively compete and dislodge the Europeans from this sector. This was not only due to their ability to reach the remote areas but also due to the fare cuts that they enforced.

In short, the history of Asian business enterprise in Kenya during this period is a story of success and fortune making. Even though not all Asians made fortunes in Kenya, many did because of their self-discipline and sacrifice. "The typical Indian trader worked for long hours and operated on a very small profit margin. His success in business can be attributed to his frugal habits and propensity to save."[8] Many of them operated businesses deep in the interior and faced all sorts of difficulties such as poor transport, hostility from the local communities, wild animals, theft of merchandise, and diseases. However, many succeeded, such as A. M. Jeevanjee. There was practically no part of the country with a potential for business that the Asians did not venture into. In rural areas, they provided markets for farm produce such as maize, beans, and potatoes. In essence Asians played many different economic roles in Kenya, especially the Oshwals.[9]

Asians were also keen on farming. The Foreign Office initially was interested in pursuing Asian farming and allocated 1,000 pounds for initiating the scheme. Unfortunately, due to politics within the colony, only 308 pounds was provided to the pioneer settlers.[10] Initially, these farmers were given other assistance, such as oxen, seeds, and hoes, to start farming. The farmers were under the direction of Mr. D. D. Waller, the acting paymaster of King's African Rifles in the treasury department. Mr. Waller was relieved of his duties and given the responsibility of being in charge of the Kibos project. Under his direction, the project for supporting the farmers was implemented and the farmers began experimenting with many crops. They initially cultivated crops such as maize, rice, cotton, sesame (simsim), chiraco (green grams), linseed, red pepper, and ceara rubber. Rubber was grown by Indersigh and Munish Ram of plot no. 15 Kibos.[11] Asians also ventured into animal husbandry; they kept cattle and even asked the colonial government for commonage in free grazing areas. They also kept other animals. For instance, Agit Singh was the first Asian to have a pig farm at Kibos.

Besides this, they also excelled in agriculture. This was because of the experiences gained from the agricultural project that some were involved in. For example, they invested successfully in sisal plantations. Muljibhai Madhvani started sisal estates in Kenya that were very successful. Other Asians settled in Mazeras in the then-Ukamba Province and grew fruit. This agricultural prowess was because most of the Asians who migrated to Kenya were peasant farmers back home; most of the Shahs, Patels, Lohans, Jains, and Hindus of Gujarat and Kutch came from agricultural villages.

The economic contribution by the Asian community during the colonial period in East Africa was broad-based. Many of the activities of Asians in Kenya, Uganda, and Tanganyika were entrepreneurial, including wholesale and retail trading and transport industry. Asians took initiative at Kibos despite the lack of support from the colonial government.[12] Needless to say, the Asians excelled; some 70 percent of household plots in Kibos were cultivated, compared with less than 10 percent in the fertile highlands dominated by Europeans.

Asians also contributed to the growth of the sugar industry in East Africa, especially in Uganda and Kenya. These two countries consistently had enough sugar between them during the colonial period and did not need to import it. For example, Uganda had a consistent surplus of production over consumption between 1955 and 1963 ranging between 10,000 and 50,000 tons while Kenya had a persistent shortfall during the same period.[13] However, the economic policies put in place by the British colonial administration affected the East African countries in terms of distribution. It seems that in Uganda the Asian sugar industry did not face stiff competition from settlers and by the 1920s had large sugarcane plantations and processing plants established by Asian entrepreneurs such as N. K. Mehta and M. Madhvani. Moreover, sugarcane farming was prone to do well and had an advantage over cotton and coffee since it found an already expanding local market[14] and did not have to depend on exports to distant markets. This was due to the new diets adapted by Africans as new food crops were introduced in the colony such as tea and coffee. Be that as it may, in Kenya, the Asians initially were involved in trade and transport and did very little in agriculture

Asians were involved in minimal agriculture quite early; many railway workers from India (coolies) who had some experience in farming in India were encouraged to remain in East Africa as settlers. The initial numbers of the Asians involved in agriculture was small; about thirty Asians acquired land at Kibos. This was not because the Asians lacked interest in farming but rather because the land available to them was not suitable for agriculture compared with the white highlands. The European monopoly in settling the highlands was highly contested by the Asians and became a major part of what came to be known as "the Indian question." And so for three decades after the Asians started farming, despite lack of adequate support, the Kibos Indian settlement grew steadily and served as an example of a small scale Asian agriculture. This changed during the World War II period; the colony adopted the Swynnerton Plan of 1954, which impacted Kenya's agriculture. Although in Uganda the industry progressed rapidly, the same could not be said of Kenya because of so many issues that affected the cane farmers such as difficulties in getting land. The colonial land policies led to a constant struggle by Asian farmers as they tried to succeed in the industry within a hostile colonial state.

COLONIAL LAND POLICIES: STRUGGLE FOR LAND

As Asians sought to expand their agricultural activities, one constant source of frustration was the land policies of the colonial state. Land policies enacted by the colonial state were based on many assumptions and misconceptions. For example, not all land was free or not in use at all since there were buffer zones between communities and also some areas were left for pasture or left fallow to restore fertility, as most Kenyan communities practiced shifting cultivation. Moreover, by the 1890s most communities in Kenya had suffered from rinderpest and smallpox, which had reduced the population of both livestock and people. The colonial masters believed in an erroneous idea of empty land. This misconception was based on stories told by early travelers who described totally unoccupied lands, but these travelers did not understand African land usage.[15] This was one reason why the land issue became controversial during the colonial period.

With conquest and control of Kenya, the colonial administration started straight away to appropriate land. For instance, when the Foreign Office took over control of the colony from the IBEA Company on July 1, 1895, land grabbing took root. This was because in the same month, Sir Arthur Hardinge, the British consul general at Zanzibar, became commissioner of the East Africa Protectorate, and land was made cheap. For example, the commissioner could easily give certificates to people to hold and occupy land for a term not exceeding twenty-one years. Furthermore, we find that as early as 1897, land within the railway zone was appropriated under the Indian Land Acquisition Act of 1894.[16]

Moreover, land policies were constantly reviewed to accommodate the wishes of the colonists. When the Indian Land Acquisition Act was amended on December 29, 1897, a new set of regulations gave the commissioner of the East Africa Protectorate the power to sell freehold land within the sultan's domain.[17] The 1897 act was also designed to secure land for public works such as construction of rails and roads. Using Indian land laws as a basis, an order-in-council was consequently enacted the following year that put the railway zone under the commissioner's purview, giving him more power. What followed was extensive land alienation in 1903 with the coming of the South African white settlers to Kenya. The Indian Act was later modified and allowed the commissioner to grant a person a certificate authorizing him to occupy land for a term not exceeding ninety-nine years.[18] Claim over land was by proof of titles, and given the fact that the Africans by then had no titles, their land could be alienated. The Kibos-Muhoroni area was also affected by this regulation. Therefore, in Kenya, right from the onset of the colonial rule, land decisions were in the hands of the commissioners who could make injurious decisions not only to the African people but also the Asians.

Moreover, due to the policy of making the colony self-sufficient, early commissioners pressured the imperial government to allocate land to European settlers. For example, Sir Arthur Hardinge was not satisfied by the 1897 regulations, and so in October 1898 he forwarded a settler demand for freehold or the 999-year lease to the Foreign Office for perusal and consideration. These views were then incorporated in the 1901 order-in-council that gave the commissioner power to dispose of unoccupied land; this was designated as crown land. The 1901 order-in-council authorized the commissioner to sell land, grant leases, or dispose of land. The specific terms were laid down in the Crown Lands Ordinance of 1902.

The actual colonization of land by Europeans was facilitated by the new commissioner to the East Africa Protectorate, Sir Charles Eliot, who came to the protectorate when a definite land policy was badly needed. The colonial government felt the need to prevent land speculation, especially along the railway, but had to be careful not to lock up areas in the hands of few individuals. Consequently, the need to prevent speculation and the need for the colony to be self-sustaining proved difficult to balance. This paradox made it hard to have a specific land policy.[19] The settlers and concessionaires wanted freehold grants, but the Foreign Office was reluctant to allow speculation; yet it wanted to attract settler capital. Since the colonial government had erroneously believed that the state could depend on the settler economy, the state therefore worked toward making large acreages of arable land available to settlers. For instance, Eliot in 1902 reiterated that the 1897 regulations were unsuitable and that the settlers' demand for freehold titles was proper. He became convinced that the highlands were suitable for permanent European settlement, comparing them to New Zealand and certain parts of Australia. He was determined to introduce the necessary legislation. Using powers granted to him by the 1901 order-in-council, Eliot drew up a notice to permit the sale of agricultural land and the lease of pastoral land. The land issue became more complicated, but rather than face hostility from Eliot and the settlers, Lord Lansdowne, the secretary of state in the Foreign Office, approved the notice.

In 1903, Eliot drew up further rules for the lease and sale of pastoral land. For instance, leases up to 10,000 acres were to be rented at one anna per acre (an anna being an old Indian currency, sixteen of which equaled one rupee). These rules were published, but the Treasury in London was wary and urged caution to safeguard against speculation. Furthermore, the Colonial Office also urged the Foreign Office not to allow the sale of land to pastoralists. Thus, in December 1903, Lansdowne sent a dispatch to Eliot asking him to take the matter into consideration. However, Eliot ignored the advice and went ahead and made a controversial offer to the two South African white settlers, Robert Chamberlain and A. S. Flemmer, of freehold grants of 10,000 acres each.[20]

Eliot indeed championed the white colonization of land in the East Africa Protectorate. He wanted more settlers to come to the protectorate and was not content with limited publicity; so he printed a notice describing land regulations, the climate, and shipping facilities, and he asked the Foreign Office to distribute the copies in Britain and the colonies. Following this advertisement, applications were received from America, South Africa, Europe, Australia, and New Zealand by mail and even cable. Not everybody supported this process. For example, W. D. Ellis, an official of the Colonial Office, rejected Eliot's and the settlers' position and emphasized that the East Africa Protectorate should not be turned into another Natal.[21]

However, it was not until later in 1903 that the Foreign Office realized that alienation of land was becoming quite problematic, especially the alienation of Maasai land. Still the settlers continued to push for more grants and were agitated by the possibility of a Jewish settlement. Throughout the year 1904, the settlers were adamant and continued to pressure for more land. The influx of more South African settlers to the protectorate made matters worse. The settlers believed that if the protectorate was transferred from the Foreign Office to the Colonial Office they would gain, since they were of the view that the Colonial Office would be more sympathetic to their case. Be that as it may, there were also changes taking place in London that culminated into the protectorate being transferred, together with Uganda and British Somaliland, from the Foreign Office to the Colonial Office on April 1, 1905. Unfortunately the land problem could not be easily solved.[22]

The Colonial Office then took over responsibility for land policy, among other issues. The change brought no quick solution; it seemed that the Colonial Office was also to blame for the tensions in the East Africa Protectorate. This was because the Colonial Office introduced legislation based on the laws passed by the governments in Australia and New Zealand. These laws, which were designed to prevent the accumulation of land and enforce the subdivision of large estates, introduced a policy that was not successful in Kenya since it was not adhered to due to pressure from settlers. Confusion became a significant aspect of Colonial Office land policy because the office was against speculation as a matter of policy. Enforcement of the regulations proved difficult.[23] More problematic was the stance adopted by the settlers who opposed land proposals and the implementation of policies by the Colonial Office. The settlers were supported by Belfield, who opposed a land tax and rejected the ninety-nine-year term, demanding a longer term of 999 years.[24] Yet the settlers were against the same terms being extended to the Asian farmers of Kibos; this in itself made it difficult for the Asian cane farmers to get land even in least desirable areas.

While the Asians continued to struggle against the colonial state's policies, the major objective of the Colonial Office was to look for a workable land policy. The Colonial Office refused to give in on the question of land tax and anti-dummying. Hence, A. C. Tannahill, a land ranger in the East Africa Protectorate land department, was given the task of solving this issue. He reviewed the land policy in the protectorate and recommended 999-year leases to the delight of the settlers, with revision of rent every thirty-three years; he strongly recommended land allocation through the auction system. Tannahill believed that there should be no limit to the accumulation of land as long as the lands were developed.[25] But even this was not obvious; white settler land was not developed at the same rate as Asian farms were.

Moreover, there was not enough arable land in Kenya for everyone who wanted it. So the establishment of settler farms in the most fertile areas of the white highlands generated a lot of tension and competition between the settlers, Asians, and Africans. For example, while the Asians were excluded from the white highlands, the Africans were also denied land rights under the Western form of tenure; moreover, Africans were forced to vacate lands they occupied and were pushed to the reserves, which generated a lot of tension among the three races in Kenya.[26]

That being the case, an East Africa land committee was set up to review Tannahill's recommendation. The Colonial Office committee agreed that the lease years be reduced to ninety-nine years. The secretary of state accepted these recommendations and ordered Governor Sir Henry Belfield to prepare a new land ordinance. Belfield, a graduate from Oxford, was the governor of Kenya between 1912 and 1917. Although he was initially not a favorite of the settlers when appointed, he was later to become their chief advocate. For instance, he influenced the passing of the Crown Lands Ordinance of 1915.[27] Belfield did not want a new ordinance that would reverse the leases to ninety-nine years, arguing that the drafting of the new ordinance would take time; for he was in favor of settler agency.[28] Due to his desire to please the settlers, Governor Belfield insisted that it was not possible to prevent dummying and illegal transfers of land, and was supported by the settlers.[29] He attacked the Colonial Office land policy and put pressure on the Colonial Office to yield to his demands. It seems that Belfield was not influenced by the settlers on his arrival to the protectorate but rather had accepted settlers' views long before he left London.[30] This in itself was a paradox, given the fact that Belfield was chosen by the Colonial Office since the protectorate needed a compromise over the land policy. The Colonial Office, tired of wrangles with no solutions by the successive governors, was in a mood to compromise over land policy and had erroneously believed Belfield was the man for the job. Instead Belfield became the settlers' crusader.

Therefore, Belfield continued to make changes, and by August 1913 the last restrictions against land accumulations were removed. The Colonial Office was forced to accept the principal of an open market in land for Europeans. By February 8, 1914, a Crown Lands bill was completed, introduced to the Legislative Council (LegCo) and eventually sent to London. It was not all smooth sailing in London. The Colonial Office was forced to scrutinize the bill because of protests from some Liberal MPs. Edmund Harvey, a strong parliamentary champion of African land rights, objected to the bill's inclusion of the lands occupied by Africans, or reserves, in the Crown lands. Harvey felt this inclusion was theft since as Crown land it meant the colonial state could take the lands. He objected to this power being given to the colonial state.[31] The bill was also contested by the coast communities since it jeopardized earlier transactions made under Islamic law. However, Belfield ignored the protest and the bill was passed into law in 1915, taking advantage of the fragile situation due to the war. The Crown Lands Ordinance of 1915 thus favored Belfield's views and the demands of the settlers. The Colonial Office nullified its own policies partly because of the fear of antagonizing the settlers with the onset of World War I since their support was badly needed. Moreover, other issues were at play. The Colonial Office gave up the struggle to impose the workable land regulation due to the unwillingness of Belfield and the settlers to compromise and because of the governor's ill health.[32]

The Crown Lands Ordinance of 1915 was a complete departure from the earlier Colonial Office principles. For example, leases were to be for 999 years, and there was to be no restriction on transfers once the full purchase price had been paid unless the transfer was between different races, in which case the governor would veto the transfer. Moreover, the policy that barred accumulation was abandoned, and there was no restriction on bidding except for the fact that conditions of auction could state if non-Europeans were allowed to bid. The 1915 ordinance was a victory for the settlers and the colonial state officials who supported them; it demonstrated the inability of the Colonial Office to impose an unpopular policy.[33] This policy was biased against Asians since while Asians held leases of ninety-nine years, European settlers were granted 999-year leases. Moreover, the Asians were initially restricted as to the amount of land they could hold per grant. For example, they could not hold more than 100 acres per grant. This applied to those who settled before 1915.[34]

Despite this restriction for Asians on the grounds that there was not enough arable land, the Colonial Office still decided to put in place a soldier settlement scheme for war veterans. To achieve this, a land settlement commission was appointed by Belfield in 1917, chaired by the attorney general J. Barth. This commission finished its work late in 1918 but did not recommend the settling of Asians or Africans who also had contributed toward the war effort (for example, many Indian soldiers died at the battle of Tanga). The scheme was implemented by Governor Sir Edward Northey, who replaced Belfield in 1919. Northey set out the terms for eligibility for the scheme and clearly stated that they must be British subjects. He knew that this claim would be hotly contested by the Asians; hence, he categorically indicated that applicants must be of "purely European extraction engaged in active service [sic] present war."[35] This not only curtailed the Asian quest for land but also showed how the colonial state was hostile to Asian farmers.

Northey did not want the ex-Indian soldiers to be settled in the colony; he believed that the European farming was sufficient. Stressing this point, Northey insisted that European farming had "greater advantages to the protectorate itself and to the local natives than by importation from another portion of the empire of immigrants of another race."[36] Moreover, Northey rarely consulted the Colonial Office in the execution of the scheme. The Colonial Office, as early as 1919, warned him against taking land from the African reserves for the implementation of the scheme, regardless of the fact that the Land Settlement Commission had suggested that land be taken from the Kikuyu reserve. Despite the warning, Northey allowed land of approximately 100 square miles to be alienated from the Nandi reserve.

Therefore, the postwar period did not change the situation in regard to landholdings by Asians and Africans. Moreover, the land issue became complicated when in 1921 a high court judgment declared that traditional African land rights were nullified when Kenya became a colony in 1920. This rendered Africans tenants of the Crown on their own land.[37] However, the world economic situation forced the colonial state to make other economic changes in the colony. In 1922 the colonial state implemented the dual policy under Governor Sir Robert Corydon. The intentions of the policy were good since it encouraged the development of both white farmers in the highlands and Africans. The dual policy was meant to spur the economy so as to protect the colony from the economic problems caused by the depressed economy after the war. The Colonial Office supported the policy for the same reason. However, the policy ignored the Asian farmers who already had made tremendous strides in Kibos-Muhoroni. Although the dual policy was meant to support even the Africans, it favored European production. For example, the findings of the Ormsby-Gore Commission of 1924, chaired by Sir William Ormsby-Gore, reiterated this fact. Although the commission supported the dual policy, it was critical of the fact that little had been done to improve African agriculture. Most importantly, it drew attention to the fear that Africans had over the loss of their lands; after it heard the views presented to it by the Africans and the missionaries who demanded the protection of native lands, the Ormsby-Gore Commission championed this fact.[38]

Therefore, in March 1927, the secretary of state for the colonies, Leopold Amery, gave instructions to the new governor of Kenya, Sir Edward Grigg, to report on the progress of the reserves. He was to specifically recommend ways to secure them permanently, that is, to have the boundaries of the reserves fixed. The Africans' fear of losing their land led to a bill being introduced in the LegCo in 1928. The bill stated clearly that land taken from an African reserve for public use had to be replaced by land of equal value. Further, it stated that the local native council had to be consulted before any such alienation occurred. This bill was eventually passed into an ordinance, the Native Lands Trust Ordinance of 1930, amid settler opposition.[39] It is worth noting that when the bill was tabled in the LegCo, despite the ferocious uncompromising stand of the European elected members, it received unequivocal support from the elected Asian representatives though very few in the LegCo. Furthermore, leases for any land taken from the African reserves became limited to thirty-three years.[40] Things seemed to move in favor of Africans with the changes at the Colonial Office and takeover by Lord Passfield in 1929. The security of African lands seemed promising. Lord Passfield reiterated his support for security of African lands in his two papers of June 1930 that came to be known as the "Passfield Pledge," in which he promised to ensure security of the reserves as provided in the Native Lands Trust Ordinance of 1930.[41]

Nevertheless, the Colonial Office wanted the protectorate to be settled by "big men," colonists with sufficient capital to develop farming or plantation industries and to employ African labor in pursuit of their self-sustaining policy. Suffice it to say, the ideal colonists in this respect were white settlers like the Grogans and the Delameres, and so long as one had capital the Colonial Office was not concerned on how much land any one of these individuals or groups obtained.[42] This was indeed a contradiction of its earlier policy. The 1915 ordinance deprived the Africans of all land rights and disinherited them. At the same time, the colonial state prevented the Asians from acquiring suitable agricultural lands even though there were Asians who had enough money to buy land if they were allowed to. The hostility of European settlers and the colonial state made it difficult for Asian cane farmers to acquire suitable land. Their situation was made worse by lack of adequate representation in the LegCo to influence decisions on land issues.

The land struggle in Kenya involved not only Asians but also Africans whose land had been alienated. The land problem was so difficult that the Colonial Office felt that the appointment of a commission was the best way to solve the land problem in Kenya.[43] Critics of this argue that this was basically Britain's idea of not tackling a problem directly. Nevertheless, the Kenya Land Commission was set up in 1932, chaired by Sir William Morris Carter, a former chief justice in Uganda who was believed to have had firsthand knowledge of African land problems. The commission was entrusted with the problem of investigating land issues in Kenya, but the membership of the commission was heavily contested as some members were accused of being stakeholders on land; hence they were not impartial. Moreover, neither Asians nor Africans were represented on the commission. Africans were denied representation on the grounds that the colonial state feared that an African could not represent views of all Africans but only the interests of his tribe. Perhaps contrary to this, an African was also denied a place on the commission because he might defend his views well and successfully challenge the land policies.[44]

Although the Kenya Land Commission was constructed to investigate African land claims and solve the problem, it worked to the contrary because other terms of reference had been included in an attempt to appease the settlers. For example, the commission was to define the area generally known as the "highlands" in which persons of European descent were to have a privileged position according to the Devonshire White Paper of 1923. This meant the entrenchment of special European land rights. This not only was a curious task for the commission but it also interfered with the commission's duties in determining African rights.[45] It was therefore not surprising when the Kenya Land Commission submitted its final report on July 7, 1933, it dismissed African views as extravagant, unreliable, and biased. The commission acted in total disregard of the "Passfield Pledge," as discussed earlier. The policy on alienation was not adhered to fully. Despite the fact that the Kavirondo Taxpayers Welfare Association petitioned the commission to compensate the Luo, whose land had been alienated to Asians at Kibos, the Luo farmers got nothing.

The commission published its report in 1934, and its recommendations were biased since its main objective was not fairness to Asians or Africans. Rather it assisted the government in achieving their goal of supporting settler land accumulation by eliminating the basis for African land claims and grievances.[46] The Asians also were ignored. This explains why the Asians were relegated to the lowlands of Nyanza at Kibos. It appears that settler politics in Kenya, and in Britain's Colonial Office, impacted land decisions and policies within the colonial government. Not surprisingly, dissatisfaction among Asians and Africans continued.

Moreover, other pro-settler policies were enacted following the recommendations of the Kenya Land Commission. For instance, the Kenya highland order-in-council of 1939 came into effect on March 1, 1939. It entrenched European privileges in the white highlands and defined the boundaries of what was to be the "Native Lands" as those set out in the Native Lands Trust Ordinance. These lands were to be administered as per the provisions of the Native Land Trust Ordinance. However, the rights of an individual were undefined and not secure. These "Native Lands" included the various lands reserved for African usage and included temporary reserves, the native settlement area, and communal reserves.[47]

Following World War II, another land policy was enacted since the Department of Agriculture wanted a change in the land tenure system that was meant to revamp agriculture in Kenya. In its report of 1952, the department was in favor of the enclosure of land in African areas and the granting of individual rights of occupancy to land. The department saw that proper land use and development could only take place after enclosure and individual ownership was established, and it criticized African land tenure systems, arguing that the land could not be cultivated intensively under them. Therefore, a program of land tenure reform, entailing consolidation of fragments into large holdings, was emphasized.[48] This led to the implementation of the Swynnerton Plan of 1954, which advocated for the amalgamation of fragmented lands into economic farming units. The latter could facilitate the application of sound intensified farming.[49] The plan also emphasized the use of extension services to transform agriculture through provisions of field officers or agricultural experts to help the farmers. The plan also recommended security of African land tenure through land registration and acquisition of titles; this process created a market for land, which became a commodity for sale.

Furthermore, the colonial state's lack of competency in understanding the ecology of different areas was evident when it gave rights to the district commissioner (DC) to prevent people from moving from one location to another. For example, in the Kano plains near Kisumu, in places such as Ahero and parts of Kibos, the communities were used to practicing transhumance for a long time. This was a way of life shaped by environmental factors such as wet and dry seasons of the area. During the dry season, the people moved near the water points for the herds, while during the wet season they moved to the hills to avoid the overflowing of rivers (e.g., the Nyando) and flooding of the plains.[50] So to prevent the people from moving to higher grounds during floods or not allowing them to return after the floods was a problem. It was practices like this that made it difficult for the Luos to embrace the consolidation program as quickly as was expected.

Suffice it to say, colonial land policies were marked by confusion and were contradictory and insufficient. The policies in most cases ignored the Asian interest while at that same time they were not committed to safeguarding African land rights. The lack of a firm decision on the part of the Colonial Office and the arm twisting by settlers undermined any good intentions, leading to racially biased policies that generated a lot of tension among the three races, especially between Asian cane farmers and European settlers.

THE COMPETITION FOR LAND BETWEEN ASIANS AND EUROPEANS

Land in Kenya was an emotional issue and involved struggles by various groups. For instance, the Asians too wanted a share of good fertile land as British subjects and also due to their efforts in World War I. However, as early as 1903, Eliot instructed the land office that no grants except of small plots were to be made to Asians between the Machokos road and Fort Ternan.[51] This was the beginning of the official exclusiveness of the white highlands. This racial segregation of land grants had been Eliot's desire.

Moreover, European settlers began to arrive in the country after 1903, mainly from South Africa; these were mainly the British and the Boers who moved from Natal, and their numbers continued to increase in the following decades.[52] Eliot became the architect for exclusive white settlement of the highlands and tried all ways possible to relax land acquisition laws in order to encourage settlers from South Africa and Europe. He assured the settlers from South Africa that all parts of the highlands that were suitable for colonization would be preserved for Europeans. He made it clearly known that no non-white would be allocated land between Kibwezi and Fort Ternan. Toward this end, Eliot got active support of the settlers led by men like Lord Delamere and Col. Ewart S. Grogan, who themselves acquired a lot of land in the colony.

Eliot's successors, Sir Donald Stewart (1904–1906) and Sir James Hayes Sadler (1906–1909), also upheld the exclusive land policy on the racial basis. Their argument was that the highlands were too small to be allocated to any other race apart from the Europeans. In October 1904, Stewart appointed a land committee to address the issue, which was incidentally chaired by Lord Delamere. The committee erroneously recommended that there was plenty of land in the lowlands where the Asians could settle and that there was no reason for Asians to hold land in the white highlands.[53] Suffice it to say, the transfer of the protectorate from the Foreign Office to the Colonial Office in 1905 seemed to swing things in favor of the settlers. For example, the new Liberal secretary of state for the colonies, Lord Elgin, in a dispatch in July 1906 to the newly appointed Governor Sir James Sadler, asserted, "it would not be in accordance with the policy of His Majesty's Government to exclude any class of his subjects from holding land in any part of a British protectorate; but that, in view of the comparatively limited area in the protectorate suitable for European colonization, a reasonable discretion will be exercised in dealing with applications for land on the part of the natives of India and other non-Europeans."[54]

This policy was further reaffirmed in 1907 when Governor Sadler, at the instigation of Lord Elgin, established an advisory land board. Due to settler influence, the board discouraged any form of Asian settlement of the highlands. Elgin made this position clear in another dispatch to the governor in March 1908 when he stated that "it is not consonant with the views of His Majesty's Government to impose legal restrictions on any particular section of the community, but as a matter of administrative convenience, grants in the upland area should not be made to Indians."[55]

The policy of not issuing land to Asians in the white highlands was effectively sealed by the enactment of a new land act in December 1912 by Belfield. The act put restrictions on transfers of land in the protectorate to people who were not of European origin, obviously targeting first and foremost the Asians who seemed to have enough resources to buy the land, and also to some extent the Africans.[56] Asians opposed the new land regulation. Their main argument was that it was irrational for them to be denied land in the highlands because they had equal rights as subjects of the British empire. However, as the protectorate became increasingly identified as a white settler colony, the Asians faced other restrictions as well. The most notable one, which also affected their farming, was in regard to immigration. The Asians challenged such biased regulations arguing that as British subjects they should enjoy the right of protection against deportation from any British territory such as Kenya. Yet throughout the protectorate efforts were being made, especially by the settlers, to restrict Asian immigration. This could partly be attributed to the Asian quest for land, among other issues.

As early as 1907, the governor had alluded to this in a report which stated that "There is a tendency amongst the white settlers in the uplands to keep the Indian, not only out of the uplands, but out of the country altogether. The spirit is akin to that prevailing in Natal and elsewhere and is due to the fact that the white cannot compete in the least with the Indian shopkeeper for supplies of provisions and articles in the daily use or as a petty trader."[57] It seems that perhaps the white settlers opposed Asians partly due to fear of possible competition for land and also in entrepreneurship. Indeed, what the settlers did in Kenya was something they had learned to perfect from their experiences in South Africa. The white settlers, as was the case in South Africa, viewed the Asians as competing with them rather than as complimenting them by contributing to the development of the colony.

The Asians did not allow such ideas to be implemented without a fight for what they deemed were their rights as subjects of the British empire. Drawing support both locally and from the Asian diaspora, members of the Mombasa Indian Association, founded in 1900 by L. M. Savle, Allidina Visram, and the two brothers A. M. and T. M. Jeevanjee were able, on several occasions, to petition on emerging critical issues like the alienation of the highlands to the whites. For example, in early 1905, the association managed to successfully petition the acting commissioner, Frederick Jackson, who then pledged that no favoritism would characterize the development of the protectorate. However, their joy was short-lived as the administration appointed a commission to re-examine the land ordinances. The commission recommended that there was a very small area in the protectorate suitable for European settlement and as such the area could not accommodate both the Asians and Europeans. Therefore, the area should be reserved only for the white settlers.[58]

The Asian community responded to this stance by holding a meeting in April 1906 to demand equal rights. Reacting to the Asian initiative, settlers too held a meeting, in June 1906, and reaffirmed their position that the highlands were a preserve of the Europeans; and they passed a resolution that "This meeting pledges itself to offer the most determined opposition to the attempts of British Indians to be allowed to take up land in such highlands, and insists upon the Imperial Government reserving for European settlement the only part of the protectorate suited to Europeans, and to which they have the exclusive right by reason of the deficit in protectorate expenditure defrayed from contributions by the British taxpayer, and by reason of a great number of the white settlers having come into the country on the faith of its being reserved for Europeans."[59] This resolution was important as it helped point the direction the history of the protectorate was taking as the settlers seemed to be favored by the colonial authorities, although there was no clear-cut policy on the seizure of the highlands by the white settlers.[60]

Nevertheless, Asians became more vocal and organized in their fight for rights, which included land rights. In 1906, for example, the Indian Association of Nairobi was formed under the chairmanship of Allidina Visram, who had shifted his business base to Nairobi. The Nairobi association carried on the struggle. Later in the same year, led by A. M. Jeevanjee, the association sent a delegation to London where they hoped to solicit for support from the larger Asian diaspora community and organizations like the All-India Islamic League, which had its headquarters in London. This group also met with officials of the imperial government. This visit would become the first of a series led by Jeevanjee that helped to popularize the Asian cause, not only in Kenya but also in East Africa generally. It also was brought to the attention of major newspapers like the *Chronicle,* which carried the story of Asian struggle for land in Kenya.

Moreover, Asian protest continued afterward in various ways. The next notable development during this period was the formation of East Africa Indian National Congress (EAINC) on March 7, 1914. Its president was T. M. Jeevanjee. One of the most significant resolutions of the congress was the reiteration of the "fundamental principle right of the Indians lawfully settled or residing within the protectorate to complete and full equality of treatment before the law." EAINC pledged "to oppose by all means any attempt at encroachments upon its rights."[61] This stance was greatly challenged by the settlers and the colonial state.

When Northey came to the colony in 1919, he felt that the Asians should be excluded from the highlands. However, they could be allowed to acquire a small area in the lowlands. The Asians, on the other hand, refused to discuss the selection of land in the lowland regions, which could be made available to them; they also did not want to discuss details of demarcation of residential and commercial areas of townships and challenged the principle of segregation. The Asians felt disregarded, as the colony of India had contributed toward the war efforts in manpower, military equipment, and financial resources. For example, India sent soldiers and officers to fight overseas who numbered about 210,000. This group was composed of army corps who served in East Africa, Egypt,

Cameroon, Mauritius, and Somaliland.[62] In the East African campaign alone, 46,906 Indians served, of which 826 were officers, 33,633 other ranks, and 12,477 noncombatants who served as porters and servants; they also provided equipment and supplies valued at $1,485,000.[63] The Asians felt the need to be considered for land like others who contributed toward the war efforts. Although the Europeans in East African Protectorate (Kenya) played down the Asian role in the war, the Indians both in the colony and back home applauded the soldiers' role and asserted that "Indian Soldiers, Indian blood, and Indian resources saved East Africa for the British empire."[64] With a volatile situation at hand, the Asians, with the help of the India Office, tried to put pressure on the Colonial Office to come up with a policy that would be beneficial to them.

However, the Colonial Office could not dictate policy to the adamant colonial state whose policy was that the highlands would continue to be reserved for Europeans only and that neither grants nor transfer of land to Asians in the white highlands would be permitted.[65] This issue was strongly contested by Asians, and it was consequently moved to the Colonial Office. The issue was eventually put to rest by the Devonshire White Paper of 1923 in which the official policy of segregation of the white highlands was accepted, even though urban segregation was abandoned. The Asian farmers had no choice but to accept the settlement in Nyanza. In fact the settlement and starting of Asian farming in Kibos had coincided with the colonial state's takeover of the Eastern Province of Uganda and the eventual restructuring of western Kenya administratively. This shift of the boundary was due to political and economic factors that impacted Nyanza Province greatly and had a great effect on Asian settlement there.

THE COMPETITION FOR LAND BETWEEN ASIANS AND LUOS

It was not only the Asians and Europeans who were competing for access to good land in the area. The native Luo were terribly disadvantaged by the colonial state. Their troubles began with the completion of the Uganda Railway in 1901 that changed the administrative landscape of the Nyanza Province. Not only did it stimulate the economy but it also enhanced the administration of the colonial state since accessibility to the interior was now easy. Furthermore, it opened up the landlocked Ugandan protectorate. Despite this, the completion of the railway also created more problems, in that it left the state with a debt of £5.5 million.[66] Nyanza Province was initially known as Kisumu and had three administrative districts, namely Elgon, Kisumu, and Ugaya. The province experienced political restructuring when on April 1, 1902, Lord Lansdowne issued directives for the Eastern Province of Uganda to be retransferred to Kenya. This was to keep the railway under a single administration.[67] Moreover, there was a great push to make the railway pay partly for the monies spent in its construction and also to help attain self-sufficiency, which actually encouraged the colonial administration to allow Asian settlement at Kibos. Furthermore, for administrative purposes and effective management of the Kenya-Uganda railway, the east shore of Lake Victoria and most parts of the fertile Rift Valley highlands were incorporated within the East Africa Protectorate, until 1948.[68] This railway would play a crucial role in the development of the sugar industry in colonial Kenya. However, the settlement of Asians at Kibos led to a struggle for land with the indigenous people, the Luo of CN.

To understand the struggle for land between the Asians and the Africans, it is important to revisit the indigenous land systems in precolonial Kenya. The Luo occupied the lands that were alienated by the colonial government to the Asian farmers as a way to discourage them from pursuing the white highlands. This was an unfortunate situation because the colonial state put the interest of the indigenous people last; they were prepared to evacuate the Luos rather than share the exclusive white highlands with Asians. On the other hand, these lands in Nyanza were acquired by Luos through conquest during their migration and settlement during precolonial period. This gave them rights of occupancy. The tenure systems in precolonial Kenya varied from one area to another in accordance to the customary law of the tribe.[69] However, the land tenure system in precolonial Kenya has been viewed as backward and wasteful due to fragmentation, which resulted from the practice of rotation of subsistence crops during this period. Due to the fact that there was plenty of land, a farmer would open up a new piece of land when the other one was exhausted; leaving it fallow so that it could regain its nutrients and be used again after some time had passed. Also, new lands were opened up due to the risk aversion. When this process was repeated over time, it led to a situation whereby the farmer could usually end up with a number of small fragmented pieces of land scattered over several miles.[70] This situation made it very difficult to supervise these farms. Moreover, indigenous economic practices were quite dynamic and such a system of tenure therefore guaranteed unlimited access to land, which was the major means of production. Generally land was communally owned in most parts of Kenya, Nyanza being no exception, making it not easy to dispose of or sell.

These lands had been acquired through rights of first occupancy after clearing or by conquest or initial hunting rights.[71] The custodian of the land was often the individual clan, which was comprised of a close-knit kinship group. Therefore, rights to land were defined by membership of a kinship group. This system put in place measures that guarded against landlessness by making it difficult for a person to trade it without the consent of the kin. Land was seen as a form of wealth belonging to the entire clan, and as such everyone was expected to protect it. Land was

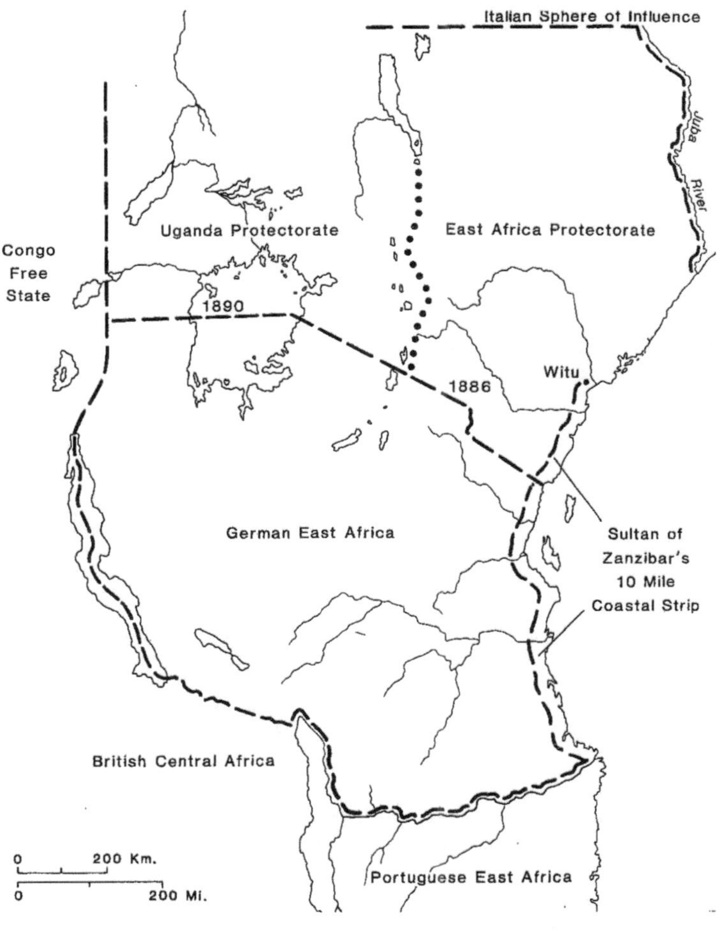

Figure 2.1. Partition of East Africa to 1895 Adopted from Maxon, *East Africa: An Introductory History* (2nd Edition), 132. Used with permission.

perceived as a form of communal wealth of a particular tribe or clan and had to be safeguarded.[72]

The Luo, being a patrilineal society, ensured that a person could gain access to land through inheritance of ancestral land or by clearing virgin lands. Inheritance was based on communal lineage rights and not individual rights. Sons inherited the father's land. The amount of land a man expected to inherit depended mainly on the size of the area occupied by the clan and family. However, with the growth of the population and the continual process of subdivision of the land through inheritance, the holdings became increasingly insufficient to meet the needs of those who wanted to farm land. Clans also had tenants-at-will who obtained the use

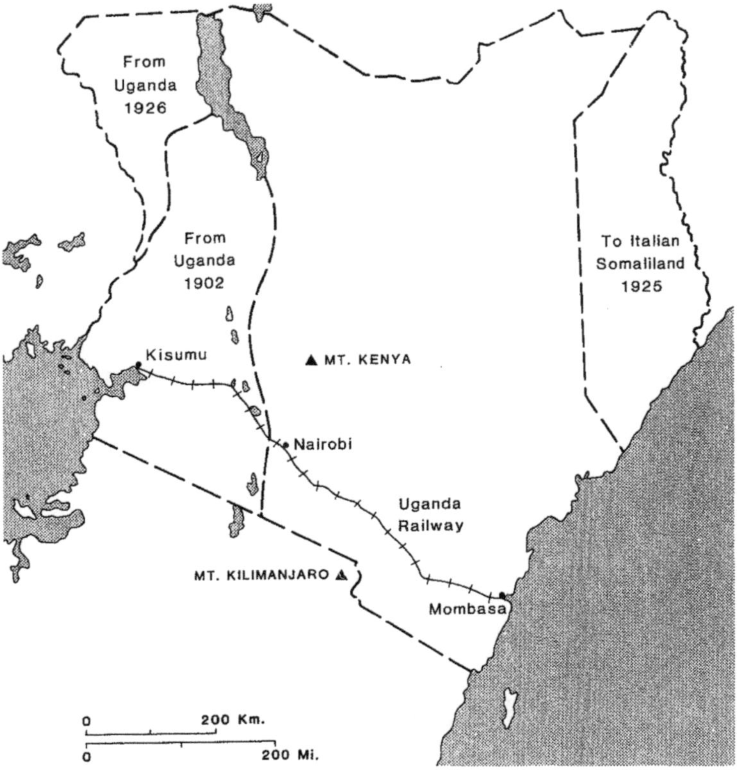

Figure 2.2. Colonial Kenya Adopted from Maxon, *East Africa: An Introductory History* (2nd Edition), 156. Used with permission.

of land through clan elders, after keen observation or on recommendation by a clan elder. In most cases the stranger had a distant relative in the specific host clan, as land was held by all. Therefore, land, an important factor in a traditional productive system, was governed by specific, but unwritten, rules. That being the case, it was not going to be easy for the Asian sugarcane farmers to get land from the indigenous Kenyan community who lived in the area where the colonial state was willing to settle the Asian farmers.

With the establishment of colonial rule, the entry of non-natives such as Europeans and Asians into Nyanza Province was of much concern to the native people. Land was alienated and the traditional agricultural system was interfered with. For example, the kinship mode of production slowly started to ebb away. A new concept of land ownership, with labor as a commodity, was introduced, which impacted heavily on the indigenous production. For instance, in Kibos the colonial state had alienated land for Asian settlement by 1903. During this period the people in the area had no means to contest the situation successfully due to the colonial regulations put in place that stated that each tribe could only claim land that it had occupied in precolonial times. If this could not be proven, then the land could be alienated without compensation. Moreover, by 1930, Nyanza experienced population growth, and so the Luos of Kibos-Muhoroni, through the Kavirondo Taxpayers Welfare Association, claimed that the land settled by the Asians belonged to them under the traditional system of land tenure, despite the fact that it had been alienated under the Crown Lands Ordinance of 1902. Therefore, in 1933 the Kavirondo Taxpayers Welfare Association submitted a memorandum during a hearing of the Kenya Land Commission that demanded compensation for the land alienated for Asian settlement.[73]

Unfortunately, the Kenya Land Commission did not make any recommendation for the compensation to the Luo community at Kibos. Yet the Luo claim over the land was justified since it had not been empty land. The area, being on the edge of Luo land, though not farmed, was used for grazing purposes and also acted as a buffer zone between them and the Nandi people. Moreover, by allocating this land to the Asians it prevented the two communities from having a natural area for expansion should their populations increase. Despite the fact that CN experienced very little alienation of land before 1930, Africans were opposed to any type of land alienation in the province. Yet Asians were attracted to the province not just for sugar growing but also due to the perceived opportunity it could offer in trading African produce. These opportunities encouraged many Asians to move to Nyanza Province.

Arable land was very limited in Kenya as a whole, and so it was not surprising when the Africans received very little help from the colonial state. This was due to the fact that the state had initially believed that the peasant economy could not make the colony self-sustaining. Moreover, there was dissatisfaction between the Asians and the Africans who felt that the colonial state was not helping them enough. On the other hand, the Asians were receiving preferential treatment, such as being given land and trading licenses. The racial strain was also as a result of what the Africans perceived as the monopoly of wholesale and retail trade in the province. The Africans also became frustrated at being outwitted by the Asians in trading. It was this feeling of letdown by the colonial state that led the Kavirondo Taxpayers Welfare Association to start cooperative self-help groups in the 1920s, particularly in grain milling, which competed effectively with the Asians in the processing of the local produce.[74]

Nevertheless, the Luos had no capital to rival the Asians in trade or in establishing the sugarcane farms. Moreover, mistrust existed between the Africans and the Asians. The latter were perceived as people who came to make money, accumulate it, and go back to their country. This dislike was evidenced when the community established the Luo Thrift Corporation partly as a rebellion against the Asian monopoly in trade in Nyanza and generally against the colonial state, which had ignored African entrepreneurship.[75] Due to such competition, the settlement of Asians at Kibos was not welcomed openly by the local community since by extension the Asian settlement led to loss of land that was a buffer zone and was used for grazing. Despite such setbacks, the Asians remained committed to agriculture and embarked on sugarcane farming.

THE BEGINNING OF ASIAN CANE FARMING

The start of Asian commercial sugarcane farming was not easy. Due to the policy put in place to exclude the Asians from the white highlands, the British fostered only a small-scale settlement of Asian agriculturalists in the colony. The area where the Asians were allowed to farm was Kibos, which is a railway stop in the lowlands north of Kisumu town, an area that was deemed unsuitable for European settlement.[76] Moreover, the government restricted ownership of land by Asians even in this area. While the European settlers could be granted thousands of acres, Asians were initially only limited to 100 acres per grant.[77] The few who settled at Kibos took to farming. This agricultural initiative began in earnest in 1903; the very first pioneers were twelve Asian laborers who got between 5 and 15 acres.[78] A few months later, they were joined by other ex-railway workers, and the number increased to thirty farmers. The sizes of the farms also increased, from between 50 to 100 acres. Most of these farmers were poor; hence, they were given seeds, farm tools, and a little financial assistance, which was to be repaid within a stipulated time. The first thirty pioneers who were granted land at Kibos-Miwani in 1903 acquired the farms under freehold or on a 99-year lease. The majority of the pioneer farmers were Sikhs (Jats), and about 60 percent of the farms they took up were on freehold, while the rest were leasehold of 99 years. By this time there was already a sizable population of Asians in the country.[79]

The thirty pioneers were helped to settle by Mr. D. D. Waller, as earlier discussed. He directed the surveying of land and the allotment of the farms. These early Asian farmers started by experimenting with various crops, as stated earlier. The pioneer Asian farmers saw the potential of cane farming and began to grow sugarcane intensively at Kibos in 1906. By 1910, many of the Asian farmers who had settled in the area were applying for more acres of land to carry out cane farming, although the governor had already ruled that no more than fifty acres should be allocated to Asians.[80]

However, the project suffered when, shortly after he initiated the promising scheme, Waller was transferred to Kisumu to pick up judicial duties; hence, he had limited time to supervise the scheme. This caused C. W. Hobley, assistant deputy commissioner in charge of Nyanza Province, to ask for a full-time supervisor; unfortunately, these requests were ignored. Although the Kibos settlement started from a modest beginning, sugarcane farming became popular as more Asians began to acquire land in the area. Over the next three decades, sugarcane farming at Kibos grew steadily, despite the lack of sufficient government support. This experience revealed the resilience and productivity of Asian small-scale farmers and the Asian agency in the Kenyan economy. The Asians demonstrated their efficiency in land use: out of the total 10,935 acres held, they cultivated 7,645 acres, approximately 70 percent of the holding.

This differed greatly from the European farms in the white highlands, which had only 5 percent of their land under cultivation.[81] This poor performance by the white settlers was due to the fact that many had never farmed before and were quite inexperienced, coupled with their reluctance to do manual work on their farms, which made them to be totally dependent on African labor.[82] These views were shared by Mr. V. Liversage, the colonial government's agricultural economist, who believed that the European farmers were grossly inefficient due to lack of knowledge of farming practices. Compared with the Asian cane farmers, the European farmers in the highlands were, before 1939, unable to use their land effectively; on average a settler farmer with about 2,600 acres of land was only able to bring about 269 acres under cultivation, a mere 10–12 percent of the total land.[83]

On the other hand, the number of Asian farmers growing cane at Kibos and Muhoroni increased by the mid-1940s. This settlement of Asians at Kibos would eventually lead to the introduction of commercial sugarcane farming in colonial Kenya. The first sugarcane was cultivated by Umar Din in Muhoroni. The crop soon became popular among other pioneer Asian settlers when they realized that sugarcane could thrive in this area. And so by 1906 most pioneer Asian settlers were cultivating the crop. From this modest beginning, the Asian agricultural settlement at Kibos expanded, as many Asians began to acquire more land in the Kibos-Muhoroni area. Most of the farmers by 1947 had bought their farms at government auctions or from previous holders and paid twenty cents per acre rent to the government.[84] Though the pace was slow, there were about eighty farmers whose farms were divided as indicated in the nine areas stipulated below and who owned a total of about 46,000 acres of land. These Kibos-Muhoroni Asian farms included holdings in the areas of Kibos, Muhoroni, Miwani, Chemelil, and Kibigori. The farms were divided categorically into nine areas as follows:

- **Area 1:** *Miwani Sugar Estate, which was owned by Devjibai K. Hindocha, Ltd.* This area had 3,615 acres of cultivated cane with a capacity of producing over 9,000 tons of white cane. The estate did not produce jaggery normally.

- **Area 2:** *The area was owned by nineteen Asian farmers and one European.* These farms were located about four miles west of the Miwani sugar factory. It covered approximately 2,500 acres, all of which was under cane cultivation. However, the area also produced jaggery and utilized up to fifty acres toward this course. Area 2 had a capacity to produce approximately 4,230 tons of white sugar.
- **Area 3:** *Asian Farms.* This area was situated at Kibos, west of the railway line. The area was owned by thirty-three farmers, and the estimate of the total acreage under cane production was 2,580. The area produced also jaggery and did not have an adequate plantation trolley line to transport cane for the production of white sugar. The area under cane cultivation had a capacity of producing 2,400 tons of jaggery.
- **Area 4:** *Miwani Estates: Asian Farms.* This area was settled by about fifty-six families of Asian farmers as early as 1903. The land was not so rich but was situated near the factory and better served by communication. The area had 1,300 acres of cane intended for the production of white sugar, approximately 1,300 tons.
- **Area 5:** *Kibigori Asian Farms.* This area had a total of twelve farms with a total acreage of 710, with 500 acres of cane under cultivation. It had a capacity of 600 tons of jaggery. The area had smaller farms and lighter soil on hilly land and was less suitable for cane.
- **Area 6:** *Chemelil Asian Farms.* The area consisted of nine Asian farms with a total of 510 acres of cane, but most was used to produce jaggery. These farms lacked any road communication with Kibigori or Muhoroni; hence, the railway was the only viable means of transport.
- **Area 7:** *Chemelil Estates—owned by Miwani Sugar Mills.* The area covered approximately 2,000 acres under cane cultivation. During the 1947–1948 period, this area produced 1,540 tons of white sugar out of 1,000 acres, and the other 1,000 acres was used in the production of 1,330 tons of jaggery.
- **Area 8:** *Muhoroni Asian Farms.* The area consisted of six Asian farmers and one Arab cane farmer. It had a total of 150 acres of cane and had a capacity to produce 1,080 tons of cane. All of it went to producing jaggery.
- **Area 9:** *Muhoroni Sugar Factory and Estate.* The farm was owned by several Asians who produced cane for white sugar.[85]

The Asian settlement at Kibos-Muhoroni shows not only that a substantial acreage of land was alienated but also that farms were well organized into areas. However, the area where the Asians settled had already been occupied by the Luo people before the advent of colonialism, especially the Kibos-Muhoroni area; and so the alienation of that land to the Asian farmers dislodged some of the Luo from their land.

CONCLUSION

The Asians who migrated to Kenya attempted a wide variety of business ventures, including wholesale and retail trading. Asian entrepreneurs penetrated the transport industry so successfully that they not only broke the European monopoly but dislodged them from the business by offering low transport rates and penetrating into the interior where the European buses and lorries had not managed to reach before. The Asians were able to do this by reconditioning used cars and using alternative routes that were not competitive in order to reach Africans in the remote reserves.

Those Asian settlers who focused on agriculture originally experimented with assortments of crops, such as maize and chili, and also the rearing of animals. They also were engaged in processing industries and made a variety of commodities such as tannin extracted from wattle trees and soap and oil processed from the coconut trees at Mombasa.

Most significant, however, was the farming of sugarcane. The Asian community initiated the sugarcane farming industry in Kenya in 1906, encouraged by the strong local demand. By 1910 farmers were seeking access to more land to bring under cultivation. But colonial land policies proved to be a major obstacle to their growth. Not only did the Asians have to buy land only in the Kibos region but they were severely limited in the number of acres they could purchase and were limited to ninety-nine-year leases rather than the 999-year leases offered to Europeans.

Colonial land policies like these caused decades of racial conflict in the area. European farmers were extended very favorable terms and support, despite the fact that the land they were granted was severely underutilized, sometimes having only 10 percent of the land under cultivation compared with up to 70 percent of Asian land. The Europeans often had little experience with farming and their farms performed poorly in contrast with the Asian farms as well.

But the racial tension between the Asian and white settlers was not the only matter of concern. As Britain encroached further into Kenyan territory, Kenyans lost land and subsequently people were pushed off their land into "reserves." Some Asians were allocated land that belonged to the indigenous Luo community, who protested the alienation of their land by the colonial state. The Asian settlement generated tension and land controversy among the indigenous community. The Luo community made efforts to have their voices heard but were ignored by the Kenya Land Commission and were not even compensated for land alienated to Asians.

In the midst of these tensions, the Asian farmers nonetheless established their farms in the Kibos-Muhoroni area in the 1900s. Their entrepreneurship, especially among the Sikhs, led to the establishment of the Asian agricultural industry in Nyanza and eventually the beginning of commercial sugarcane farming in Kenya, which were categorically divided into nine areas. Such an initiative was undertaken despite the hostility of the colonial state that denied Asians land in the fertile highland, which in essence showed the lack of political will by the colonial state to support the Asian agency in sugarcane farming in colonial Kenya and the wider colonial economy.

NOTES

1. Gregory, *South Asians in East Africa,* 189.
2. Mangat, *A History of the Asians in East Africa*, 55.
3. Gregory, *India and East Africa,* 67. Deroda Shamji Harji became a leading building contractor in Kisumu.
4. Bhatt, "A History of Asians in Kenya," 57–58.
5. Gregory, *South Asians in East Africa,* 55.
6. Gregory, *South Asians in East Africa,* 272.
7. Gregory, *South Asians in East Africa,* 125.
8. Bhatt, "A History of Asians in Kenya," 69.
9. Zarwan, "Indian Businessmen in Kenya," viii.
10. McCormack, *Asians in Kenya: Conflict and Politics*, 26.
11. Kenya National Archives (KNA): PC/NZA/3/22/1/5 Sugar General.
12. Gregory, *South Asians in East Africa,* 241.
13. Charles S. Frank, *The Sugar Industry in East Africa* (Nairobi: East African Publishing House, 1965), 2.
14. Maxon, *East Africa*, 192.
15. McGregor W. Ross, *Kenya From Within: A Short Political History* (London: Cass Publishers, 1968), 42.
16. M.P.K Sorrenson, *Origins of European Settlement in Kenya* (Nairobi: Oxford University Press, 1968), 49.
17. F. H. Okoth-Ogendo, "African Land Tenure Reform," in Heyer et al., *Agricultural Development*, 154.
18. Ross, *Kenya From Within*, 42.
19. Sorrenson, *Origins of European Settlement*, 44.
20. Sorrenson, *Origins of European Settlement,* 65.
21. Sorrenson, *Origins of European Settlement,* 85.
22. George Bennett, *Kenya: A Political History* (London: Oxford University Press, 1963), 18.
23. Sorrenson, *Origins of European Settlement*, 94.
24. Robert Maxon, *Struggle for Kenya The Loss and Reassertion of Imperial Initiative 1912–1923* (London: Associated University Presses, 1993), 56.
25. Sorrenson, *Origins of European Settlement*, 135.
26. Rita Breen, "The Politics of Land: Kenya Land Commission (1932–1933) and Its Effects on Land Policy in Kenya," Ph.D. diss., Michigan State University, 1976, 17.
27. Robert Maxon and T. Ofcansky, *Historical Dictionary of Kenya*, 2nd ed. (London: Scarecrow Press, 2000), 31.
28. Maxon, *Struggle for Kenya,* 58.
29. Dummying is the buying of land under someone else's name. In Kenya it was common practice for settlers to buy land using their children's names.

30. Sorrenson, *Origins of European Settlement*, 132.
31. Maxon, *Struggle for Kenya*, 62.
32. Maxon, *Struggle for Kenya*, 63.
33. Sorrenson, *Origins of European Settlement*, 141.
34. KNA: PC/NZA/3/22/2-16, Kibos Indian Settlement 1910–1916.
35. Maxon, *Struggle for Kenya*, 143.
36. Maxon, *Struggle for Kenya*, 162.
37. Bennett, *Kenya: A Political History*, 54.
38. Maxon, and Ofcansky, *Historical Dictionary*, 213.
39. Maxon and Ofcansky, *Historical Dictionary*, 191.
40. KNA: PC/NZA/1/25, Memorandum, Copy of Petition by Kavirondo Natives to the House of Commons.
41. Priscilla M. Shillaro, *A Failed Eldorado: Colonial Capitalism, Rural Industrialization, African Land Rights in Kenya and the Kakamega Gold Rush 1930–1952* (New York: University Press of America, 2008), 27.
42. Sorrenson, *Origins of European Settlement*, 147.
43. Breen, "The Politics of Land," 52.
44. Breen, " The Politics of Land," 60.
45. Breen, " The Politics of Land," 62.
46. Breen, " The Politics of Land," 113.
47. Ministry of Agriculture Animal Husbandry and Water Resources, *African Land Development in Kenya* (Nairobi: Government Printer, 1962), 233.
48. Colony and Protectorate of Kenya, *Department of Agriculture Annual Report 1956* (Nairobi: Government Printer, 1958), 2.
49. Roger J. M. Swynnerton, *A Plan to Intensify the Development of African Agriculture in Kenya* (Nairobi: Government Printer 1954), 9.
50. Maxon, *East Africa*, 59.
51. Bennett, *Kenya: A Political History*, 13.
52. Their numbers had reached 3,175 in 1911 from 506 in 1901. See Mangat, *A History of the Asians in East Africa*, 97.
53. Stewart to Lyttleton, August 14, 1905 as quoted by Gregory in *India and East Africa*, 76.
54. Gregory, *Indian and East Asian*, 77. Sadler had initially served as a commissioner of the East Africa Protectorate from 1905 until his appointment as governor.
55. Bennett, *Kenya: A Political History*, 24.
56. Bennett, *Kenya: A Political History*, 79, 93. See also Bhatt "A History of Asians in Kenya," 81.
57. As quoted in Bhatt, "A History of Asians in Kenya," 82.
58. Bhatt, "A History of Asians in Kenya," 79.
59. As quoted in Bhatt, "A History of Asians in Kenya," 80.
60. Sorrenson, *Origins of European Settlement*, 159–175. Sorrenson points out that there was no official statement on the alienation of the highlands until the Elgin Pledges of 1908.
61. KNA: PUB.1/1/1/6, East Africa Indian National Congress Papers-Resolutions of the Annual Meeting in 1915.
62. Gregory, *India and East Africa*, 146.
63. Gregory, *India and East Africa*, 155.
64. Gregory, *India and East Africa*, 152.
65. Bennett, *Kenya: A Political History*, 13.
66. C. C. Wrigley, "Kenya: The Patterns of Economic Life 1902–1945," in Vincent Harlow and E. M. Chilver, eds., *History of East Africa*, Vol. 1 (Oxford: Oxford University Press, 1965), 209–210.
67. Maxon and Ofcansky, *Historical Dictionary*, xvi.
68. Maxon and Ofcansky, *Historical Dictionary*, 186.
69. *African Land Development in Kenya*, 234.
70. *African Land Development*, 235.

71. Robert Maxon, *Going Their Separate Ways: Agrarian Transformation in Kenya 1930–1950* (London: Associated University Presses, 1993), 24. For more discussion on communal ownership of land in Kenya; see also Gunter Wagner, *The Bantu of Western Kenya with Special Reference to Vugusu and Logoli, Vol. II: Economic Life* (London: Oxford University Press, 1956), 75–84.
72. Norman Humphrey, *Liguru and the Land: Sociological Aspects of Some Agricultural Problems of North Kavirondo* (Nairobi: Government Printers, 1947), 19–20.
73. Fearn, *An African Economy*, 93.
74. Atieno Odhiambo, "Economic Mobilization and Political Leadership," 143.
75. Atieno Odhiambo, " Economic Mobilization and Political Leadership," 145–155.
76. Gregory, *South Asians in East Africa*, 240.
77. Ross, *Kenya From Within*, 314.
78. Osamba, "The Forgotten Minority," 3.
79. Sorenson, *Origins of European Settlement*, 37.
80. KNA: PC/NZA/3/22/1/9, Kibos Asian Settlement 1910–1912.
81. Gregory, *South Asians in East Africa*, 241.
82. Bruce Berman, *Control and Crisis in Colonial Kenya: The Dialectic of Domination* (London: James Currey, 1992), 133.
83. Berman, *Control and Crisis*, 134.
84. KNA: DC/CN/1/3/1, CKDAR, 1947.
85. KNA/ AK/11/50, Report by W. M. Lyne Watt on Sugar Growing Areas, August 1947.

THREE

Asian Sugarcane Farming in Colonial Nyanza

Asian settlers began farming in the Kibos-Muhoroni area of Nyanza Province in the first decade of the 1900s. This chapter discusses in detail how these farmers initiated the sugarcane industry in Kenya and how that industry grew before, during, and after World War I. It reviews the first landholders, the size of the early farms, and how license agreements and the location of farms hindered their attempts at growth. Of particular importance in this regard was the establishment of the first sugar mill in Kenya, the Victoria Nyanza Sugar Company. It examines the Asians' initiative in sugarcane growing over the years, particularly how their numbers, landholdings, and sugarcane farms increased in acreage alongside European settler farming in the area.

The chapter also addresses the technical assistance given to the farmers in the interwar period by the colonial state, after years of neglect. The agricultural offices carried out experimentation in cane varieties as well as farming techniques. These experiments were quite important and despite the problems that Asian farmers faced—such as environmental difficulties, lack of capital to buy machinery, and disease—the sugar industry was able to overcome the challenges by 1950 to become a vibrant industry in colonial Kenya. Of particular importance to its growth was the entrenchment of Asian capital in the manufacturing of sugar after World War II.

ASIAN AGENCY IN KENYAN ECONOMY: SUGARCANE FARMING

The beginning of Asian farming in Kenya was a gradual process. The pioneer growers started farming in earnest after European settlers had begun to cultivate in the white highlands. These European settlers acquired land much earlier than the Asians. By 1902 there were about thirty European settlers in Kenya, and they did not want the Asians to be given land. The opposition was not only based on racial prejudice but also fear of possible competition; the white settlers did not want Asians to benefit from possible land speculations.[1] Due to pressure from settlers, Asians had a difficult time getting land, while European settlers were offered generous grants by Eliot. Later, Eliot advocated for Asians to be settled at Kibos, as stated in the earlier chapter. Many of the pioneer Asian settlers of Kibos were ex-railway workers. After the completion of the railway, some were offered land along the railway line between Kibos and Muhoroni.[2] The land occupied, as discussed in the preceding chapter, was not empty; hence, some Luos were displaced with minimal compensation. About thirty-seven villages in the Kibos area and another sixty in Miwani were alienated to the Asians.[3] The number of these Asian farmers increased a little, from thirty farmers in 1903 to thirty-four in 1907, when four more Punjabi farmers were able to join the Kibos settlement after the visit by Waller to India in an effort to recruit more farmers. This was as a result of the recommendation by Sir Edward C. Buck, the secretary of the Department of Revenue and Agriculture in the government of India, who had visited the protectorate earlier in 1905. He advocated for more Asians to go to Kibos because he was impressed by the work of the pioneer settlers. The four Punjabis who were able to come from India acquired 600 acres.[4]

This trend of more colonization of the lowland continued. For example, in 1912, twenty-one more Asian farmers were able to acquire land in the Kibos-Muhoroni area through government auction near Muhoroni Township. Each obtained fifty acres; hence, about 1,051 acres was added to the Asian holdings. During the same period, another twenty-one Asian farmers rented farms at Muhoroni. All this attests to the fact that Asians were increasingly becoming interested in farming in the region. The result was that by 1913 there were approximately seventy-five farmers in the Kibos-Muhoroni area with a total acreage of 2,623.55.[5] This acreage was greatly increased when, during the same period, the London-based Asian company Messrs Imhazali & Sons purchased a large estate, the Kisumu Rubber Estate, at Kibos. The latter contained 6,122 acres of land. This Asian family struggled to grow rubber.[6] Eventually, the total acreage of Asian farms, small scale and large scale, amounted to approximately 8,734 acres in 1913.[7] From the initial number of twelve pioneers in 1903, by 1913 there were approximately thirty-five Asian farmers settled in the Kibos settlement scheme as indicated in table 3.1.

Table 3.1 shows that as early as 1913 there was one European, a George Matthew, among the cane farmers, who later sold his farm to Bresler and Starr. The table also shows that the terms of occupancy were not the same for all landholders. For example, apart from freehold and

Table 3.1. Kibos Indian Settlement and Land Acreage, 1913

Farm No.	Farmer	Acreage	Type of lease	Price/ Rent
1	Kharka Ram & Labu Ramm	39.40	Freehold	Rs. 200/=
1A	Kharka Ram & Labu Ramm	60.67	Freehold	
2	Vali Hasham & Co.	50.0	Freehold	Rs. 100/=
3	Chowdri Ram Jowaya	25.00	Freehold	Rs. 50/=
4	Mokanjee	50.00	Freehold	Rs. 100/=
5	George Mathew	45.5	Freehold	—
5A	George Mathew	58.0	Freehold	—
6	Jagat Singh & Surean Singh	54.9	Freehold	Rs. 210/=
6A	Jagat Singh & Surean Singh	50.9	Freehold	
7	Gujar Ram	35.2	Freehold	Rs. 70/=
7A	Malawa Ram Probdial	51.0	Freehold	Rs 100/=
8	Harriram Probdial & Co	50.0	Freehold	Rs. 200/=
8A	Harriram Probdial & Co	50.0	Freehold	—
9	Veharibhay C. Patel	50.0	Freehold	Rs. 200/=
9A	Veharibhay C. Patel	50.0	Freehold	
10	Pala Singh	30.84	Leasehold 99 yrs. from 1-1-09 -rent	Rs. 18/=
11	Jagat Singh & Surean Singh	88.2	Leasehold 99 yrs. from 1-6-10	Rs. 29/=
12	A.A. Visram & Jan Mohamed Giga	50.00	Freehold	Rs. 100/=
12A	A.A. Visram & Jan Mohamed Giga	50.00	Freehold	Rs. 100/=
13	Malawa Ram Probdial	24.50	Leasehold 2 yrs. from 5-2-11	Rs. 17/=
14	Gulam Moyud Din	50.00	Freehold	Rs. 100/=
15	Mohamed Din Alla Jawaya	25.00	Freehold	Rs. 50/=
16	Gujar Ram	50.00	Freehold	Rs. 50/=
17	Gujar Ram	50.00	Freehold	Rs. 50/=
18	Abdul Majid & Chetu	100.50	Freehold	Rs. 200/=
21	Dewa Singh & Sodager Singh	7.60	Leasehold 99 yrs. from 1-1-09	Rs. 15/=
22	Karaka Ram & Labu	7.0	Leasehold 99 yrs. from 1-1-09	Rs. 15/=

23	Koloo Ram	14.40	Leasehold 99 yrs. from 1-1-09	Rs. 15/=
24	Bhai Dan Devi	79.00	Leasehold 99 yrs. from 1-1-09	Rs. 27/=
26	Koloo Ram	12.0	Leasehold 99 yrs. 1-1-09	Rs. 15/=
27	Koloo Ram	36.4	Leasehold 99 yrs. from 1-1-09	Rs. 19/=
39	Jamna Dass Karamsi	54.4	Leasehold 99 yrs. from 1-6-10	Rs. 23/=
40	Gurma Singh & Buta Singh	50.0	Leasehold 99 yrs. from 1-6-10	Rs. 22/=
41	Gobind Ram & Koloo Ram	50.0	Leasehold 99 yrs. from 1-6-10	Rs. 22/=
42	Jagat Singh & Surean	44.0	Temporary occupancy for 2 yrs. from 5-2-11	Rs. 21/=
43	Inder Singh & Munshi Ram	50.0	Leasehold 99 yrs from 1-11-10	Rs. 22/=
44	Ladha Ram	50.0	Leasehold 99 yrs. from 1-6-10	Rs. 22/=
45	Sadhu Singh	50.0	Leasehold 99 yrs. from 1-6-10	Rs. 22/=
46	Malawa Ram Probdial	50.0	Leasehold 99 yrs. from 1-9-10	Rs. 22/=
47	Jagat Singh & Surean Singh	14.60	Freehold	Rs. 30/=
48	Messrs Imhazalhi & Sons (Kisumu Rubber Estate)	6,122.0	Leasehold 96 yrs. from 5-1-10	Rs. 920/=

Source: Kibos Indian settlement, KDAR 1912-13, KNA: DC/CN/1/5/1.

leasehold, there was also a temporary occupancy of two years in Kibos. This was an exception since this type of tenancy was mainly found at Muhoroni. Even the leaseholds were different. Whereas almost all farms were granted leases of ninety-nine years, the Kisumu Rubber Company was granted a lease for ninety-six years, starting in January 1910. The lease came with specific conditions. The company had to spend approximately 1,600 pounds for the first five years; 165 acres were expected to be cultivated in the first year, 330 acres in the second year, and 496 acres in the third year. Although this farm was engaged in growing rubber as the main crop, other crops were planted as well. For about ten years, the Kisumu Rubber Company experimented with rubber, but in vain.[8] Even taking into account some absent landowners, it is not clear why such stringent measures were part of the Asians leases and not the European leases, since many European farmers had large estates as well and were absentee landlords just like the Imhazalhi & Sons. Asians also owned other farms; these farms ranged between 100.5 acres to 7.0 acres. Not all Asians got fifty or more acres.

There were also farmers with multiple plots of land. For instance, Jagat Singh and Surean owned both plots number 42 and 47 in the same area while others owned several strips of land in different parts of the Asian settlement. For example, a farmer could have one strip in Kibos and another at Miwani with holdings varying from fifty acres to about eight hundred acres. A case in point is Maganbhai Patel, who owned land at Kibigori and Miwani. In these types of farms, some of the land was held on yearly or two years occupancy from speculators.[9] The farmers mostly lived on their farms and worked it. In some cases, a partnership was put in place whereby one partner worked the farm in Kenya while the other resided in India or did business in Nairobi or Kisumu. For instance, Dhanwant Singh, who owned a farm at Miwani, was also an attorney at Kisumu. Satbachan Singh was an inspector of police, while Maganbhai Patel, who had several farms as earlier stated, was also a straddler. He had a flourishing business in Kisumu as a moneylender.[10] Furthermore, some Asians settled at Muhoroni by 1912 but under very different conditions. The Muhoroni farmers were only initially granted farms on a two-year occupation license that began on August 10, 1912. These farmers were twenty-one in number, as indicated in table 3.2. It is not clear how and when, if at all, the licenses were renewed.

Table 3.2 indicates that even if one farmer had 48 acres and another farmer had 51 acres, a difference of three acres, both paid the same annual rent of Rs. 22. Rashid Khamis, who had 55 acres, paid Rs. 23. The tenure on this settlement was based on a temporary occupation of two years. Some of the farmers who had acquired land in Kibos also rented farms at Muhoroni, for example Allidina Visram. However, there was no large variation of acreage of the farms in Muhoroni as was found in the Kibos settlement.

The location of the Asian sugarcane farms and the few European farms was important to the growth of the industry. The farms were situated to the north in the Nandi District, stretching from the foot of the Nandi escarpment toward the Kenya-Uganda railway line. To the south, the farms bordered West Kano, Kisumu, and East Kano locations of CN and also Kericho District. These farms were situated conveniently to the main railway stations from Muhoroni to Kisumu. The location of the farms was very appropriate since the rail line facilitated transportation not only of cane from the farms to the factory but also other products such as fertilizers and farm tools as well as labor. The location of most of the cane farms near the railway line made it easier for farmers to move from one farm to the other and also facilitated supervision by the agricultural officers. The location of these farms near the Kenya-Uganda railway

Table 3.2. Muhoroni Indian Settlement Farms Granted on Two Years' Occupation License on August 10, 1912

Farm No.	Farmer	Acreage	Price/Rent
2	Fateh Chand	50.00	Rs. 22/=
3	Allidina Visram	51.6	Rs. 22/=
4	Hasham Jamal & Co.	48.1	Rs. 22/=
5	Hussein Abdul Rassul	51.3	Rs. 22/=
7	Rashid Khamis	55.00	Rs. 23/=
8	Mohamed Bux	50.00	Rs. 22/=
9	Alibhai Kassim	50.00	Rs. 22/=
10	Mohamed Majid	52.8	Rs. 22/=
11	Mohamed Kassim	51.2	Rs. 22/=
14	Sheik Noor Din	50.0	Rs. 22/=
15	K.M. Bhogaita Bros.	50.0	Rs. 22/=
16	Moladad Dolatakha	48.7	Rs. 22/=
17	Ava Dan	51.3	Rs. 22/=
18	Sultani Ram	50.0	Rs. 22/=
19	Abdul Wahid	51.3	Rs. 22/=
20	Gulam Mohamed	50.0	Rs. 22/=
21	Jalal Din	52.1	Rs. 22/=
Total		863.4	

Source: Muhoroni Indian settlement, KDAR 1912-13, KNA: DC/CN/1/5/1.

was supposed to have made the sugar industry economically viable because the colonial government was expected to accord the farmers the much-needed access to reasonably price rail freight transportation. The farms were strategically located as shown in figure 3.1.

Figure 3.1 shows how the farms that were within easy reach of the railway benefited from accessible transportation services. Moreover, the location of the farms near the railway was found to be very convenient during the harvesting period. Unfortunately, some of the farms were situated along the rail line but were far from the main factory at Miwani; they included farms in Chemelil and Muhoroni. Despite the availability of the railway, cane farming remained expensive because of transportation cost; the colonial state did not take advantage of the strategic location of the Asian sugarcane farms to boost the industry through negotiation with the Kenya-Uganda railway for subsidized transportation costs. The unavailability of low-cost rail freight transportation would continue to affect the sugarcane farmers for some time.

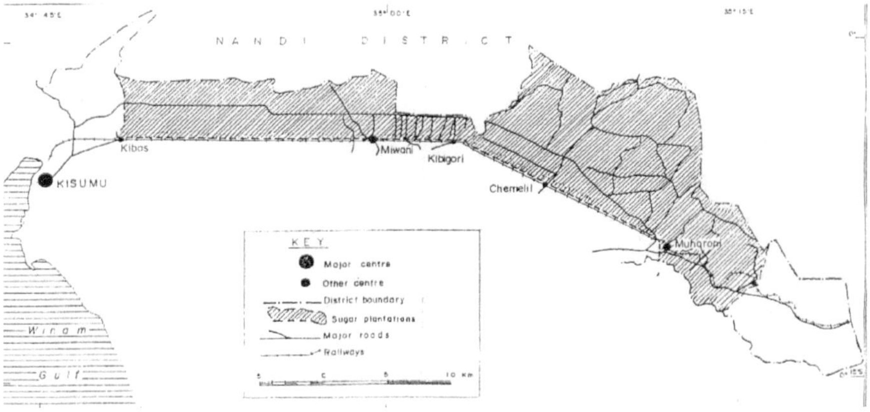

Figure 3.1. Location of Asian Sugar Plantations Kisumu, Survey of Kenya, 1979

ASIAN SUGARCANE FARMING IN COLONIAL NYANZA: THE PRE-WORLD WAR I PERIOD, 1903–1913

The Asian farmers by 1913 had started growing sugarcane, experimenting with a variety known locally as Uba, originally from Natal in South Africa. By the time the Asians started cane farming, Uba had long been planted in Kenya;[11] contrary to common belief, the Asians did not introduce sugarcane there. However, when the Asians started cultivation, they introduced other varieties and commercial farming not only in Kenya but also in Uganda, and through experimentation with different varieties they revolutionized cane farming in Kenya completely. Early on the farmers relied on any information and knowledge from India, and the colonial state encouraged the cane farmers. As early as 1913, Indian farmers in Kenya began to experiment with a sugar variety called CO, which was a variety imported from Coimbatore, a renowned sugar research station in India. There were many types, such as CO 221, CO 290, and CO 421. The imported seedlings of these varieties were distributed to the farmers who, apart from using trial and error, were assisted at this early stage by the office of the provincial director of agriculture, who not only encouraged them but was the driving force behind the early experimental trials on the Asian farms.[12] With such support, the Asians, who had the majority of farms, embarked on growing cane at both Kibos and Muhoroni.

There were a few European sugarcane farmers in the Kibos-Muhoroni area as well. Almost all were found at Muhoroni, starting as early as 1911, although later some settled at Miwani, Kibigori, and Chemelil. It is difficult to ascertain how many acres they had. The Europeans who settled in Muhoroni by 1911 were just as few as the Asian farmers in that area. For instance, whereas there were twenty-one Asian farmers in Muhoroni, there were approximately twenty-six occupants in the European farms as indicated in table 3.3.

Table 3.3 illustrates how European settlements at Muhoroni, unlike the Asian farms, had names; this was the custom for European farmers in Kenya. These were big estates owned either by individuals or companies. Unlike the Kibos-Muhoroni Asian farms, a majority of the owners were not the occupants who worked the land. For instance, out of the twenty-three European landowners, only five occupied their respective farms, the other eighteen occupants were not farm owners. This partly explains why the total acreage put under cultivation on European farms was far below that of the Asian farms. Moreover, the data shows clearly that whereas there were exclusive settlements for Europeans, when it came to Asian farms, some European settlers were found in areas designated for Asian settlement as was the case around the Miwani-Muhoroni area.[13] This was not applicable for Asians in the white highland; colonial state practiced double standards whenever it was deemed to benefit the European settlers. Be that as it may, other European farms were later purchased, such as the Drumeen Estate that was growing sugarcane and had its own jaggery factory. Also, Nottidge and Allen got a lot of land in Kisumu District, for example, land registration number L R No. 3976 comprising of 1,539 acres, L R No. 5977 of 2,339 acres, and another parcel of land, L R No. 3978 of 1,960 acres.[14] These European cane farmers would pose as competitors to the Asian cane farmers on issues dealing with labor recruitment as will be discussed later.

SUGARCANE FARMING DURING THE WORLD WAR I PERIOD, 1914 TO 1923

Generally the European settlers got land more easily than the Asian farmers. As mentioned earlier, the 1915 Crown Land Ordinance was a major boost in European farming in Nyanza as elsewhere. Due to the persistent pressure from the settlers, the outbreak of World War I in 1914 was a boon to them since the Colonial Office feared to antagonize them partly due to the German threat that was eminent from the neighboring territory of Tanganyika with the onset of the conflict. The settlers took advantage of the situation and had their way; and the land bill was passed.[15] This greatly affected the Asian farmers whose leases were not changed by the ordinance.

Despite this unfair treatment, Asian farmers' resilience helped them; forged ahead, and had planted 621 acres in sugarcane by 1916. During this period, the area under cane continued to increase and the cane that was produced was sold in the form of jaggery, which found a ready local market.[16] During this period there was minimal production probably due

Table 3.3. European Settlers at Muhoroni, 1911

Farm No.	Name of Farm	Name of Owner	Name of Occupier
642		Hale	Hale
449		I. Clerk	W.B. Saul
450	Pendet Estate	Capt. S.H. Charrington	Capt. S.H. Charrington, N.D. Allen, W. Wilmont Eardley
451	Koru Farm	C. Harvey	C. Harvey, J.P. Kotze
452	Fair Garden	Play Fair and Gardner	Mr. & Mrs. G. Hampson
453		Capt. O'Brien	Mr. & Mrs. J. Klopper
454	Soba Estate	Duirs & Wheelock	E. Brown
455	Mbogo Vale	P. Guy	E.E. Hutchins & C.P. Hutchins
456	Tinderet Estate	Seth Smith & Tetley	E. Tarlton
458		Smith & Faucus	H. Hannay
458A	Usone	M.A. Harbord	Mr. & Mrs. M.A. Harbord
459		Dunman	J.A. Randall
460	Maraba Estate	J. Mckinnon	H.M. Harries, Mr. & Mrs. Bezedenhout, Child & Brother-in-law
1461	Thousand Springs	Stocker	O.C. Harries
1462		F. Fletcher	E.A. Phelps
1463		E.A. Webb	A. Cook
1464	Red Hill	J.M. Penruddock	J.M. Penruddock
1465	Mberere	B.F. Webb	B.F. Webb & J. Mortishaw
1466		Wessells	Geldenheuse
1467	Chemalich Farm Songhor Estate Kepchera Estate	1. Montgomery	
1474	Kamassio Height	F. De Cergit	W. Cockburn
1475	Mount Seret	J.W. Wright	E.T. Behn
	Mtete Valley	F. S. Clarke	F.C. Burmeister

Source: List of European Residents, Muhoroni District, KDAR 1912-13, KNA: DC/CN/1/5/1.

to the lack of adequate farming knowledge in cane production as Asian farmers were still at an experimental stage with the crop. Their output was further affected due to the fact that there were no machines established for milling sugar. This led to minimal production since the cane that was produced was turned into jaggery (coarse sugar). The small machines that the farmers had were oxen-operated crushers, which consequently determined the acreage under cane production. Before the establishment of the Victoria Nyanza Sugar Company, the Asians mainly cultivated a small amount of sugarcane that was all turned into jaggery. They had to supplement their farming by growing other crops such as maize.[17]

On the other hand, European farming was also expanding as settlers continued to increase in number and acreage of landholdings, especially at the end of the war that saw the implementation of the soldier-settlement scheme in 1919. This scheme added 2.5 million acres to the settlers' holdings, part of which were found in Chemelil, Kibigori, Muhoroni, and Koru. The European farmers who were allocated farms under the scheme within the said areas grew wheat, coffee, sisal, and maize. The soldier-settlement scheme was another example of the contradictory nature of the colonial state policy. On the one hand, the state was supporting the Asian cane farmers. On the other hand, due to settler politics, Asians were restricted as to where they could obtain land and were not themselves included in the soldier-settlement scheme, despite the fact that many Asians fought in East Africa. The colonial administration was focused on giving the best land to European settlers only. For example, Governor Sir Edward Northey, in his address to the Convention of Association in August 1920, was reported in the newspaper to have said: "by the time this whole settlement scheme [the soldier-settlement scheme] has been readjusted, i.e. that farms which had not turned out to be what it was hoped they were would be replaced by something better, there should not be a farm left worth having."[18]

Such statements were criticized by the government of India. In a rejoinder, the Indian government sent a dispatch to the secretary of state for India criticizing the land policy. The communique noted that not only were the Asians locked out of the white highlands, but their access to land in the lowlands which the colonial state had initially rejected as being unsuitable for Europeans because of the bad climate, was limited because some land there had also been alienated to the Europeans.[19]

Even Europeans from other countries did not find it easy to get land in Kenya. This was seen when in 1919 the colonial state offered land for sale for the production of sugarcane through the official Gazette, General Notice number 920. While the land could have been offered to the Asians, it was open to anyone to bid on. Consequently, tenders were invited for the purchase of the lease of an area of 671.6 acres near the Kenya-Uganda railway between Kibigori and Kibos. The terms of the lease were ninety-nine years, effective September 1, 1920, with an annual rent of £44.15. This amount could only be paid at the office of the Crown Agents for the Colonies in London or at the land office in Nairobi. While the settlers protested against Asians being permitted to bid for land in that area, an Italian national bid and the application was declined. This led to uproar and a petition filed by A. C. Cavicchioni, the Italian consul, on behalf of the Italian government against the treatment of their nationals. Cavicchioni criticized the terms of the sale, which required one to be a British subject to bid for land. He specifically singled out the divergent treatment between British nationals and foreigners, arguing that it was in opposition to the principle in section 5 of the General and Final Act of the Congress of Berlin of February 26, 1885.[20] Regardless of the protestation, the case was dismissed by the Colonial Office on grounds that there was a misinterpretation of article 5 of the Berlin Act. Consequently, Governor Northey was advised to dispose of the land according to the powers accorded to him under the Crown Lands Ordinance of 1915.[21] This event showed the discriminatory land policies that were rampant in colonial Kenya; they denied fair access not only to Asian cane farmers but to any non-British farmers.

The postwar era was a period marked by the confusion and contradictions of colonial policies but also by the tremendous efforts made by both Asians and the Europeans farmers. For example, the Asian cane farmers were not discouraged by the biased soldier-settlement scheme; although they complained bitterly, they continued to make improvement in cane farming. This was because by 1919, the Asians of Kibos-Muhoroni had put 800 acres under cane cultivation; this was up from 621 acres in 1916. Furthermore, due to challenges that they faced such as poor infrastructure, they organized themselves and formed organizations such as the Kibos Planters Association, formed in 1920. Meanwhile, the Europeans settlers at Miwani, Kibigori, Chemelil, and Koru also started to plant sugarcane and sisal, with some focusing on one crop while others planted both.

Meanwhile, the colonial state changed its policies, and efforts were made to support sugarcane farming in its infancy stage. By 1920 experimentation with sugarcane had already started in Kibos, and many farmers were becoming more interested. During this period, trials were conducted on varieties such as the Sealy seedling, Uba, B.3922, and B.1528. The last two varieties had poor results. Progress in the expansion of cane farming was slow, as seen by the negligible increase to only 640 acres that was cultivated and planted with Uba and the Striped Ribbon variety, known locally as *kampala*.[22] On the other hand, Europeans who had settled in the area also by mid-1921 had started cultivation and managed to put 691 acres under cane.[23] Furthermore, during the same period, land estimated to be 6,000 acres was alienated near Kibos for sugar production and plans were already put in place for the construction of a sugar mill that would enhance production. By then it was believed that the sugar manufactured would be enough to meet the local demands of the colony.

Unfortunately, before the farmers in Kenya, both Asians and Europeans could be firmly established in agriculture, the colony faced economic problems in 1921 caused by the abnormal rainfall pattern, especially in Nyanza Province; most crops, such as maize and sugarcane, were affected due to a prolonged wet season. The situation was made worse by the fall in prices for agricultural goods. Despite these problems, the sugarcane industry continued to expand. By 1921, the first mill, the Victoria Nyanza Sugar Syndicate (the forerunner of the Victoria Nyanza Sugar Company) was formed and engaged in clearing and planting of approximately 1,000 acres of cane.[24] This initiative saw an increase in area under cultivation because the cane farmers could sell their cane to the factory. A year later, the area under cane increased to 3,787 acres.[25] This improvement was to continue, and there was a substantial increase when the acreage under cane increased from 3,787 to 4,193 in 1923. The expansion of the sugar industry was partly facilitated because of the ready local market.[26] It soon became clear that the industry was viable.

SUGAR PRODUCTION DURING THE INTERWAR PERIOD, 1924–1938

There was marked progress in sugar production when in 1923 an Australian pioneer settler, Mr. G. R. Mayers, who came to Kenya in 1920, bought land at Miwani and opened the Victoria Nyanza Sugar Company near Kibos. This was a major boost to the Asian sugarcane farmers, as the company started to manufacture crystallized sugar. Previously all cane was turned into jaggery, which limited the profitability of cane farming and limited the acreage put under cane. The company grew its own cane, but not enough to supply its needs. Since it did not have a large estate, it depended on cane brought from the Asian farmers. For example, the new company contracted approximately 75 percent of the Asian farmers to supply it with cane. To attain this goal, the company was able, after applying to the authorities concerned, to lay side trolley lines as far as Kibos. This enabled the sugarcane farmers to deliver cane to the factory. Due to these efforts, the area under sugarcane plantation increased by 1,050 more acres, from 4,193 in 1923 to 5,243 in 1924.[27] As cane farming became popular, more farmers started to plant the crop. Unfortunately, the farmers in Chemelil and Koru still did not have good access to the mill because of poor infrastructure in their area.

In the meantime, the growing of sugarcane increased among the Asian farmers because the prospects now seemed good.[28] This was as a result of the expanding local market and the colonial government's support with various experimentations and improved farming methods. Furthermore, by 1923, the area under cane cultivation increased due to the fact that there was a reduction in the price of sugar in the local market. It was partly because of the establishment of larger mills like the Victoria

Nyanza Sugar Company. Since it produced in bulk, the cost of production during the period was lower.[29] This produced lower prices and created increased demand among consumers.

Moreover, by 1925 Asians had managed to acquire more land with total holdings of 12,000 acres.[30] Consequently, within only two years there was a lot of improvement in sugarcane farming since the Victoria Nyanza Sugar Company was able to acquire a large estate at Miwani and had approximately 4,500 acres under cane by 1925. The total area under cane by 1925 was 6,556. Production also increased. For instance, during the 1924–1925 season, the infant factory was able to manufacture approximately 2,891 tons of crystallized sugar.[31] Although there was adequate rainfall in 1926, which made sugarcane to do well and enhanced the industry's performance, there was very minimal increase in acreage under cane. Only 191 more acres were added, which brought the total area planted with cane to 6,747 acres.[32]

Nonetheless, during the same year, another European company, Messrs. Nottidge and Allen, acquired land at Muhoroni. The owners were John Thomas Ripley Nottidge and Thomas Allen; the two had initially settled at Lumbwa. At Muhoroni they planted approximately 800 acres of cane. By 1932 Nottidge and Allen acquired more land in K-L district, about 5,838 acres.[33] Both factories had milling machinery and were able to produce excellent grades of sugar; this found a ready, expanding market since the Africans had started to add sugar to their diet.[34] Due to the remarkable improvement in the industry, the Asian farmers started experiencing delays in the delivery of their cane to Victoria Nyanza Sugar Company. This was because the company had to cut its own cane first before accepting any cane from the Asian farmers. Their situation was compounded by the fact that the Nottidge and Allen mill only crushed their own cane. This meant that the Asian farmers had only one mill to receive their cane. The growers at Muhoroni suffered more since the distance to Miwani was far. Moreover, between 1922 and 1926, a lot of attention was focused on European cane farming with the entrance of European cane farmers on the scene, Victoria Nyanza Sugar Company at Miwani and Nottidge and Allen at Muhoroni. The Asians wanted to acquire machinery to mill their own cane, but colonial policy made permission to acquire the machinery difficult for Asian farmers, although the two European companies had no restriction in importing machinery for their cane. For example, the Muhoroni Asian farmers were denied a permit to import a milling machine for fear of possible competition to the young factory of Nottidge and Allen. It seemed that the lack of political will by the colonial state to help the Asian sugarcane farmers was a great impediment to the growth of the sugar industry in colonial Kenya.

Regardless of that, the sugar industry continued to expand. By 1927 there was little increase in land under cane cultivation. The area grew only from 6,747 to 6,811 acres in 1927. Furthermore, the Nottidge and

Allen Company also purchased more land toward the end of the year and planted an additional 450 acres. The greatest increase in sugarcane cultivation was in 1928 when European farmers cultivated 9,403 acres while the Asian farmers, who by then were forty in number, owned approximately 4,217 acres. They brought under cultivation another 3,693 acres of cane.[35] This brought the total acreage under cane to 13,096.[36] During the same period, two other small mills were established, one near Nairobi at Ruiru and the other at Ramisi near Mombasa. Such expansion of the industry with the establishment of the two mills clearly indicated the potential of the industry, which was poised toward meeting the consumption needs of the colony. Demand increased partly because of the increase in consumption of sugar by Africans, who had seen an increase in their purchasing power.[37] However, cultivation during this period was beset by challenges due to the mosaic disease outbreak in the province, as will be discussed in a later chapter.

European cane farmers also made progress by 1929. Their acreage under cane greatly increased. For instance, there was an increase of 1,758 acres, bringing the area under cane to 11,161 acres. Together with the Asian farm holdings of 8,820 acres of cane, the total of acreage under cane increased to 19,981 acres, compared with 13,096 the previous year.[38] Asian farms saw an increase of 5,127 acres on their holdings. This was partly because the number of Asian farmers increased from forty in 1928 to forty-eight in 1929.[39] This was a huge increase despite the Depression that the country faced.[40] This led to a great increase in sugarcane production whereby 39,900 tons of cane were cut and taken to the Victoria Nyanza Sugar Company. Nevertheless, greater losses were incurred than was normal because of the mosaic disease that invaded the farms. The 1928–1929 seasons were difficult for the cane farmers. Therefore, in an effort to help the farmers, the colonial state changed its stance and put in place heavy protections in the form of custom duties and preferential railway freights in favor of the farmers in the colony, including the sugar producers, so as to help them survive the difficult economic times.

The sugarcane farmers had no time to recover as they were hit again by another spate of locust invasions in 1932. During this period the agricultural officers embraced new methods to fight the infestation whereby farmers were taught of the use of smudge fires as a protective measure against locusts.[41] Due to these difficulties, there was a decrease of the area under cane cultivation in 1931. The area decreased to 12,363 acres, down from 13,493.[42] Consequently the production of sugar also decreased. However, growth was steady for both Asian and European sugarcane farmers as was witnessed by the output of the sugar produced by the two factories from 1925 to 1932, as indicated in table 3.4.

The data in table 3.4 show that despite the missing 1928 records, sugar production increased in this period. For instance, Nottidge and Allen produced 560 tons in 1926 despite the fact that the factory was closed for

Table 3.4. Performance of the Sugar Companies, 1925–1932

Year	Output (in tons)	
	Victoria Nyanza Sugar Co. (Miwani)	Messrs Nottidge and Allen (Muhoroni)
1925	2,833	—
1926	6,570	560
1927	7,302	847
1929	6,200	310
1930	7,450	700
1931	3,500	250
1932	2,833	310

Source: K-LAR, 1926, KNA: DC/CN/1/4/1.

three months to allow installation of new machinery.[43] The poor performance reflected by the output of 1929 in both factories was due to the unfavorable weather conditions during the year. Not only was the rainfall erratic it was also insufficient in many months. To add to this, there was an outbreak of mosaic disease that destroyed crops and caused serious losses.[44] Meanwhile, the decrease during the 1931–1932 season was due to the effects of a locust infestation that ravaged the area. It should be noted that the output by Victoria Nyanza Sugar Company included cane from the Asian farmers. During this season, the performance was commendable because the farmers had to contend with other problems such as disease, locusts, and inadequate labor.

Because of the progress and zeal shown by the Asian farmers and the need for land for expansion, the lowland issue was revisited in 1925. Kenya Asians had shown interest in the exclusive settlement in the Kibos area. This led to the Colonial Office to institute the Standing Emigration Committee in 1924 to look at the possibilities. The members of this committee included Asians such as Sir Hari Singh Gour, Chaundri Muhammad Ismail, and Ramchandra Roo, among others. Unfortunately, the committee was compromised because political rather than economic opinion prevailed, and the committee members objected to the idea of reserving the Kibos area exclusively for Asian farmers on the grounds it might be misconstrued by Asians to mean acceptance of the principle of segregation.[45] Therefore, after much discussion and exchange of views with London and the government of India, as well as pressure from Asian politicians back home and the East African Indian Association, the idea was shelved completely. This meant that the Asian sugarcane farmers were not going to get land easily in the lowlands, yet the highlands remained closed to them.[46] The farmers who had initially been supported by the colonial state now felt betrayed by the government. What the Asian farmers needed was land; although they complained to the Kenya Land Commission, their requests were ignored.

Be that as it may, the Asians continued to look for land from Europeans who were ready to dispose of their farms in the region. This led to the development of a very interesting phenomenon. During the 1920s and 1930s, Asian absentee landlords and land speculators became very common. Many purchased land from Europeans around the Kibos and Chemelil areas. For example, N. J. Desai, an insurance broker in Kisumu, purchased large farms in Kibos and Chemelil. These absentee Asian landlords then leased the farms to others. By 1928 there were forty-eight Asian farmers in the Kibos-Miwani area alone, each with approximately 200 acres under cane cultivation. Yet out of these farmers, 60 percent were absentee landlords. Most were not Sikhs.[47] The relentless pursuit for farms continued. In 1930 the 4,500 acres belonging to the Walji and Sons estate (part of the former Kisumu Rubber Company) was leased to Asian tenants at an annual rent of 50 cents per acre.[48] The tenants planted sugarcane since the farms were very near the Victoria Nyanza Sugar Company (5 kilometers away). The subdivision of the Walji estate brought the total number of Asian cane farmers to fifty-six in the Kibos-Miwani area alone. They occupied 15,400 acres of land, out of which 9,120 acres were under cane cultivation.[49]

In 1931, the Asian farmers suffered some setbacks. The spread of striga weeds and a locust infestation reduced the total area under cane cultivation almost by half, to 4,470 acres.[50] Furthermore, the landholdings also decreased to 14,614 acres. This was because one of the farmers sold out his land, reducing the number of farmers from fifty-six to fifty-five.[51] The locust infestation contributed to low sucrose content in the cane, which had a ripple effect as the farmers received smaller proceeds from the cane especially given the depression in world prices.

During this time the farmers responded by putting 700 acres of maize under cultivation. The situation continued till 1932, when the total area under cane for the whole district was reduced again to 12,088 acres compared with 12,363 acres under cane in 1931.[52] Many took to growing of cotton, maize, and pulses.[53] The frustrations of these years led the Asian farmers to diversify more. Therefore, by 1932, the acreage under maize increased to 1,942 acres, further reducing the area under cane cultivation.[54]

In the meantime, cane farmers continued to suffer due to the world depression that affected the sugar industry; producers not only continued to struggle to maintain the home market but also were disadvantaged due to threats posed by the rapid increase in production from Uganda, which then dumped its excess sugar into Kenya. Nevertheless, the European cane farmers made progress after the poor performances experienced over the previous two years; they were able to increase the area under cane cultivation. The total area increased from 12,088 acres to 12,704 acres during the 1933–1934 season.[55] On the other hand, Asian cane farming during the same period seemed to have declined. Not only did their number not increase but also their landholdings dropped significantly, from 14,614 acres in 1931 to 10,935 acres in 1933.[56]

Fortunately, this changed in 1934 as a rigorous campaign was undertaken to eradicate the striga weeds that destroyed most of the crop in the district. The experimentation with imported varieties from India's Coimbatore station also encouraged the farmers.[57] The Asian farmers increased their holdings and some bought more land when the government auctioned two farms in Kibos. These farms, one of which belonged to Major Carnegies, were bought at Shs. 23,700 and Shs. 21,700, respectively. Hence, the total Asian landholdings by 1934 were approximately 24,000 acres.[58] These farms were not bought individually; rather the Nyanza Indian Farmers' Association bought the farms for their members who had to reimburse the funds.

The prices of these two farms were exorbitant at the time, given the fact that European settlers got land cheaply. However, the fact that the Asian farmers were willing to invest their money for land meant that they were resilient and hoped to benefit from cane cultivation. Unfortunately these hopes were dashed when the Victoria Nyanza Sugar Company's managing director, B. Y. Abel, decided to arbitrarily cancel some of the contracts due to the Great Depression that led to the fall in the price of sugar. The company took its own cane first, and the farmers were left in a desperate situation, with only some small amount of cane received from some farmers. The situation at Kibos-Muhoroni was so bad that fifteen out of the fifty-six farmers could not pay the rents and leases of their farms.[59] Furthermore, there was a threat from Uganda, which was producing more sugar. These are some of the factors that led to the company declining cane in 1930. As a result, many of the Asian farmers suffered from anxiety about how to pay for their farms. The situation was dire, and it forced the Nyanza Farmers' Association (also known as the Nyanza Indian Farmers' Association or NIFA) to petition the government for an extension of payment. Such a situation brought to light issues of sustainability of peasant production; the fact that small-scale farmers had their contracts revoked by the Victoria Nyanza Sugar Company indicates otherwise.

However, during the same period, the European farmers in Nyanza were successful growing coffee and sisal. The expansion was partly due to the boom of World War I when prices for the settler produce were lucrative. Not only did their production increase but it was diversified as well, with the farmers growing cereals such as maize and wheat.[60] However this situation was to change when the world depression impacted Kenya in 1929 and the prices of commodities crumpled. The prices of maize, groundnuts, and sesame fell by 50 percent, while those of coffee fell by 40 percent. This effect was so bad that a number of European settlers went bankrupt, and many more would have been hard hit, had it not been for the state's intervention through the provision of rebates, tariff concession, and the creation of the Land and Agricultural Bank in 1931 that offered loan subsidies on exports. Furthermore, they were also given low railway rates for their cargo to the port of Mombasa.[61] It was unfortunate that such assistance was not accorded to the Asian sugarcane farmers.

Therefore, the period between 1930 and 1932 proved to be very challenging to both European settlers and the Asian cane growers; not only were the prices bad but various calamities also set in. These were years of depression where no marked progress was witnessed, especially in Nyanza, which was invaded by locusts that destroyed crops leading to the infamous famine that was known locally as *Nyangweso.*[62] As a result, European settlers and Asian cane farmers were affected and suffered adversely. The most affected cane areas were Songhor and Muhoroni. The effect of depression, drought, and locusts was evident in the total sugar production. The area under cane was reduced from 13,493 acres during the 1930–1931 seasons to 12,363 acres in 1931–1932 seasons.[63] As a result, the Victoria Nyanza Sugar Company output dropped by more than a half, from 7,457 tons of white sugar produced in 1931, when it purchased 49,000 tons of cane from the Asian farmers at the cost of 26,769 pounds, to 3,000 tons of white sugar in 1932.[64]

This poor performance was due to the fact that throughout the year, the prices of sugar remained extremely low, which affected production. For the following year, 1933, the production of Victoria Nyanza dropped further to a mere 2,836 tons, while the Muhoroni Sugar Company, which had managed 709 tons in 1930, produced only 250. The Asian small-scale farmers were badly affected, and they responded by starting to operate small shops on their farms. The situation was so bad that those that resided at Kibigori resorted to selling fuel to the Victoria Nyanza Sugar Company just to make ends meet.

Similarly, the European farmers in Nyanza had mixed results; some gained while others performed poorly. For example, while the coffee farmers had a good harvest, (despite the fact that some had poor results, possibly because they had planted in inferior soil); the sisal farmers at Songhor faced immense problems. Many stopped cultivating the crop altogether due to the low market prices. Unlike those at Songhor, the sisal farmers at Koru did better, while the farmers growing maize and wheat at Lumbwa and Fort Ternan were adversely affected by the ravages of locusts, just like the Asian cane farmers. To help the farmers in such a difficult time, the colonial state started the Land and Agricultural Bank in 1931, since the Agricultural Ordinance had already been passed in 1930, which offered provisions to finance agriculture.[65] The bank was to extend loans to farmers during the depression period in which they could use their land as collateral. Unfortunately, the bank provided loans mostly to European farmers. The Asian cane growers' applications were turned down, despite the fact that they too had farms that could be used as collateral.[66] Once again the hostility and lack of political will by the colonial government to support the sugarcane farmers stifled the industry in colonial Nyanza.

The colonial government's bias in favor of European farmers continued to hamper the economy. The government not only gave subsidies, especially to coffee farmers, but also offered incentives in the form of the Guaranteed Minimum Return (GMR). Again, these rescue measures were not made easily available to the Asian farmers. Initially Asian applications were denied on flimsy grounds such as that sugarcane was not an annual crop. Those colonial officials who most vehemently opposed extending such help to Asian farmers were C. G. Lloyd and the chief executive officer of the Board of Agriculture, E. F. Martin.[67] As the struggle against the colonial state continued, Asian farmers had to contest the decision since the depression, bad weather, and calamities like locust infestations affected both the Asian cane farmers and the European farmers alike.

Be that as it may, the Victorian Nyanza Sugar Company made strenuous efforts to recover and replanted cane that had been destroyed by locust the previous year. By 1932 the company planted sugarcane on both of its estates at Miwani and Muhoroni, amounting to 8,000 acres. This performance was, however, affected by fluctuating world sugar prices. The severe impact of the low prices led to drastic measures to be taken by producers in Kenya in an effort to remedy the situation and ensure stability of the local prices in the colony.[68] The low sugar prices and the expansion program undertaken by the company most likely influenced Victoria Nyanza Sugar Company to reduce its intake of cane from the farmers. The company found itself in a very difficult situation for, while expanding, it could not on the other hand encourage the small-scale farmers to plant cane for sale to the factory. Nevertheless, the small-scale farmers had also started to recover and by 1934 had enough cane, though they faced problems of disposing their crop. Their problem was compounded by the fact that they had no machinery to mill sugar. And on seeing that their cane was not going to be easily accepted, a deputation led by Shams ud Deen met with the governor, Sir Joseph Byrne, and requested access to machinery to mill sugar. The Asian farmers were eventually granted a license to import one.[69] However, there was delay as they organized to get funds.

At the same time, many complaints had been made against Mr. Abel, the manager of Victoria Nyanza, especially by the Nyanza Indian Farmers' Association. He was not only accused of not taking cane from farmers but also of discriminating against those who complained about how they were treated by the company. The manager himself claimed that the allegations were not true. Nevertheless, despite the farmers' complaints, for some of them, their cane was not received. However, contrary to what the farmers claimed cane was delivered to the factory and bought from the farmers as shown in table 3.5.

Table 3.5 shows that cane was received from both contracted and other farmers, and during good seasons more cane was obtained from the two groups of farmers, for example in 1930–1931 and 1931–1932. Furthermore, the cane obtained in 1931–1932 had a difference of approximately only 10,000 tons to that of the factory estates, while during 1930–1931, Victoria Nyanza Sugar Company had 50,389 tons of cane and still went ahead and purchased 49,042 tons from the Asian farmers, both contracted and those without contracts, although some farmers complained that their sugarcane was not received. It seemed likely that the tonnage fluctuated depending on the financial performance of the factory. The situation was further aggravated by the fact that the factory also faced problems, such as the dumping of sugar in the country, which affected how many extra tons could be received from local farmers.

Nevertheless, the sugar industry continued to grow in the following years. Both Victoria Nyanza Sugar and Muhoroni Sugar companies had fully recovered and registered good performances after 1934. For instance, Victoria Nyanza Sugar Company, holding 9,000 acres by 1936, produced 11,180 tons of white sugar, up from 5,051 in 1934. Muhoroni Sugar also made headway after initial bad years. The company under-

Table 3.5. Sources of Sugar Produced by Victoria Nyanza Sugar Co., 1924–1934

Year	Total Sugar Manufactured (in tons)	Estate Cane (in tons)	Planters under Contract (in tons)	Other Planters (in tons)
1924–25	1,385	14,774	—	5,296
1925–26	4,048	34,199	—	18,188
1926–27	8,018	55,712	12,214	28,523
1927–28	5,700	45,299	16,266	13,972
1928–29	5,723	41,436	15,568	23,978
1929–30	6,216	36,979	19,158	16,335
1930–31	7,498	50,389	24,455	24,587
1931–32	3,498	39,409	18,099	11,201
1932–33	2,821	32,011	6,858	3,493
1933–34	5,051	49,294	11,479	10,221

Source: Extract from the Report of the Commission of Enquiry into the Sugar Industry, Victoria Sugar Company, Ltd–Miwani, 1934, KNA: PC/NZA/3/2/34.

took an aggressive expansion and intensive planting program when it installed a system of irrigation as a measure to counteract drought, having gone through drought conditions in the early 1930s. The irrigation, facilitated by water from the Nyando River, produced encouraging results. Meanwhile, the small-scale farmers also increased production. The sugar produced during this period found an already expanding local market because the consumption rate had increased from 9,100 tons during the 1932–1933 season to 13,100 tons during the 1935–1936 season.[70] The increase in consumption and the ready market seemed to have motivated the farmers to grow more cane. Consequently, those at Kibos and Chemelil formed the Nyanza Farmers' Cooperative Society (NFCS) in 1936, which marketed their jaggery to England and managed to get good prices for it at double the price it fetched locally. Through the society, the farmers also managed to get free rail transport of their cargo to the port of Mombasa. The following year, on their own initiative, they were able to collectively purchase 20,000 acres from European farmers and managed to increase the cane acreage by an additional 7,000.[71]

Meanwhile, there was also improvement in the European farming within the region. The European farmers increased production of sisal in their farms, partly because they were assisted by the Sisal Industry Committee, which was appointed in 1934 to look at the interests of the sisal growers.[72] Furthermore, there was increase in world prices. For example, fiber prices increased weekly from May to December of 1935, and the market continued to expand in the United Kingdom. The lucrative world prices were a factor that influenced the European farmers to increase sisal production. The overseas market in sisal remained steady throughout the 1936 period.[73]

During 1936, coffee farmers still languished in financial stress due to poor harvests and poor prices. Many farms in the Songhor area closed completely. These were the effects of the depression that had left the farmers struggling. For instance, the Barham Investment Trust Estate, formerly Kitchener's African Estate, suffered seriously. The owners could not cope; out of frustration they uprooted the coffee trees, discharged the laborers, and abandoned the farm with a few caretakers. Other farmers included Mr. W. Maxwell who also gave up his farm at the foot of Nandi hills, while Mrs. D. A. Irvine gave up farming altogether and left the district.[74] This was a difficult period for many farmers. The slump in the world market forced about 400 out of the 2,000 Kenya settlers to give up farming, and many abandoned their farms.[75] These problems led to the colonial state's intervention to save the European farming community from total collapse. And so in 1936, in an effort to help the farmers, the Farmers' Conciliation Board was formed with the aim of assisting farmers who were experiencing financial problems. Approved farmers were allowed a grace period of five years to pay up their loans. The board also provided loans on the condition that the farms would be supervised by the Agricultural Board. To qualify for loans, the applications had to be approved by the local agricultural committee. Though helpful to European farmers, the board, just like the Land and Agricultural Bank, did not offer any meaningful help to the Asian cane farmers. For instance, less than 10 percent of the Asian farmers got the financial assistance.[76] It was interesting that the colonial government always took initiatives to put in place measures to rescue the European farmers while the Asian farmers were left to their own devices.

By 1938, other small mills were also manufacturing sugar. During this period there were a total of five factories. Apart from the Victoria Nyanza Sugar Company at Miwani, there were also others; Muhoroni, Sukari Company at Ruiru near Nairobi, Manoni, and the Kenya Sugar Company-Ramisi at the coast.[77] These factories were an indicator of how promising the sugar industry in Kenya could be.

SUGARCANE FARMING DURING THE WORLD WAR II PERIOD, 1939–1949

The sugar industry prior to World War II was in a fragile state, and the depression period left many farmers struggling. The European farmers, however, got more assistance from the colonial state to deal with the problems of the 1930s than did the Asian cane farmers. Even with such preferential assistance from the government, European farming in the Nyanza region continued to decline. For example, the coffee farms at Koru, Fort Ternan, and Songhor reduced their acreage under cultivation. The European farmers faced so many difficulties that by 1939 they were unable to cultivate more than 10 percent of their land. On average, farms of approximately 2,600 acres only managed to effectively cultivate 269 acres.[78] However, this changed during the Second World War. The war led to increased food production to meet the demands of feeding troops and prisoners of wars. The colonial state introduced the Increased Production of Crops Ordinance of 1942, which facilitated the provision of short-term credit and also the Guaranteed Minimum Return (GMR) for all annual crops grown on European farms. Under the GMR, European acreage under maize and wheat increased.[79] Asian agriculture also expanded tremendously; for instance, there was a huge shift as the Asians began to manufacture sugar and cotton.[80]

The expansion was evident in 1942 when an Asian miller, Premchard Raichad and Company, bought the estate belonging to the Nottidge and Allen Company at Muhoroni (often called Muhoroni mills, approximately 4,800 acres in extent) and started processing sugar.[81] This was a relief for the cane farmers in Muhoroni-Chemelil who previously had to transport their sugarcane to Miwani or were forced to produce jaggery instead. The Asian farmers also started to grow maize because it fetched better prices and the colonial government offered incentives for its cultivation. Their efforts were encouraged when the Board of Agriculture encouraged farmers to form agricultural committees for easy dissemination of information among other services. Thus the Asian farmers organized themselves and formed committees in their areas.[82] Such committees included the Kibos Planters Production Sub-committee, which had forty members; the Miwani-Kibigori Production Sub-committee, with sixteen members; and the Chemelil-Muhoroni Production Sub-committee, which had ten members. The subcommittees did a lot to improve farming on Asian farms. They not only reviewed past performances and made plans for the coming season; they also worked together with the department of agriculture in an effort to procure loans to the farmers.

During this same time European farming expanded as the world prices of sisal went up. This led to an increase of cultivated acreage on the European sisal estates at Chemelil and Koru. With this expansion, the colonial state assisted the farmers further by enacting the Increased Production of Crops Ordinance of 1942; this was helpful as it provided farmers with financial assistance to break new land. The Asian cane farmers complained that the breaking grant was by no means sufficient. For example, the farmers were given Shs. 40, which was enough to plough once, but cane needed to be ploughed three times. Still, the assistance was timely because the Ethiopian campaign and North African theater created a continuous demand for foodstuffs. These put pressures on farmers to not only produce more toward the war effort but also feed prisoners of war.

The war burden stimulated production by all sectors: African, European, and Asian.[83] However, by 1945 production decreased and the five factories in Kenya were able to produce only 8,187 tons of sugar. This low production was due to the severe drought that affected cane in the early months of the year.[84] The poor performance continued into 1946, when a huge drop was registered, from 8,187 tons of sugar to only 5,760 tons from the five factories.[85] This was caused by the poor weather conditions of the previous two years and the prevalent labor shortage experienced during the year, which affected harvesting.

Meanwhile, European agriculture was reinvigorated when the colonial state came up with policies and established institutions to solve the problems faced by European farmers. For example, a 1946 development plan was adopted to intensify the use of land. To assist the European farmers further, technical training was started at the Egerton College of Agriculture in Njoro. The training, based on diversification, stressed the importance of mixed farming, which included the grade cattle.[86]

However, the CN cane farmers had a major breakthrough on June 16, 1947, when an Asian businessman from the Kakira Sugar Works Company in Uganda, an experienced planter and miller named Devjibhai Hindocha, bought the whole of the Victoria Nyanza Sugar Company's estate and factory. He renamed it Miwani Sugar Mills. Hindocha then started expanding the operation and infused life into the moribund factory by undertaking three important steps. First, the production of the mill increased as he acquired more land. Within a short period the Miwani Sugar Mills had three plantations: the Miwani estate with 9,300 acres, Chemelil estate with 4,252 acres, and Kibigori estate with 543 acres under sugarcane. Second, the company also stepped up production when it reached an agreement with the neighboring cane growers for the factory to absorb their cane. This was a relief to the farmers who initially had been complaining about such issues with the former Victoria Nyanza Sugar Company. Third, during the takeover period the factory had inherited shortages of rails and rolling stocks for the tramway that made it difficult for farmers to deliver their cane. Therefore, to ensure that cane reached the factory, the Miwani Sugar Mills introduced more trolley lines to transport cane.

All these efforts made the farmers intensify their cane growing. Many sought loans from Asian moneylenders. Before 1945, it was not easy for Asian cane farmers to access loans. However, by late 1940s, two Asian banks opened in Kisumu: the Standard Bank of India and the Bank of Baroda. These became the major source of financing to the Asian farmers, which led to expansion of the sugar industry; the number of Asian farmers now increased to eighty-three with a total acreage of 22,071 by 1947. Furthermore, there was heavy rainfall and adequate labor in 1947, which helped the farmers to increase the acreage under cultivation.[87] Most of these cane farmers were found in Kibos. This area had the most land under sugarcane cultivation. Comparatively, there were few farmers at Chemelil, yet they managed to have a lot of land under cane; about 4,335 acres, while the farmers at Kibigori planted another 3,990 acres of cane.[88] The increase in the total acreage under cane was partly as a result of Miwani's aggressive policy of expansion, which in turn influenced the farmers to expand as well.

During the same period, the Miwani Sugar Mills, eager to expand, took another initiative when in 1948 it offered the colonial state part of its land near Ngeta River, three kilometers from Miwani, for the purposes of building another research station. The purpose of the station was to carry out research on various crop varieties in order to come up with a variety that would be not only high yielding but also compatible with the ecology of the area. In response, the government not only established the station for trials, as will be discussed later in the chapter, but also appointed an Asian sugar expert, Sampura Gill Singh, as an agricultural assistant to help the farmers by instructing them on good farming methods. He also carried out investigations on sugar varieties.[89] Due to his work as well as favorable weather conditions and the improvement in the procurement of labor, there was marked improvement in cane production, as the total acreage rose to 13,289 of planted cane.[90]

THE EXPANSION OF SUGARCANE FARMING IN THE POSTWAR PERIOD, 1950–1963

In the postwar period, economic growth was registered generally in the agricultural sector in the colony, including in the sugar industry. There was a change in agricultural policy that brought about the tremendous development. The new policy encouraged African agriculture alongside European farming.[91] In Asian cane farming, progress came from the experimentation into improved farming methods, including the adoption of better varieties.

The farmers did have challenges in 1950, however. The poor short rains led to a severe drought. The situation was no better once long rains arrived since the rains destroyed the Asian maize in Kibos-Muhoroni area, due to waterlogging caused by poor drainage. Furthermore, the rainfall was erratic, and the areas around Kibigori and Chemelil experienced drought that also led to failure of the maize crop.[92] Although the cane farmers suffered from the effects of waterlogging, their cane crops were not adversely affected compared with the maize. Many cane farmers were able to cultivate more land since they got help through the use of a machinery pool as a coping strategy. Through such initiatives, many were able to use tractors to plough land, and hence the number of oxen used was greatly reduced. More acres were cultivated, especially by large-scale farmers and factory estates. Furthermore, experimentation had intensified during this period as farmers started to plant the CO varieties that led to an increase in production.

In 1952 the preliminary results from varieties planted in 1950 were reported. The results indicated that the CO 421 was a better variety than CO 281 for the area, as will be discussed in the latter part of this chapter. Such information was crucial to the cane farmers for economical farming. Hence, most farmers changed over to planting of CO 421 by 1952.[93] Other trials were conducted during the same year on the possible use of factory-made long-term fertilizers that employed the use of nitrogen, potash, phosphate, molasses, and filter mud press from cane. Other experiments were done on the use of herbicides to control weeds, especially in young cane.[94] During this period there was marked improvement in cane production, despite the problems of waterlogging. The acreage under maize also was increased, mainly to feed the labor on Asian farms despite the fact that the area was not suitable for maize growing due to poor drainage during the heavy rains.[95]

On the whole, there was marked improvement in Asian farming, and the Asian farmers were able to produce approximately 129,602 tons of sugarcane in 1953. But increased production brought its own problems, as the Miwani Sugar Mills began to refuse cane, just like its predecessor, the Victoria Nyanza Sugar Company, had done when production was too high. Complaints had started as early as 1951, and by 1952 it was apparent that some farmers' cane was not accepted. A prime example was provided by the case of Milwant Singh and Makan Singh. These two farmers were behind the calls for another factory at Kibos, and it seems that they were intimidated by the Miwani Sugar Mills Company, which refused to accept their cane, a situation which made Milwant Singh to sell his farm.[96]

Despite this situation, the Miwani Sugar Mills estate alone was able to harvest in 1955 approximately 97,656 tons of cane, and the mill bought 33,124 tons more from the Asian farmers. This almost doubled in the next three years, for by 1958 the farmers produced 234,952 tons of sugarcane.[97] Due to the good performance of the Asian cane farmers, approximately six European farmers also began to grow sugarcane in the Miwani-Muhoroni area. For example, a Mrs. Martin grew sugarcane at her Miwani estate and also allowed her farm to be used for experimental purposes. Some of the European farmers who took to cane farming by the mid-1950s had initially been cultivating sisal. This shift to cane was partly due to the poor world prices of sisal then.

Needless to say, the intensification of agriculture continued as was attested by the colonial state's implementation of the Swynnerton Plan in 1954, which was meant to enhance commodity production. Although meant to revamp African and European farming, the fact that it promoted consolidation of farms and granted individual land titles helped the Asian sugarcane farmers, who could use their land as collateral to secure loans. Therefore, the colonial state's interventions during the depression and post-depression period and the coming of the Asian manufacturers in the industry led to the expansion of sugarcane growing and production. Furthermore, there was a great increase in the number of farmers: by 1960 there were approximately 100 small-scale and large-scale farmers holding approximately 46,000 acres. Meanwhile Miwani Sugar Mills also continued to stabilize its production with an average output of approximately 20,000 tons of sugar annually. This stability continued until 1960, as indicated in table 3.6.

The data in table 3.6 clearly indicate that the sugar industry was growing. For example, from 1955 the increase was steady, with the greatest increase in 1959–1960. Such an achievement could not have been possible without the efforts of the farmers and the government interventions through technical support and better policies. It is evident that whenever the colonial state supported the sugarcane farmers, there was a marked progress in the industry. By 1961, the sugar industry had improved and the Miwani Sugar Mills was at the peak of its performance; the company started to manufacture refined sugar, as will be discussed in the next chapter.

Table 3.6. Miwani Sugar Mills Production, 1953–1960

Year	Sugar Produced (in tons)
1953	10,587
1954	8,554
1955	12,005
1956	14,465
1957	16,706
1958	12,693
1959	19,015
1960	21,582

Source: The East African High Commission, *An explanation of the trends and Prospects of Sugar Production and Consumption in East Africa, 1950–1970*, 5.

AGRICULTURAL ASPECTS OF SUGARCANE GROWING IN COLONIAL CN

The Asian sugarcane farming in CN developed over the years from a humble beginning of only a handful of pioneers with a few acres to over eighty farmers by 1960 with much acreage. The industry became a successful one by the mid-1950s with Asians as the main growers and manufacturers of sugar. This growth was due to the resilience and determination of the Asian farmers who learned new techniques and adopted new varieties of cane despite the challenges that they faced such as poor ecological conditions, killer weeds, a lack of machinery, and a dearth of adequate financing.

The beginning of commercial sugarcane farming started when farmers experimented with the locally known variety, Uba, as well as some varieties from India. Initially all Asian planters worked on small-scale farms. However, as more farmers took to farming, three different types of farms emerged: large scale, small scale, and nuclei farms belonging to factories. Small-scale farms comprised fewer than twenty acres.[98] Large-scale farms had twenty acres or more, with an average of fifty acres. For example, Mohinder Singh, who to date still owns 500 acres of land at Kibos, of which 300 acres were under cane cultivation, had initially inherited 100 acres from his father Nihla Singh, a former Kenya-Uganda railway worker. His father had 200 acres at Miwani by 1922 and another 300 at Kibos. In 1964 the farms were divided among the three brothers—Mohinder Singh, Ranjit Singh, and Arjit Singh—when their father left for India.[99]

Nucleus farms included sugar factories such as Victoria Nyanza Sugar Company, Muhoroni Sugar Mills, and later Miwani Sugar Mills, which all had their own estates where they grew cane. These were relatively large estates. For example, between 1951 and 1952, the Miwani nucleus estate planted 4,800 acres of cane compared with the 2,000 acres planted by individual farmers. By 1953, the Miwani estate cultivated 5,000 acres of cane while individual farmers cultivated 2,500, and by 1960 the Miwani estate had acquired approximately 10,000 acres, stretching from the Nandi escarpment to the railway line. Other large farms were found among the few European cane growers in Koru-Muhoroni-Songhor area. These farmers had initially planted a considerable acreage in anticipation that Uganda sugar entrepreneurs would expand and build another factory near their farms. Since most of them were too far from Miwani to send their cane, much of their production was turned into jaggery.[100]

Moreover, sugarcane in Nyanza was grown on these different-sized farms under a wide range of rainfall conditions, varying from 25 inches up to 100 inches, though an even distribution of 40 inches was essential. Thorough cultivation of soils was essential for sugarcane, plowing to approximately a depth of three feet. The main reason for this was to increase soil aeration and water-holding capacity.[101] This was important to the Kibos-Muhoroni sugarcane farms since they were basically rain-fed. During cultivation and opening up of new lands, Asian farmers used oxen-driven ploughs to break the land, which proved to be an arduous task, mostly done by African laborers. Occasionally the breaking of the land was done by contracting European farmers who had tractors. The cost of a tractor was out of the reach of most small-scale Asian farmers; hence, they devised a coping mechanism whereby they pooled their resources together by paying £20 each, which enabled them to rent the tractors from European farmers collectively. Some farmers felt that this slowed the process since the land had to be worked upon in turns. Due to the nature of the soil in the area, mechanized farming required deep cultivation.

Most farmers preferred a chisel-type cultivator with teeth at 10–11 inches spacing. One such model was IHC, model DYRR5FH, which had a lengthened, strengthened frame able to carry an additional tooth on each end with couplings attached to the tractor to allow free movement on axes and transverse to the direction of movement, lifting and clearing the ground. The IHC cultivator was efficient on black soil and functioned just like the Caterpillar G series.[102] Some farmers used M.D.V. tractors, others, like Dhanwat Singh, used a Fordson tractor that cost him Shs. 31, to plough an acre in 1949. This was not economical for the farmer. Furthermore, the tractors were problematic during the rainy season due to the heavy soil; many got stuck.[103] Those who could not afford tractors had to use the oxen ploughs on their farms. In this case, it was found that in a 100-acre farm, approximately 100 oxen were kept for cultivation. Moreover, since the soil was hard, the small-scale farmers were not often in a position to prepare as fine a seedbed as was required for sugarcane sowing.[104]

Sugarcane farming also involved other types of land preparation, such as leveling to make the land uniform, avoiding potholes, for good, even draining. Hence a uniform platform was important.[105] Land was then harrowed, which was mostly done by African labor. On previously used land, the farmers skimmed out old cane stumps and chopped them down before a tractor or plough could be used. This was common in small-scale farms. Before planting it was necessary to plough land almost three times and harrow it twice in order to attain the required uniformity of surface.[106]

During cultivation, the land had to be ridged along the contour to prevent erosion and conserve moisture. For large-scale farms and factory estates, a high clearance tractor was used after experimentation had been done at the Kibos estate in 1950. However, this tractor was very expensive when first imported; at the price of Shs. 29,115 (1950) very few farmers could afford it.[107] Moreover, the high clearance tractor was normally used after the land was broken by a heavier tractor. As a result of these varied conditions and difficulties, Asian farmers experienced problems during land preparation since many were poor, small-scale farmers who found it difficult to buy machinery. It was not until 1949 that the colonial state made efforts to solve the problem by allocating six tractors to Amar Singh and M. Baktara Singh of Kibos to help them in farming. This was far from efficient given the fact that by the 1940s there were about fifty-six Asian cane farmers in the area. Having the six tractors was still not conducive to better plowing.

Whatever the means used in land preparation, planting followed. Cane can be classified in two types. One was the noble canes that were usually high in sucrose content but required high rainfall conditions and fertile soils. These were not found in CN. The second type was the slender canes with strong roots and great ratoon. These slender canes adapted easily to poor soils and had high fiber content; hence these were the canes found at Kibos-Muhoroni. This type of cane could further be divided into two: hard cane such as Uba, which was drought resistant and mainly used for jaggery, and soft canes such as the CO varieties, which were good for sugar production.

In Nyanza, sugarcane normally had two seasons of planting, either in March or October-November. However, the October-November season was preferred since growth took place during the long rains where the plants made good progress for almost six months, although some farmers felt that planting could be done throughout the year.[108] The method of planting varied according to the size of the farm. Small-scale farmers preferred two methods of cultivation, pre-planting and inter-row cultivation, especially on sandy soils where waste material was left where it fell instead of removing it into an alternate row. This was found to be helpful since it not only helped limit labor costs but also helped in the conservation of moisture while also preventing soil erosion. Cane was planted in rows and sections separated by infill roads that were used to control fire, to transport inputs, and to make harvesting easier.[109]

However, the large-scale farmers and factory estates preferred flat planting in lines of six feet apart, which was most suited for mechanized cultivation, while some used ridge planting, which proved to be popular in swampy areas. This type of cultivation was especially found in the Miwani sugar estate. The method of planting adopted by farmers also varied according to the nature and condition of the soil. When planting cane, small pieces of stem known as setts, each about one foot in length and preferably from a plant crop, were used. However, only healthy young setts were to be used. The setts were then generally planted in furrows opened by a mould-board plough at a distance of between four and six feet. The cane setts were then placed flat, touching each other in the furrows, and then covered with soil. When planted in an irrigated farm, the setts were to be covered with three to four inches of soil. However, since Kibos-Muhoroni farmers relied on rain-fed farming, the setts were placed deeper. To plant an acre of cane in this way, one or two tons of setts were required.[110] Growth was noted from the eleventh day after planting if temperatures were high enough for approximately one month.

The use of manure was also very important in cane production. Farmers could use either natural or chemical fertilizer. These could be obtained from farmyard manure, green manure, and chemicals such as sulphate of ammonia and potash. Other farms also used molasses and filter press mud. Proper fertilization with sulphate of ammonia required three to six cwts of sulphate of ammonia per acre. Fertilization with farmyard manure required cooperation with dairy farmers: molasses and cane tops from the sugarcane farms were useful to the dairy farmers as feeds, while animal waste from the dairy farms could be used as fertilizer by the sugarcane farmers. This practice was very common among the Asian cane farmers at Kibos. Moreover, sugarcane did not need to be rotated with other crops. For example, the practice of "resting" sugarcane land under other crops like sunflower, which was quite prevalent by the 1940s, proved not worthwhile as far as fertility of the soil was concerned.

Cane farmers also had to deal with the problems of weeds. It was necessary to control the growth of weeds from the early stages until the crop had grown sufficiently to cover the rows. Farmers had to hoe the land four to five times and weed at least twice during each sugarcane crop until the cane reached maturity, which in most cases was between twenty-two to twenty-four months for crop cane and twenty to twenty-two months for ratoon when there was an even distribution of sugar in the stem. Small-scale farmers weeded manually whenever necessary. Most large-scale and estate farms, such as Miwani, used chemicals to control weeds, especially the most stubborn weeds like *Digitaria scalarum*.

The most common chemicals used was an agroxone hormone weed killer as a pre-emergence treatment after the field was ridged. The chemical was applied using one to four gallons per acre, diluted with water to make forty gallons of spray.[111] After this process, cane had to be planted immediately or soon after. Fernoxone was used as a pre-emergent at three pounds per acre and YF2618 also a pre-emergent at twenty-seven fluid ounces per acre. Fernoxone was also used with P.C.P. oil (two pounds per acre) as a post-emergence application.

The first pre-emergence spray of the chemicals was done three to seven days after planting while the second pre-emergence spray was done four to six weeks after the first application. Post-emergence spray had to be done as soon as the grass weed emerged.[112] The sprays did not exceed thirty gallons per acre. However, when using hand spray it was found necessary to use up to eighty gallons per acre to get uniform coverage.[113] All these treatments made sugarcane farming an expensive venture for poorer farmers.

During the growth season some cane flowered while others did not; some flowered sooner than others. Flowering of cane was the stage when the vegetative growth changed to reproductive activity. Farmers had to ensure that before such changes took place, a sufficient number of mature nodes had appeared in the cane. The flowering of cane was not good since it stopped the growth, though the sugar content increased for about four months afterward.[114] Flowering was due to heavy soil and favorable weather, as was the case in Lall Singh's farm at Kibos. Yet on this particular farm, the indigenous cane, Uba, did not flower under the same conditions as did the flowering varieties of CO 312 and CO 271. Moreover, these flowering varieties, being very hard, were resistant to pests, wild animals, and drought.[115] As such, they were highly recommended to be grown in the area.

Harvesting was done when the cane was fully ripe. Determining factors included rainfall, the variety, and whether it was a plant crop or ratoon. This could be determined by the appearance of new young shoots in the cane stool of over eighteen months of age. During this period the tops were cut just above the node and removed. The length of the top depended on weather conditions. For example, in the drier period the length would be shorter than in wet weather since the plant continued to grow. Cane was then fired, especially Uba, which was difficult to strip. Firing was done to remove dangerous insects and animals like snakes and also to remove the leaves. This method of firing, however, destroyed the fertility of the soil and was later abandoned.

After firing, the blades or *pangas* were sterilized to avoid the transmission of diseases, and then cane was cut with as close to the ground as possible since the lowest portion of the stem had the highest sucrose content.[116] The dried materials of the crop after harvesting were then collected and put in alternative lines of cane so as to allow inter-culture in other rows. In this way, the dried waste material was left to decay and mix with soil to increase fertility; it was also found to hasten the germination of the ratoon crop, and it preserved moisture in the soil for longer periods. After cutting, the stools were allowed to grow again (creating a ratoon crop). It was usual to harvest a plant crop and two ratoon crops, and sometimes even a third ratoon was harvested. The first ratoon crop was referred to as the "money crop" since, in most cases, it paid off the cost of the crop.[117] However, it was normal to notice a decline of about 10 to 20 percent in each successive ratoon yield. For example, in a non-irrigated, well-managed farm, cane would yield as follows in 1946:

- 16–20 tons of cane per acre from plant crop
- 12 tons of cane per acre in the first ratoon
- 10 tons of cane per acre in the second ratoon[118]

Although some poor farmers attempted to harvest even a third ratoon, this did not yield much. Therefore, it was necessary to replant after the second ratoon. An incidence of harvesting a fourth ratoon was rare in Nyanza, as opposed to India, and was not recommended by agricultural officers such as Mr. Sampura Singh Gill.

Once the cane was cut, it was transported to the milling factories. It is clear from this account that cultivating sugarcane was not cheap, especially for the small cane farmers as illustrated in the 1946 estimates. The 1946 estimates were chosen because during that year the cane farmers had problems. It was a dry season and the production was adversely affected by labor shortages.[119] These were frequent issues that the cane farmers dealt with. Furthermore, the war had ended and so demand had decreased. Therefore, 1946 provides a good case study of typical challenges for Nyanza farmers. It is clear from table 3.7 that, even after the first harvest, farmers had to spend more money with each consecutive ratoon crop.

Table 3.7 indicates that cane grown by 1946 was expensive to the small-scale farmers; the estimates given above were just for one acre, but, on average, most of the cane farmers had fifty acres. Additionally, cane cultivation involved a lot of essential tasks such as plowing, harrowing, and harvesting. All of these tasks required labor costs as well. Despite the challenges, the Asian farmers persisted and were determined to improve on cane farming.

Fortunately, cane growing picked up again in the following year, as there was heavy rainfall in 1947. The short rains fell during the early part of the year and continued until the long rains, which resulted into a period of nine months of good weather conditions. This was good for cane growing, but the rains destroyed maize on Asian farms due to flooding.[120] There was also adequate labor during the year, as such production increased. Moreover, it was during this time that the new management of

Table 3.7. Estimated Cost of Cultivation of Sugarcane per Acre, 1946

		Tasks	Amount (in shillings)	
1st Crop	(a)	Clearing and de-rooting	140.00	per acre
	(b)	To plough three times, harrowing and ridging by tractor	90.00	per acre
	(c)	Price of seed 1½ ton per acre at 10/- per ton	15.00	
	(d)	Transport charges of seed for one acre	5.83	
	(e)	Labour charges for planting	5.33	per acre
	(f)	Hoeing four times at 8/- per acre	32.00	
	(g)	Hoeing with cultivator and ploughing	8.00	per acre
	(h)	Cost of harvesting the cane at 2/- per ton on the basis of 15 tons per acre	30.00	
	(i)	Laying line, etc.	9.00	
	(j)	Supervision and other charges	20.00	
		Total	355.00	per acre
1st Ratoon	(a)	Cost of clearing trashes	4.00	per acre
	(b)	Cost of ploughing with oxen and harrowing	15.00	
	(c)	Cost of hoeing with plough and planter	8.00	
	(d)	Cost of hoeing with hand three times	24.00	
	(e)	Cost of harvesting the cane at 2/- per ton on the basis of 12 tons per acre	30.00	
	(f)	Laying lines, etc.	9.00	
	(g)	Supervision and other overhead charges	20.00	
		Total	110.00	
2nd Ratoon		Same as 1st Ratoon	110.00	

Source: Estimated Cost of Sugarcane Production, 1946, KNA: AK/11/50.

Miwani Sugar Mills started under Hindocha brought improved relations between the factory and the cane farmers.

TECHNICAL ASPECTS, STATE SUPPORT, AND THEIR IMPACT IN COLONIAL SUGARCANE FARMING

Asian cane farming developed over time, although during the period up to 1930, some of the farmers especially the small-scale farmers experienced difficult times. However, the period of the 1940s and 1950s was characterized by a marked increase in productivity, due to the resilience of the cane farmers coupled by the changes in colonial policies that were put in place to meet the needs of production during World War II. By the 1950s the sugar industry was transformed and solidly established, in part due to the colonial state's provision of technical support for experimentation and testing. Such experimentation began in the 1920s but reached their peak in the 1950s when major trials and experimentations were undertaken, covering almost all aspects of sugarcane farming from trials of the sucrose content of different varieties, spacing of cane, growth on different soils, and irrigation. It was these trials that transformed sugarcane farming in colonial Kenya. Without the technical support the industry would probably not be where it was at independence. Once the colonial government changed its stance toward the sugar industry, it not only imported varieties from many countries but also invited sugar experts from different countries to advise the farmers. This led to the growth of the industry as more technical support was availed to the cane farmers.

The colonial state's support for experimentation was significant. For example, the state put procedures in place for the safe import of different varieties. Before importation, the cuttings had to be inspected and certified in their originating homes. Furthermore, an authorized official of the department of agriculture of the country of origin had to certify that there was no evidence of the presence of any insect pests or plant disease before the varieties were shipped to Kenya. The cuttings, once imported, were then railed to Kisumu. Upon arrival they were inspected by the PAO before being sent to the government experimental station at Kibos. The importation permit was valid for only six months, and if any contamination or disease was found, the varieties affected had to be destroyed immediately. Trials were meant to investigate major issues such as the use of different fertilizers and weed killers, methods of cultivation, such as spacing, ridging and irrigation, and, finally, investigation of the possible use of molasses and filter press cake as fertilizers.[121]

Prior to 1930, the main commercial variety at Kibos-Muhoroni was CO 281. However, small amounts of CO 290, CO 421, and Uba were also cultivated. By 1948, the colonial state was fully involved and supported farmers by facilitating variety trials that were undertaken by the department of agriculture. It was these trials that revealed that CO 421 was a better variety for the Kibos-Miwani area than CO 281 in terms of yield and sucrose content. Moreover, it was found that CO 421 did not bend easily on the farm from the effects of strong winds, which was a desirable characteristic should mechanical harvesting be employed.[122] The trials made it possible for farmers to have many varieties of sugarcane from which to choose from. This was important because farmers were very careful with the varieties of crop that they planted; hence, the need to experiment on different varieties so as to improve production. Consequently, in 1947, the Miwani Sugar Company embarked on a project of cane variety trials. The seeds were obtained from the farms of Kala Singh and Kisham Singh of Kibos and some from Badian Singh and Bharmant Singh of Miwani. This initiative was followed in 1948 when the government started its own sugar research station at Kibos, as indicated earlier. The aim of the station was to try and identify disease-resistant varieties that were suitable to the ecology of the area and of high yield potential. The work was supervised by an AAO (sugar) known as Sampura Singh Gill.[123]

These experimental farms were important, for they in turn supplied cane varieties to other plantations such as Turney Sugar Estate in Bukoba and Nachingwea Plantation in Tanganyika.[124] The Kibos Station also worked together with the Scott Laboratory in Nairobi where the plant pathologist, Dr. Nottrass, was based, whose technical expertise was often sought. The resulting provision of improved seed varieties was a great push by the colonial state. Many of the trials were done at the research stations or on individual farms before being released to farmers. To get better results, many tests were done on different varieties from different countries in the world. By the 1950s, the farmers were well informed and had many varieties to choose from, apart from Uba, as indicated in table 3.8.

Table 3.8 shows the development of sugarcane research in colonial Nyanza over the years. It started with only about three varieties in 1913 and later made great progress. The data demonstrate how the colonial government tried to improve the yield of sugarcane through experimentation to identify the most suitable varieties for the farmer. These varieties were from many different research stations in the world. The experiments provided an enormous wealth of knowledge that was found to be necessary to improve sugarcane farming in Nyanza. Other varieties were introduced later in the early 1960s, such as CO 617, which was ready for harvest after only seventeen months and did not need intensive weeding; the newest variety today is CO 808.[125] Varieties used at Kibos-Muhoroni came from the following sources by 1953:

- Sugarcane Research Station, Canal Point, Florida, United States
- Sugarcane Research Station, Hawaii, United States

Table 3.8. Different Varieties Imported and Grown at Kibos-Muhoroni by 1953

Variety	Origination
M 171/30	Imported from Mauritius
M 134/32	Imported from Mauritius
MF 240	Imported from Mauritius
CO 281	Imported from Coimbatore—Ratoon canes very thin. Resistant to infection, drought resistant, not suitable for small scale farmers, soft—favorite for wild pits and monkeys, has weeds until it is harvested, need chemical weeding.
CO 290	Imported from Coimbatore—Does better in rich soils at the foot of Nandi hills
CO 312	Imported from Coimbatore—Better yields, especially during first year of cutting
CO 408	Imported from Coimbatore
CO 411	Imported from Coimbatore
CO 421	Imported from Coimbatore—Known locally as "manywele," drought resistant, resistant to mosaic disease, high sucrose content, and best cane
CO 419	Imported from Coimbatore and had better yields
CO 271	Imported from Coimbatore, mostly favorable for making jaggery production
CP 807	Imported from Canal Point—Florida
POJ 2878	Imported from Jamaica.
POJ 2961	Imported from Jamaica.
PR 803	Imported from Puerto Rico
B 34104	Imported from Barbados
B 37161	Imported from Barbados
B 4098	Imported from Barbados
B 3254	Imported from Barbados
N CO 334	Imported from South Africa
N CO 339	Imported from South Africa
N CO 349	Imported from South Africa
N CO 310	Imported from South Africa —good yields
N CO 79	Imported from Tanganyika
N CO 291	Imported from Tanganyika—originally from South Africa
N CO 330	Imported from Tanganyika—originally from South Africa
CP 34/79	Imported from Hawaii
CP 34/120	Imported from Hawaii

CP 36/105	Imported from Hawaii
CO 331	Imported from Coimbatore
CO 441	Imported from Coimbatore
CO 453	Imported from Coimbatore

Source: Varieties Grown at Kibos-Muhoroni area, 1953, KNA: AK/11/50.

- Sugarcane Research Station, Coimbatore, India
- Asian Institute of Sugar Technology, Kanpur
- B. W. 1 Central Sugarcane Breeding Station, Barbados
- Sugar Research Station, Mauritius
- Sugar Research Station, Mt. Edgecombe, Natal, South Africa
- Agricultural Experiment Station, Mayaguez, Puerto Rico[126]

The state's support improved sugarcane cultivation as Asian farmers received more attention from the colonial state, which had in the initial years ignored the agricultural potential of the area. The colonial state was now keen on improving the sugar industry, and so the experiments were not only confined to the Kibos-Muhoroni area but also were carried out in other cane-growing areas within the Nyanza sugar-belt such as the Miwani-Kibos area.

SUGARCANE VARIETY TRIALS MIWANI-KIBOS AREA

These trials were conducted on individual Asian and European farms where a farmer would be issued with various varieties and other necessities such as fertilizers and weed killers. In some of these trials, the farmer had one section of land that was tested with a new method or variety while the rest of the farm would be worked on normal basis and the data from the two sections (e.g., yield) compared. The trials on individual farms were given the code 29 to differentiate them with those on research stations. For example, the trial at Kala Singh's farm was planted in 1950 and the first ratoon was harvested in April 1954, while the second ratoon was harvested in October 1955.

The data in Table 3.9 show that there was a large reduction in yield of the first ratoon compared with the plant crop because of the drought experienced during the dry season of 1952–1953. However, by comparing the plant crop with the second ratoon, it is clear that the mean yield reduction was only 8 percent. As such the second ratoon was as good as the plant crop. It is also evident that the CO 331 and CO 213 varieties produced the highest calculated yield of sucrose per acre of all the tested varieties.[127]

These same varieties were also tested during harvesting when the canes were stripped of leaves and various differences were noticed. For

Table 3.9. Trials Comparing Yields of Plant Crops and Ratoon Crops Planted in 1950

Variety	Plant Crop	1st Ratoon	2nd Ratoon	Total Tons/ Acre
CO 331	58.76	33.42	51.09	143.27
CO 213	58.10	25.07	31.93	115.10
CO 434	45.29	28.23	36.52	110.04
B 3337	41.21	24.80	40.71	106.72
CO 419	48.10	19.41	36.79	104.30
CO 408	43.81	19.64	35.80	99.25
CO 285	40.91	19.35	35.95	96.21
CO 281	31.07	22.26	40.14	93.47
CO 421	41.72	18.93	30.93	91.58
CO 417	34.98	19.76	32.68	87.42
CO 432	38.34	16.25	26.82	81.41
MF 2/40	35.59	17.37	28.38	81.41
POJ 2747	26.96	15.95	29.41	72.32
CO 244	33.18	13.16	25.42	71.76
B 2257	22.17	11.15	24.15	57.47
POJ 2714	20.33	10.37	21.64	52.34
Mean	38.77	19.69	32.02	—

Source: Sugar Variety Trials, Kala Singh's Farm, Miwani-Kibos Area (29/2/2), KNA: AK/2/6.

instance, observations were made on cane length, thickness, and number of canes per stool and lodgings with the following results: Varieties with largest canes were CO 331, CO 408, CO 432, POJ 2747, CO 244, CO 281, CO 417, and CO 434. Varieties with the thickest cane were CO 419, CO 421, and POJ 2714. Varieties with the thinnest cane were CO 244, CO 408, and CO 432. Varieties with the most cane per stool were CO 213, CO 421, CO 244, CO 285, CO 432, and CO 331. Varieties with least cane per stool were POJ 2714, POJ 2747, MF 240, and B 3257. Varieties with most lodging were CO 434, CO 417, MF 240, CO 281, and CO 244. Those with the least lodging were B 3337, CO 421, CO 213, CO 408, and CO 432.[128]

These trials were important as they made it possible for the cane farmers to be better informed and enabled the growers to choose the most profitable variety of cane. The farmers participated in many trials to get better results. Many offered their farms for experimentation. Further trials were undertaken; for example, sixteen varieties were tested at Kala Singh's farm. Trials were undertaken in different soils, such as the upper red soil of the Miwani Estate and lower black soil. These trials were meant to determine which soil was most conducive to cane production, as indicated in table 3.10.

Table 3.10 shows the yields in red soil of various varieties. The results from this particular trial provided important information for farmers whose farms had red soil since they were able to know the best variety to plant. Another experiment was conducted concurrently using the same varieties on the black soil for comparison. This second experiment indicated that CO 331 gave a significantly higher yield than any other variety as was the case in the first ratoon harvest on the upper red soil. (See table 3.10.) However, the significant mean fall in yield was attributed to a ratoon-stunting disease that was prevalent in the area.[129] Moreover, the comparison between the yield grown in red soil and black soil indicated that sugarcane did better in black soil. The information from these trials was shared among the farmers and many adapted and grew the varieties that had successfully been tested. This greatly improved the sugar industry. For example, Kibos alone had approximately sixty-eight different varieties grown by farmers who had different types of soil. The technical support given to farmers by the colonial government enabled them to have enough information to improve their farming.

Apart from the tests that determined the suitability of varieties to the ecology of the area, the sucrose content of sugarcane was important to cane farmers. Therefore, many trials were also carried out on experimental farms, not only at Kibos but also at the Miwani estate, to determine such factors. A trial of sixteen varieties was done whereby planting took place in March 1953 on the sandy red loam soil typical of the upper part of the estate; it was harvested in January 1955. The results of the yields of cane together with the sucrose of the yields calculated in tons per acre are shown in table 3.11.

Table 3.10. Sugarcane Planted on Red Soil, Miwani Estate, and Harvested 1955

1st Ratoon	**Tons of Cane per Acre (Upper Red Soil)**
Variety	Yield
CO 331	31.45
CO 419	22.74
CO 421	22.44
CO 213	19.42
CO 408	19.09
Mean	23.02

Source: Sugarcane Variety Trial Miwani (29/2/5), KNA: AK/11/52.

The trial shows that CO 411 yielded the highest, while CO 453 had the highest sucrose content. The poorest cane was CP 34/79, which had both low yields and sucrose content. The data also indicated that the N CO 291 was a good cane since it had good yield and sucrose content. Such information indicated that perhaps there were other determinants for a good cane, such as sucrose content, which determined the tons of sugar produced from a particular cane. All these trials that were carried out by farmers with the help of the colonial state greatly assisted the Asian sugarcane industry in the 1950s.

Both the farmers and the colonial state continued to make improvements in cane farming. For example, another trial was undertaken of twenty-five varieties at Miwani. It was planted in September 1954 to compare with other varieties tested. This test incorporated new varieties also. The crop was harvested in April 1956, as indicated in table 3.12, which indicates the yield figures of cane and of sucrose in tons per acre.

Table 3.11. Sugar Variety Trial on Yield and Sucrose Content, Miwani Estate, 1953

Variety	Yield of Cane (tons/acre)	Sucrose Yield (tons/acre)
CO 411	52.80	4.78
N CO 291	49.01	5.13
CO 453	47.82	5.52
B 37193	47.75	4.45
B 41227	41.64	4.01
POJ 2961	40.52	4.67
N CO 79	40.28	5.26
CO 290	39.66	3.54
CO 421	39.39	3.58
CO 281	37.94	4.84
N CO 330	36.68	3.94
B 3337	34.68	3.63
CP 36/105	29.99	3.07
B 3257	29.22	3.54
CP 34/120	22.77	2.24
CP 34/79	12.06	1.29
Mean	37.64	3.97

Least significant difference = 0.05. Source: Sugarcane Variety Trial, Miwani (29/2/6), KNA: AK/2/6.

Generally the yield from this experiment was very high, and the variety CO 419 out-yielded other varieties. N CO 399, which was a new variety, also registered a high yield and had the second best sucrose content. Hence, the need for constant trials was important as new varieties were found. However, the data also revealed that while CO 331 was mostly

Table 3.12. Sugar Variety Trial on Yields and Sucrose, Miwani Estate, 1954

Variety	Yield of Cane (tons/acre)	Sucrose Yield (tons/acre)
CO 419	66.01	7.13
N CO 339	63.44	6.76
CO 331	62.39	5.68
N CO 334	62.26	6.42
CO 301	61.97	6.47
CO 434	61.00	6.18
N CO 349	59.80	6.42
CO 312	58.30	6.14
CO 411	55.02	4.99
B 41227	54.96	5.61
N CO 330	53.49	6.19
N CO 291	52.99	5.53
N CO 310	52.03	6.05
CO 281	50.75	5.26
CO 452	50.61	5.15
CO 421	48.29	5.13
POJ 2875	45.95	4.84
B 37193	45.90	4.40
CO 408	45.82	5.86
NCO 79	45.10	4.75
POJ 213	43.57	4.14
POJ 2961	43.08	4.55
B 3337	42.23	4.26
B 4362	41.09	5.32
CO 213	36.59	3.28
Mean	52.10	—

Least significant difference = 0.05. Source: Sugarcane Varity Trials –Miwani (29/2/8), KNA: AK/2/5.

favored and recommended for planting, it had a low sucrose yield. The trials indicated that the highest sucrose content cane was CO 419 and the lowest CO 213. Farmers needed such information to make economical choices since they needed cane that had good yields as well as good sucrose content if farming was to be profitable. Cane with low sucrose content often produced less sugar and hence fetched lower prices.

Furthermore, to ensure more yields, other experiments were conducted on cane spacing during cultivation to help the farmers attain good yields. A sugarcane spacing experiment was conducted at Chemelil farms using the CO 281 variety. The cane was planted in December 1949, and the crop harvested in October 1951. The experiment was geared toward studying three different rows of spacing: four feet, five feet, and six feet. When the cane was harvested in 1951, there was no significant difference in yields between the three different spacings. The first ratoon harvested in June 1953 showed no major difference either. Although there was very little difference in yield among the various rows in the plant crop, however, on the second ratoon, there was a tendency for the two wider spacing amounts (five feet and six feet) to yield more than the four-foot row. Hence, six-foot spacing was recommended to cane farmers since it was the most suitable for mechanized cultivations and in large-scale farming.[130]

EXPERIMENTS ON THE POSSIBLE IRRIGATION OF SUGARCANE FARMING

Although in the Kibos-Muhoroni area cane cultivation was basically rain-fed, experiments with irrigation had started in 1930 when, due to the severe effects of drought, Muhoroni embarked on irrigation experimentation and got good results. These initiatives were followed in 1953 when Miwani Estates started an experimental irrigation scheme for sugar. The water was drawn from the tributary of the Nyando River near Chemelil.[131] The idea was to pump water up to the top of the small hillock and then gravity-feed the ditches from there. In 1956, after the trials, Miwani Sugar Mills irrigated 1,100 acres of cane at the Chemelil nucleus estate. Another 500 acres was also irrigated at their Miwani estate, where they used boreholes.[132] However, there was no large-scale irrigation of sugarcane on individual Asian farms since the only rivers available were the Kibos, which was seasonal, and the tributaries of the Nyando, which were already used by the Miwani estate. Moreover, there were no permanent streams coming from the Nandi escarpment into the sugarcane estates. Nevertheless, the main point for experimenting with irrigation of sugarcane on Miwani estate was to see whether the low-yielding CO canes could be replaced with the high-yielding POJ canes by one to two watering during the dry season from January to March. Storage of water on the lands would have required very large dams, which were expensive and needed large capital. However, because it was a worthwhile venture, the colonial state could not ignore it.

There was a great push for irrigation during this period, which led to a deputy hydraulic engineer to visit Kisumu in December 1953 to make a preliminary assessment of the possibilities of irrigating the sugarcane in the Kibos area.[133] The engineers who were contracted by the government evaluated the Kano plains in a pilot project begun in 1954–1956. The evaluation showed the potentialities of irrigation and swamp reclamation in Nyanza Province. The engineers' report recommended that the development of the Kano plains was a priority in the development of the lake basin. The report clearly stated that it was possible to irrigate by pumping water from Lake Victoria, which could be supplemented by the gravity flow from the Nyando River.[134] The irrigation potential of the Kano plains was assessed at approximately 30,000 acres based on the possibility of two crops: sugar and rice. If implemented, this could have gone a long way to help in sugar production in Kenya and toward food sufficiency as some farmers, especially the peasant farmers in the region, could have expanded their rice farming. The report also recommended the establishment of experimental farms to study the growing of sugarcane under irrigation and the construction of pilot schemes to determine, among other things, the size of holding best suited to small-scale irrigated farming.[135]

The area estimated to benefit from possible irrigation was approximately 19,500 acres under cane and an additional 16,000 acres from the Miwani estate and the Asian farms in the Kibos-Muhoroni area where sugarcane could be grown successfully. Hence, it was possible to irrigate in total about 35,500 acres as indicated in table 3.13.

Table 3.13 shows that if irrigation were adopted, more acreage could have been planted, which could have gone a long way toward self-sufficiency in sugar consumption. Furthermore, it shows that irrigation of sugarcane could have improved productivity for Kenya as a result of additional water supplies being made available during the critical period from November to March. The irrigation would have increased production by eliminating the effects of the dry season, making it possible for the cane to grow continuously; that way, the length of the cycle required to produce three crops would be shortened from five years to four years.

Moreover, irrigation would have permitted the growth of improved varieties of noble canes, which yield 25 percent more than other varieties in the Kibos-Muhoroni-Miwani area. However, the noble canes could not be grown since they are not drought resistant.[136] It was estimated that if 20,000 acres of cane were irrigated with those varieties grown at Kibos, there would be an increase in production of 360,000 tons of cane a year, worth £540,000. However, if improved varieties of noble canes were grown, the increase in production would have been 510,000 tons of cane a

Table 3.13. Possible Acreage to Benefit from Irrigation, 1953

Estates	Acres under Sugarcane	Total Possible Acreage When Irrigated
Miwani Main Estate	7,000	9,000
Miwani Chemelil Estate	3,500	5,000
Kibos Asian Farms	6,000	12,000
Miwani Asian Farms	2,000	3,000
Kibigori Asian Farms	700	5,000
Muhoroni Asian Farms	300	1,500
Total	19,500	35,500

Source: Irrigation of Sugarcane, Nyanza Province, 1953, KNA: AK/11/51.

year, worth £765,000. For example, 60 tons from one acre in five years yielding 12 tons per acre, would produce 240,000 tons per annum, valued at Shs. 301/ton, which would be worth £360,000. This could have made cane growing very profitable. Moreover, there was also a significant difference between ordinary cane and noble cane grown under irrigation. Normal cane under irrigation could produce 120 tons from one acre in four years, which would equal 30 tons per acre per annum. Likewise, 20,000 acres would produce 600,000 tons per annum. Valued at Shs. 301/ton, it would be worth £900,000. On the other hand, irrigation with improved noble canes could produce 150 tons from one acre in four years that yielded 37 tons per acre per annum. And 20,000 acres would have produced 750,000 tons per annum, which, valued at Shs. 301/ton worth, would be worth £1,125,000.

Therefore, if irrigation could have been possible, the cane farmers could have greatly benefited. Moreover, some of the Asian farmers had already the experience of growing sugarcane under irrigation in India.[137] Although it required a large amount of capital, Miwani Sugar Mills was able to build a dam, and by 1953 it was possible to irrigate 2,000 acres at Chemelil.

It is possible that there would not have been enough water to irrigate large areas of the sugarcane plantations in the area. Assuming that the water requirements were one cusec per 100 acres (cusec meaning a unit flow of water equal to one cubic foot per second) and taking into account only the existing 1953 acreage, the water requirement would have been as listed in table 3.14. The data in table 3.14 show that during the dry weather, the flows in the rivers in the area would be very limited and far from these figures. Because of this, storage would have required the use of very large dams, which in turn required large amounts of capital.[138] Moreover, this was the most critical period (November-March) when irrigation was needed. Therefore, based on the findings, the irrigation of sugarcane in Nyanza was shelved due to these technicalities, and sugarcane farming continued to be rain-fed in the Asian farms.

Nevertheless, apart from the experimental trials by the colonial government, other initiatives were seen in terms of personnel, when experts were invited to visit the sugar zones. For example, in December 1952, Mr. Barnes, a sugar expert from South Africa, toured the Songhor-Muhoroni and Miwani areas where the farmers had 28,000 acres of land under cane. Unfortunately, not all were used for sugar, as some was turned to jaggery. Besides discussing this with the agricultural officers, he met the farmers and gave them technical advice on how to improve cane cultivation and make it profitable.[139] In his report, Barnes discouraged the establishment of a second factory in the Kibos-Miwani area on the grounds that the quantity of cane produced in the area was far from adequate to maintain a separate mill economically. Instead, he recommended that communications in the area be beefed up so as to allow farmers to deliver their cane to the existing factories.[140] However, later when Mr. P. O. Wiehe, the sugar research expert from Mauritius, visited CN at government invitation in 1956, he reiterated the need for irrigation to improve cultivation. He recommended that the improved varieties had to be complemented by irrigation to attain better results. Such information was crucial to determine if irrigation was viable, especially to the small-scale farmers, and if peasant production in sugarcane farming was sustainable at all in comparison with large-scale farming

CONCLUSION

The growth of the sugarcane farming industry during the colonial period was dogged by severe difficulties and even more persistent resourceful-

Table 3.14. Approximate Amount of Water Required for Irrigation, 1953

Estates	Continuous Flow Required	Storage for 5 Months (November–March)
Miwani Main Estate	70 cusecs	21,000 acre feet
Miwani Chemelil	35 cusecs	10,000 acre feet
Kibos Asian farms	60 cusecs	18,000 acre feet
Miwani Asian farms	20 cusecs	6,000 acre feet
Kibigori Asian farms	7 cusecs	2,000 acre feet
Muhoroni Asian farms	3 cusecs	900 acre feet
Total	195 cusecs	58,000 acre feet

Source: Irrigation of Sugarcane, Nyanza Province, 1953, KNA: AK/11/51.

ness on the part of the farmers. This chapter reviewed the growth of Asian cane farming in the early years of the twentieth century, including the number of farmers, size of farms, the varieties used, and the farm location, especially vis-à-vis the transport routes that allowed them to get the cane to mills and other markets. The farms were strategically located along the Kenya-Uganda railway, but the state failed to subsidize the transportation of cane or come up with reasonably priced rail transport, which inhibited the full potential of the industry. Although the local variety known as Uba had been grown in the area for years, it was the Asian farmers who undertook serious farming ventures and introduced the CO varieties and other techniques.

During World War I and in the years just after, the farmers faced obstacles to their growth. Transport continued to be expensive, so cane that could not be transported to mills was turned into jaggery instead. The farms themselves had no milling equipment and even had to rely on ox-drawn ploughs for tilling. However, the industry received a boost with the construction of the Victoria Nyanza Sugar Syndicate in 1921 and later the Muhoroni Sugar Mills, both owned by Europeans. By 1923 roughly 4,193 acres were under cane cultivation in the Kibos-Muhoroni area.

The industry received a further boost in the interwar period when the farmers and the colonial state embarked on an intense, thorough series of trials on cane varieties and cane farming techniques. Trials were conducted on fertilizers, row spacing, irrigation schemes, soil, and more. These trials represented one of the first serious investments that the colonial state made in the sugarcane industry and was of immense value.

These trials were of particular value because of the continuing challenges the farmers faced during these years. The weather alternated between droughts and overly heavy rainfall. There were infestations of locusts and invasive weeds. The colonial state continued to limit the Asian farmers' efforts to acquire more land for cultivation. And the Great Depression caused a decrease in prices and demand that lasted for years.

The beginning of World War II marked an upturn for the industry. The Depression ended, and the colonial state changed its policies to produce enough foodstuffs to meet the war demands. After World War II remarkable progress was achieved with the entrance of Asian entrepreneurs in the sugar manufacturing industry in 1947. These experienced sugar industrialists revolutionized the industry completely by buying more large estates, contracting farmers, and laying more trolley lines for cane transportation. In addition, the massive government program of experimentation saw the successful testing of sixty-eight varieties from various countries; sugar experts were also invited from different countries to help out. This state's positive intervention through the trials and experimentations boosted the sugar industry in the postwar period.

Although the Asian farmers had made important strides, they were still handicapped by the colonial state's preferential treatment of the European farmers. Access to land was a continual problem, as was discrimination in financial matters. For example, while the Land Bank provided loans to European farmers, it denied credit to the Asian farmers. Even as demand increased and prices rebounded, these discriminatory policies and lack of political will by the colonial state proved a serious obstacle to the industry's growth.

NOTES

1. Ross, *Kenya from Within*, 309.
2. M. F. Hill, *Permanent Way: Story of the Kenya-Uganda Railway and Harbours* (Nairobi: East African Railway and Harbours, 1949), 253–254.
3. Peter Anyang' Nyong'o, "The Development of a Middle Peasantry in Nyanza," *Review of African Political Economy* 8, no. 20 (1981): 110.
4. Gregory, *India and East Africa*, 70.
5. KNA: DC/CN/1/5/1, Kisumu District Annual Report (KDAR) 1912–13.
6. Osamba, "A Forgotten Minority," 3.
7. KNA: DC/CN/1/5/1, KDAR 1912–13.
8. Joshia Osamba, "Colonial African Labour in Asian Owned Sugar Plantations in Kisumu District, 1901–1963," M.Phil. Thesis, Moi University, 1996, 43.
9. KNA: PC/NZA/3/22/1/8, Report by Magor of a Visit to Asian Farms in the area of Kibos-Miwani-Kibigori-Chemelil, October 6–10, 1947.
10. KNA: PC/NZA/3/22/1/8, Report by Magor.
11. Colony and Protectorate of Kenya, *Department of Agriculture Annual Report (DAAR)* 1919–1920 (Nairobi: Government Printer, 1921), 126.
12. KNA: PC/NZA/3/3/22, Sugar Cane Industry, 1913.
13. Nicola Swainson, *The Development of Corporate Capitalism in Kenya* (London: Heinemann, 1980), 124. This discrepancy could have been due to the thirty-year rule placed on accessing archival sources.
14. KNA: BN/1/249, Survey and Registration Department to Acting (Ag.) Commissioner for Local Government, Lands and Settlement, July 11, 1932.
15. Robert Maxon, "The Years of Revolutionary Advance, 1920–1929," in W. R. Ochieng', ed., *A Modern History of Kenya*, 72.
16. *DAAR 1919–1920*, 5.
17. Fearn, *An African Economy*, 106.
18. *East African Standard*, August 14, 1921.
19. Indian Overseas Association, *East African Pamphlet* 6, no. 45 (1921):10.
20. KNA: AG/2/19, A. C. Cavicchioni, Royal Consul, on Behalf of H M the King of Italy, to Northey, August 20, 1919.
21. KNA: AG/2/19, J. W. Barth, Attorney General, to Chief Secretary, September 20, 1919.
22. *DAAR 1919–1920*, 126.
23. Colony and Protectorate of Kenya, *DAAR of the Year Ended 31st March, 1921* (Nairobi: Government Printer, 1922), 21.
24. *DAAR of the Year Ending March 31st 1921*,139. According to the Report, the whole area stretching from Kibos to Fort Ternan was engaged in sugarcane growing.
25. Colony and Protectorate of Kenya, *DAAR for the Year Ended 31st December1922* (Nairobi: Government Printer, 1924), 21.
26. Colony and Protectorate of Kenya, *DAAR 1923* (Nairobi: Government Printer, 1924), 15. The report indicates that the year was good to farmers as even the European wheat farms at Lumbwa performed well.

27. Colony and Protectorate of Kenya, *DAAR* 1924 (Nairobi: Government Printer, 1925), 8.
28. KNA: PC/NZA/3/2/34, NIFA to Sir Robert Brooke-Popham, September 28, 1937. This letter highlights the background of Asian farming in Kibos.
29. *DAAR of the Year Ended 31st December 1923*, 15–16.
30. Colony and Protectorate of Kenya, *DAAR* 1925 (Nairobi: Government Printer, 1926), 17.
31. *DAAR 1925*, 17.
32. Colony and Protectorate of Kenya, *DAAR 1926* (Nairobi: Government Printer, 1927), 14.
33. KNA: BV/1/249, Survey and Registration Department to Ag. Commissioner for Local Government, Lands and Settlement, July 11, 1932.
34. KNA: DC/CN/1/4/1, K-LDAR, 1925.
35. Colony and Protectorate of Kenya, *DAAR 1929* (Nairobi: Government Printer, 1930), 24.
36. Colony and Protectorate of Kenya, *DAAR 1928* (Nairobi: Government Printer, 1929), 17.
37. Colony and Protectorate of Kenya, *DAAR 1928* (Nairobi: Government Printer, 1929, 17.
38. *DAAR 1929*, 25.
39. *DAAR 1929* 37.
40. *DAAR 1929*, 25.
41. Colony and Protectorate of Kenya, *DAAR 1931* (Nairobi: Government Printer), 29.
42. *DAAR 1931*, 27.
43. KNA: DC/CN/1/4/1, K-LDAR, 1927.
44. *DAAR 1929*, 68.
45. Gregory, *India and East Africa,* 298–299.
46. Gregory, *South Asians in East Africa,* 241.
47. Wrigley, "Kenya: The Patterns of Economic Life," 214.
48. KNA: AGR/6/4308, Miwani Farms 1930–1956.
49. Fearn, *An African Economy*, 105.
50. *DAAR 1931*, 44.
51. Colony and Protectorate of Kenya, *DAAR 1932* (Nairobi: Government Printer, 1933), 18.
52. *DAAR 1932*, 32.
53. *DAAR 1932*, 18.
54. *DAAR 1932*, 18.
55. Colony and Protectorate of Kenya, *DAAR 1933* (Nairobi: Government Printer, 1934), 21.
56. *DAAR 1933*, 18.
57. *DAAR 1933,* 156. The varieties were experimented to find out if they were drought resistant, but this proved to be unsuccessful.
58. KNA: PC/NZA/3/2/34, Nyanza Farmers Association to DC K-L, October 11, 1934.
59. KNA: PC/NZA/2/2/35, Crown Lands Rent Premier, 1932–1945.
60. Tabitha Kanogo, "Kenya and the Depression," in W. R. Ochieng', ed., *A Modern History of Kenya 1895–1980* (Nairobi: Evans Brothers [Kenya] Ltd., 1989),112.
61. Berman, *Control and Crisis*, 170.
62. Atieno Odhiambo, "Economic Mobilization," 143.
63. *DAAR 1931*, 27.
64. KNA: DC/CN/1/4/3, K-LDAR 1931. See also the 1932 *DAAR,* which indicated that the total area under cane had drastically reduced to 12,088 acres.
65. Kanogo, "Kenya and the Depression," 117–118.
66. KNA: DC/KSM/1/3/123, Land and Agricultural Bank Recipients, 1932–1935.
67. KNA: AK/11/51, Proposed GMR on Sugar.

68. *DAAR 1932*, 32.
69. KNA: PC/NZA/3/2/34, Petition by Nyanza Indian Farmers Association.
70. Colony and Protectorate of Kenya, *DAAR 1936*, Vol. 1 (Nairobi: Government Printer, 1937), 16.
71. KNA: PC/NZA/3/2/34, NIFA to Brooke-Popham, September 26, 1937.
72. Colony and Protectorate of Kenya, *DAAR 1935*, Vol. 1 (Nairobi: Government Printer,), 11. The farmers also found a ready market exporting the produce as a result of the Italian-Ethiopian War, possibly for the manufacture of ropes and bags.
73. *DAAR 1936*, 15.
74. KNA: DC/CN/1/4/5, K-LDAR 1936.
75. Roger Van Zwanenberg and Anne King, *An Economic History of Kenya and Uganda* (Atlantic Highlands: Humanities Press, 1975), 39.
76. KNA: DC/KSM/1/3/123, Application for Land Rehabilitation Fund, 1952.
77. KNA: BV/6/613, Sugar Export Quota.
78. Berman, *Control and Crisis*, 134.
79. Smith, "An Overview of Agricultural Development," 122.
80. Swainson, *The Development of Cooperate Capitalism*, 125.
81. Osamba, "Colonial African Labour," 96.
82. KNA: DC/KSM/1/3/122, Meeting of Kibos Planters Association, August 1942.
83. Swainson, *The Development of Corporate Capitalism*, 112.
84. Colony and Protectorate of Kenya, *DAAR 1945* (Nairobi: Government Printer, 1946), 14.
85. Colony and Protectorate of Kenya, *DAAR 1946*, Part I (Nairobi: Government Printer, 1948), 10.
86. Alex Winter-Nelson, "A History of Agricultural Policy in Kenya," in Scott Pearson et al., eds., *Agricultural Policy in Kenya; Application of the Policy Analysis Matrix* (Ithaca: Cornell University Press,1995), 35.
87. KNA: DC/KSM/1/3/122, Agricultural production on Indian farms, 1947. For more information about the sugar research station see also KNA: AK/11/50, Memorandum of Sugar Investigation for Kisumu Area, 1948, and Colony and Protectorate of Kenya, *DAAR 1947* (Nairobi: Government Printer, 1948) on the good weather conditions during the year.
88. KNA: DC/KSM/1/3/122, Agricultural Production Indian Farms, 1947.
89. Colony and Protectorate of Kenya, *DAAR 1948*, Part I (Nairobi: Government Printer, 1950), 32.
90. *DAAR 1948*, Part I.
91. Heyer, "Achievements, Problems and Prospects in the Agricultural Sector," 11.
92. Colony and Protectorate of Kenya, *DAAR 1950*, Vol. 2 (Nairobi: Government Printer, 1953), 28.
93. Colony and Protectorate of Kenya, *DAAR 1952*, Vol. 1 (Nairobi: Government Printer, 1953), 28.
94. *DAAR 1952*, 28.
95. *DAAR 1952*, 34.
96. KNA: AK/11/51, Indian Sugarcane Farmers, Nyanza Province.
97. KNA: AK/2/7, 1953–1958, Asian Areas Food Crops.
98. Hezborn Atudo (O.I.), former agricultural manager at Miwani Sugar Mills, June 6, 2009.
99. Mohinder Singh (O.I.), a large scale farmer at Kibos, June 6, 2009.
100. KNA: DC/CN/1/4/3, K-LDAR 1931.
101. KNA: AK/11/50, Memorandum on Sugar Investigation for Kisumu Area, 1948.
102. KNA: AK/3/15, Soil Chemist Scott Agricultural Laboratories Nairobi to Director of Agriculture, January 3, 1951.
103. KNA: AK/11/50, Minutes from the 70th Meeting of Agricultural Production Sub-Committee, Muhoroni-Kibos Area.
104. KNA: AK/11/50, Sugar Industry in Nyanza Province, Miwani-Kibos Area, 1945.
105. Martin Okeyo (O.I.), Kibos Nucleus Estate Farm Manager, June 24, 2009.

106. KNA: AK/3/15, Notes on Sugar Industry, prepared by Sampura Singh Gill, 1948.
107. KNA: AK/3/15, Notes by Sampura.
108. KNA: AK/3/15, Notes by Sampura.
109. Hezborn Atudo (O.I.), June 6, 2009.
110. KNA: AK/11/50, Sugar Industry in Nyanza Province, Miwani-Kibos Area, 1945.
111. KNA: AK/3/15, Agriculture Officer (AO) (experiments) to Director of Agriculture, July 24, 1952.
112. KNA: AK/3/15, Provincial Agriculture Officer (PAO), Nyanza, to Director of Agriculture, June 23, 1950.
113. KNA: AK/3/15, Provincial Agriculture.
114. KNA: AK/11/30, Assistant Agriculture Officer (AAO) (Sugar Research) to Sugarcane Planters, Nyanza Province, September 24, 1951.
115. KNA: AK/11/30, Kibos Planters Association to Sampura Singh Gill, AAO (Sugar Research), September 28, 1951.
116. KNA: AK/11/52, Sugarcane (Saccharum Officinarum).
117. KNA: AK/11/52, Sugarcane.
118. KNA: AK/11/50, NIFA to Chairman, Agricultural Production Board, August 6, 1946.
119. *DAAR 1946*, 10.
120. *DAAR 1947*, 9.
121. KNA: AK/3/15, Department of Agriculture to AO (experiments) Kisumu, April 9, 1951.
122. KNA: AK/3/16, AO, L. J. Collings-Wells, to PAO Coast Province, November 9, 1955.
123. KNA: AK/3/15, AO (investigation) Nyanza to E. G. P. Sherwood, Department of Agriculture, Eldoret, February 27, 1950.
124. KNA: AK/3/15, Director of Agriculture to Officer in Charge, Sugar Experimental Station Kibos, October 17, 1952.
125. Mzee Hezron Kaka Muynze (O.I.), Small Scale Farmer at Chemelil, 80 Years Old; He Has Only 3 Acres of Sugar Cane, July 1, 2009.
126. KNA: AK/3/15, Assistant Director of Agriculture (Research) to PAO Nyanza, July 1, 1950.
127. KNA: AK/2/6, Experimental Farm, Kala Singh-Miwani-Kibos.
128. KNA: AK/2/216, Sugar Variety Trials, Miwani-Kibos Area.
129. KNA: AK/2/5, Nyanza Province Agriculture Annual Report, 1956–1957.
130. KNA: AK/2/6, Sugarcane Spacing Experiment Trials, Chemelil.
131. KNA: AK/11/51, PAO Nyanza to Director of Public Works, December 8, 1953.
132. KNA: AK/2/5, Progress in Improved Farming-Asian Areas.
133. KNA: AK/11/51, Public Works Department Head Office, to PAO Nyanza, December 8, 1953.
134. Bank for Reconstruction and Development, *Economic Development of Kenya* (Baltimore: The John Hopkins University Press, 1963), 80.
135. Bank for Reconstruction, *Economic Development*, 80.
136. KNA: AK/11/51, PAO Nyanza to Director of Agriculture, May 4, 1953.
137. KNA: AK/11/5, POA Nyanza to Director of Agriculture.
138. KNA: AK/11/51, B. R. C. Koch, to Member for Agriculture and Natural Resources, September 29, 1953.
139. KNA: AK/11/51, D O, C N to the Assistant Director of Agriculture Nyanza, December 11 1957.
140. KNA: AK/11/51, A. G. Dalgleish to Mr. A. Pritam, April 18, 1953.

FOUR

Transportation of Sugarcane, Production, and Marketing of Sugar in Colonial Nyanza

The Asian cane farmers struggled through the difficulties of the interwar period: locust and weed infestations, droughts and wet seasons, the Great Depression and plunging prices. World War II brought a boost in demand, and the colonial state supported invaluable experimentation in cane varieties and farming techniques. Throughout all of these years, however, basic problems related to turning cane into sugar continued to beset the farmers.

In this chapter we will examine more closely three interrelated processes that were crucial in the process of turning sugarcane into a finished product to be marketed: the transportation of sugarcane to the factories, the acceptance of the cane by the factory, and finally the processing of the cane into sugar. At each of these three stages, the farmers were hampered by discriminatory policies that supported European farmers while neglecting Asian farmers. Moreover, colonial policies were marked by confusion and contradictions as the colonial government supported cane farming in some ways but inhibited its expansion in others.

The chapter also examines issues related to the pricing of sugarcane. These include payment terms, the low prices caused by the monopolistic power of the mills, and the devastating effect of dumping Ugandan sugar in the Kenyan market.

TRANSPORTATION OF SUGARCANE TO THE MILLS

Sugarcane farming involves crucial stages that have to be done at the required time to avoid wastage. We have seen how cane farming involved various experimentation to achieve better yield. Once the cane was ready it was harvested and taken to the factory. Transportation of cane remains an important aspect of sugarcane production. During the colonial period this was a huge problem due to poor roads and infrastructure. This forced those running the transport system to raise the cost of transport, which sometimes led to post-harvest losses due to delays associated with transportation as farmers struggled to transport their cane to the factories. After World War I, the small-scale farmers transported sugarcane from their farms to the main road or railway line using oxen-drawn carts. The use of carts was preferred also by the African farmers when they started planting cane. Although the carts were an inexpensive way to take cane to the main road, during the wet season it was difficult. From the main road, cane was then either transported by road or rail.[1]

SUGARCANE TRANSPORTATION BY ROAD

There were generally three means of transporting cane. The first was the use of medium-sized lorries, especially among the small-scale farmers who found the carts problematic to use, especially during the rainy season. By late 1940s some cane farmers acquired ex-military lorries left over from World War II, which were reconditioned and used for transportation. These lorries were heavily built and far better than the carts. However, they too were uneconomical and difficult to use, especially during the heavy rainfall since many became stuck in mud. The second method involved the use of trucks, which were actually lorries that were reconditioned and made larger using metal frames so that they were capable of carrying heavier loads. In most cases, the trucks belonged to the factory that had contracted with the cane farmers. Third, there were also some Asian businessmen who were not farmers but owned a fleet of trucks that transported cane to the factories. For example, Simba Singh, who owned many trucks by the 1950s and still operates them today, had a fleet of over 100 trucks whose purpose was just to ferry sugarcane to the factories.[2]

Whichever the means used, the transport of cane to the factory was mainly the responsibility of the farmer. Those who could afford to hired the heavily built trucks, which were found to be stronger than the lorries. Of course, the longer the distance the more the farmers paid. Yet there were no loans extended to help the farmers transport cane or to buy their own truck. The lack of adequate funding shows the contradictions of the colonial state's policy that was reluctant to support the cane farmers; yet the state was fully engaged in the experimentation process. After harvesting, a farmer had to find means to deliver cane to the trolley lines from the farm to the factory, which was very problematic. As put by Mohinder

Singh, as far as transport was concerned, *itho-thoni* (you carry your cross).[3] This was because the colonial state neglected the transport needs of the Asian cane growers and many, especially the small-scale farmers, had to struggle to ferry their cane to the factories.

Needless to say, apart from hiring the trucks the farmers had also to hire drivers and loaders who were not only responsible for loading but also unloading the cane at the factory. The loaders and truck drivers were only paid after the cane was accepted by the factory. Furthermore, when the roads were impassable during the rainy season, most of the trucks were stuck in the mud, and the driver had to wait until it was removed at no extra pay. Truck drivers claimed that in most cases the cane farmers suffered. For example, when stuck in the mud, the factory would send a wedge to pull the truck out and then charged the farmers by deducting the cost from the cane payment.[4] Sometimes the trucks had accidents, especially in sloppy areas; hence the need for experienced drivers. Sometimes the trucks themselves were in bad condition, which caused the cane to fall off the trucks, which incurred losses to the farmers.

Transportation was often a major problem since the roads were in poor condition. For example, the road that accessed the Muhoroni sugar mill was in very bad shape in 1930, because it had not been repaired for two years. The colonial government's neglect of the roads directly impacted the farmers' transportation and profit. Most of the roads were completely worn out and reverted to the status of a mere track. The situation was made worse by the collapse of the bridge over the Kapchuire River at Kibigori in 1930 which by 1931 had not been replaced.[5] Moreover, the roads in Kericho, Londiani, Kibigori, and Muhoroni, though maintained by the DC, were not in good shape. It seemed that the distance from the headquarters of the Public Works Department in Nyanza and a lack of supervision might have contributed to the conditions of the roads, which greatly affected the cane farmers in the area. To alleviate the situation, the executive engineer for the Public Works Department was asked to undertake the task of supervision, which led to better conditions on these roads by 1932. The initiative saw the construction of eighteen miles of new road between Kibigori, Muhoroni, and Chemelil, which greatly helped the farmers to transport their cane.[6]

With the positive road improvements in 1932, the colonial government embarked on a wider project of improving roads in 1933 after many complaints by the farmers. The roads in the sugarcane area were murramed throughout the year. The road between Kibigori and Londiani was upgraded to an all-weather road. Also the section from Kibigori to Songhor was improved when the construction of a truck road that had stalled in 1930 was revived and widened, while the Kibos Bridge, whose construction was begun in 1932, was completed in 1933.[7] Many farmers transported cane to the factory using these improved roads. Companies also took the initiative to build roads to help in transportation of cane. By

1932, for example, the Muhoroni-Kericho road had been constructed by Mr. Allen of the Muhoroni Sugar Factory, which was then inspected by the executive engineer and found to be satisfactory.[8] The willingness of the factory owners to construct the road showed the manufacturers' belief in the potentiality of the sugar industry in the area, unlike the colonial state, which had a long history of ignoring its potential. The poor state of roads had impacted the pricing and transportation of both cane and sugar because transport was expensive and required the maintenance of cane haulage fleets.[9]

Despite these modest improvements, some farmers in Chemelil and Muhoroni still suffered because by the mid-1930s the Asian farms in the area did not have mills close at hand. Cane had to be transported all the way to the Miwani factory since the factory at Muhoroni did not accept cane from outside its nucleus estate. Therefore, the Asian cane farmers in the Muhoroni-Chemelil area had no choice but to turn their sugarcane into jaggery. This situation continued until 1942, when an Asian entrepreneur, Premcard Raichad and Company, bought the Nottidge and Allen factory at Muhoroni and started receiving cane from the farmers around the Muhoroni-Chemelil area.[10]

As the problem of sugarcane transportation to the factory persisted, the Nyanza Indian Farmers' Association (NIFA) appealed tirelessly to the government during the 1930s for the construction and improvements of roads. In 1937, they proposed that the Kisumu-Kibigori road be improved to access Asian farms, since that particular road had already been granted £5,500 by special warrant for that purpose, as acknowledged by the acting colonial secretary.[11] Unfortunately, the colonial state was not willing to assist the cane farmers and responded that there were no funds to undertake such development and even suggested that the route could be changed.[12] It was interesting to see the colonial state respond in such a manner, knowing that funds had been set aside for that purpose. It was unfortunate that the Asian sugarcane growers were treated differently than the European farmers. With the implementation of the soldier-settlement scheme, the colonial state embarked on a program that extended the railway and improved roads and port facilities to ease the transportation of the European settlers' produce.[13] European agriculture received more support in transportation, while Asian cane farmers struggled with the colonial state for much-needed road improvements and generally paid heavily for transportation. Those Asian farmers who opted for transportation by use of factory trucks had approximately one-fourth of their total cane value skimmed off as transport charges. By 1960 the charges were Shs. 1 per ton per mile.[14] Truck transportation was used by only a few, in most cases large-scale farmers. It required skilled operators and careful maintenance to avoid breakdowns. Therefore, transportation of cane by road was a major problem that the farmers had constantly to struggle with, especially during the rainy season, due to the poor condition of roads. Transport by rail was a better option.

SUGARCANE TRANSPORT BY THE KENYA-UGANDA RAILWAY

Rail transport became handy during the wet season, particularly for farms that lay adjacent to the railway line. The railway facilitated the transportation of cane in a more secure manner as rails were not adversely affected during the rainy season. Farmers who were situated far away from the line needed cane to be moved from the farms to the railway line for transportation to the factory. Some farmers opted to use other means of transport that entailed the use of rail and trolley lines that were set up by the factory. This involved the use of large wheeled metal containers for transportation of heavy goods. Such arrangements were made possible because the Victoria Nyanza Sugar Company laid its own tramway lines to transport cane, according to the Victoria Tramway Ordinance No.1 of 1922. This ordinance was introduced specifically to enable the company to convey sugarcane from the neighboring farms to the factory on trolley lines that were drawn by small locomotives. There were even private trolley lines that could be used, but to build such tramways, bills had to be introduced, first in Britain, to allow such measures to be taken.[15]

The provisions of the Victoria Tramway Bill were made through the Roads and Traffic Bill, which ensured the license for temporary construction and operation of works on road reserves. In the construction of roads and tramways, the question of land for through-pass always arose. In the case of the sugar-belt areas, recompense for the use of land became an issue. The colonial state saw that, due to the advantages afforded to the community by the establishment of the sugar factory, the company needed to be accorded free use of the public road for the construction of the tramway.[16] But the local community did not see it that way and wanted compensation for any land lost to road construction. Unfortunately, due to colonial policy, there was no compensation payable to anybody on account of right to lay trolley lines along the strips of land. Most of the trolley lines did not reach most of the farms, especially those located far from the factory.

Nevertheless, a tramway was built in 1923 by Victoria Nyanza Sugar Company, starting on its own estate, running along a line of a public road, and terminating at the Kibos railway station.[17] The fact that the tramway was built by the Victoria Nyanza Sugar Company instead of the colonial government shows how the colonial state neglected the Asian farmers while providing rail and road services to the European settler farmers. Perhaps this was partly due to the inability of the colonial state to visualize the potential of the sugar industry in CN.

According to the agreement between the Victoria Nyanza Sugar Company and the railway, the company was allowed the use of the railway siding line from the Miwani railway station to the company's nearby sidings. This was meant to be used for light trolley conveyance; thus the railway provided the engine for use of the trolley.[18] However, the railway refused to take any responsibility for accidents that occurred upon it while being used by the company. The railway mostly preferred, and allowed the company to use, a light trolley on the railway siding line during the wet season when the roads were impassable.[19] In addition, the company was to keep the siding line in good repair during its usage, and if the railway did the repairs it would recover the cost from the company. In essence, according to the agreement, the Kenya-Uganda railway administration was relieved of all responsibility. This left a huge financial commitment on the part of the Victoria Nyanza Sugar Company, but the tramway was essential to the growth of the industry. For example, the provincial agrarian development officer (PADO) advised that the tramway was a very crucial part of the whole sugarcane production industry and encouraged the company to install the trolley lines.[20] As such the company was ready to commit its finances while both the colonial state and the Kenya-Uganda railway shied away from the responsibility. The location of the sugarcane farms near the main line should have made the colonial government reach an agreement with the Kenya-Uganda railway to extend the line or could have even subsidized rail freights, but this was not the case. It seems the economic importance and potential of sugarcane farming in the area, which could have contributed toward the reduction of the import revenue, was not given priority.

Needless to say, farmers situated adjacent to the main railway used trolley lines to move their cane to the railway line from where it was transported direct to the factory. Farmers near the factory at Miwani had to cut cane and have it delivered to the trolley line to be conveyed to the factory.[21] Unfortunately, even the trolley lines were not sufficient. As an alternative, farmers were expected to allow feeder lines to be laid and maintained in their fields for transport of cane from the adjoining farms by extension of the feeder lines. Farmers also had access to portable tramlines that were connected to the nearest mainline and, when not in use, were removed and returned to the nearest mainline. This arrangement by the Kenya-Uganda railway of using portable tramlines facilitated the transportation of cane from the farms to the factory; however, it could only help those farmers whose farms were near the railway line, since the feeder lines were not very long. Furthermore, the rail charges were also not cheap. Due to such difficulties, by 1937 NIFA asked the colonial government to reduce the railway freight charged in transportation of sugarcane and jaggery intended for export; they wanted the freight to be similar to that charged on exported maize.[22] This was because the rail charges for sugarcane transportation were high. For instance, by 1944, J.

T. Moon, the senior agricultural officer (SAO) for the province, estimated that transport of cane by the Kenya-Uganda railway from Kibos to Muhoroni was Shs. 3.14 per ton for 10-ton lots while from Kibigori to Miwani it was Shs. 2.24 per ton for a 10-ton lot. Hence for a 10-ton lot of cane, some farmers paid approximately Shs. 30 for transport alone, despite having to shoulder other expenses such as the payment of the loaders.[23] These transport expenses were difficult for the farmers to shoulder, especially given the low price paid for cane and the long delay in payment for it.

These serious transport issues should have led the colonial government to intervene and use the Kenya-Uganda railway to enhance sugarcane farming in colonial Nyanza. Unfortunately this was not the case. The problem persisted and affected both the small-scale and large-scale farmers. Even the Miwani factory experienced transportation problems. By late 1948, for example, Miwani Sugar Mills started to experience problems especially in transporting cane from its estate at Chemelil to the mill. This was because although it had mature cane, cutting it all at once was not possible since the Kenya-Uganda railway could only provide eight wagons per day to transport the cane. The limitation of the number of wagons inhibited the transportation of cane from Chemelil, which was situated far from the mill, and the farmers found that it was not economical to use road transport given the conditions of the roads especially during the rainy seasons. Having experienced firsthand these transport problems, Miwani Sugar Mills embarked on an aggressive campaign of reaching out to sugarcane farmers. So in 1948, the mill extended trolley lines and took sugarcane from Asian farms up to a distance of about twelve miles from the factory. Even those who had initially produced cane for jaggery turned over one-third of their produce to Miwani to be processed into sugar.[24]

Farmers who were unable to get labor to cut their cane on time, on the other hand, entered into an agreement with the Miwani Sugar Company by which the company would cut the cane, transport it to the factory, and charge the farmers for these services. Transportation had become so problematic that the Kibos cane farmers organized themselves into seven groups to make the best use of the transport facilities that were available to them. They worked closely with the factory, which required that they arrange staggered planting of cane; this was to ensure that all the cane did not mature at the same time, which created a bottleneck in transport.[25]

Another strategy used by the Kibos cane farmers was to transport as much cane as possible during the dry season since most vehicles got stuck in the mud during the wet season. Toward such an initiative, three of the seven groups were assisted by the provision of more lorries, while the other four were accorded more trolley line extensions by the Miwani Sugar Mills. In return, the farmers would undertake the cutting of cane as soon as it matured so as to avoid idle lines and trolleys.[26] Despite the fact that both Miwani Sugar Mills and Muhoroni had their own trolley lines for transportation of sugarcane to the factories by 1947, these were not sufficient to serve all the farms. Moreover, obtaining the metal sleepers was very difficult, and the problem was compounded by the fact that the wooden ones were frequently stolen. However, it was not the responsibility of the mills to provide transport.[27] The millers argued that it was uneconomical for them to send lorries into the farmers' fields to collect cane; and they only received cane delivered to their trolley line. Furthermore, the Muhoroni mill was small and so most farmers, whenever possible, took their cane to Miwani, which had a greater milling capacity. This overreliance on the Miwani mill generated a lot of tension between the Asian small-scale farmers and the manufacturers. The cane farmers complained that the manufacturers did not always accept their cane. The mills, on the other hand, complained that the feeder rails that they constructed and trucks they provided involved a heavy financial outlay. Moreover, this was not economical since the feeder rails could receive as little as three tons of cane per day. This lack of harmony between the farmers and manufacturers affected production.[28]

Nevertheless, Miwani Sugar Mills abided by their undertaking to accept cane delivered to their trolley lines but claimed it was uneconomical to send lorries to the farmers' fields to collect cane. The company asserted in 1952 that doing so was impossible, especially during the wet season.[29] Hence, some farmers were stranded with ready cane in the field due to unavailable means of transport. For example, in 1952, Milwant Singh Sandhu could not get some ninety acres of cane from his farm to the mills. This left such farmers with the only option of converting the cane into jaggery. This in itself was problematic since once a farmer installed a jaggery mill it became difficult to take sugarcane to the factory; besides a jaggery factory was expensive, especially for small-scale farmers. For instance, by 1952, the approximate cost of installing an average size jaggery factory was £1,250.00.[30]

Due to these difficulties, it was apparent that the Asian cane farmers faced huge problems when it came to transport services offered by the colonial state. By contrast, European farmers were greatly assisted; the government assisted the settlers by improving and extending railway, road, and port facilities.[31] Even more important, the European settlers were given low rail freight rates for their wheat and maize produce. The European settlers simply had more influence on policies undertaken by the colonial state; for instance, the European farmers' freight rates were adjusted considerably when the high rates became too much of a burden for them.[32]

Such initiatives were not accorded to the Asian sugarcane farmers who struggled to deliver their produce to the factories, nor were the same freight rates accorded to the peasant farmers in colonial Kenya. European agriculture received tremendous attention by the government at the expense of Asian farmers in colonial Kenya. What affected the Asian cane farmers was the lack of political will by the colonial state as it failed to take advantage of the Kenya-Uganda railway line, yet many farms were near the line. This could have gone a long way to reduce transportation problems. Such actions (and inaction) demonstrated the contradictory nature of the colonial state's policies: the government insisted on financial self-sufficiency and yet provided widely varying degrees of support for different farmer groups, which often led to disputes among the classes.[33]

THE CONVEYANCE AND ACCEPTANCE OF CANE BY THE MILLER

During the harvesting period, farmers were also expected to pre-arrange with the factory so as to fix a program for harvesting their cane. In most cases, it was found that the factory fixed the dates and times of harvesting to fit their convenience. This was because the factory had first to harvest its own cane and then manage overlapping demands by different farmers. However, the arrangements were not fair to the cane farmers since the factory accepted no liability for any delays of harvesting dates, which were often frequent, and farmers incurred losses. Furthermore, there were conditions laid down for supplying cane to the factories. For instance, the Victoria Nyanza Sugar Company initially insisted on prior arrangements and also that the cane had to be loaded to maximum capacity to avoid weight discrepancies. Meanwhile, Miwani Sugar Mills, by the 1950s, preferred fresh cane that had been cut within the last twenty-four hours and also insisted that the cane supplied must not be burnt and that it be of CO and POJ varieties, and not Uba.[34]

Such stringent conditions did not work well for the farmers. Despite the fact that a farmer had to receive written consent before the cane was accepted, there were cases where such consent was received and later canceled. This resulted in losses incurred by the farmers. For example, between 1943 and 1944, the Victoria Nyanza Sugar Company canceled the orders of eight farmers who lost a total of 12,670 tons over that period. The gross was valued at Shs. 9 per ton: hence a loss of Shs. 114,000.[35] This posed another problem: the farmers, though willing to supply cane, were sometimes dissatisfied with the form of contract, since they were at the mercy of the company and so struggled to remain in business as the manufacturers did as they wished due to lack of supervision by the colonial state.

According to the chief chemist of the Victoria Nyanza Sugar Company, once the cane was transported to the factory, the focus then shifted to the activities at the factory where certain requirements had to be met before the cane was accepted. The quality and type of cane determined the price offered by the factory, though the price was meant to be determined by the world prices, which frequently was not the case. The Victoria Nyanza Sugar Company set regulations for the acceptance of cane delivered. For example, once the cane was transported to the factories, it was received at the weigh bridge, where it was weighed and tested to affirm that it was good for milling. The test included the sucrose content of cane, which was supposed to be at least 10 percent, and the purity of the juice was to be at least 78 percent. Any sugarcane below these minimum standards was not accepted by the factory.[36] Moreover, cane delivered to the factories in dirty condition was often rejected or sidetracked to be cleaned first and made fit for crushing before being accepted. [37]

During the process, first the cane and truck was weighed; a second weighing was for sugarcane alone. The weighing of cane was done at the factory and delivery was preferred in the evenings, since it made it easier to start weighing and testing the next day. Farmers and truck drivers had to attend when their cane was checked and weighed; thus the weighbridge, in most cases, had to be installed in a suitable place. However, some farms had weighbridges installed, especially the large-scale farmers, for fear of being cheated on the weight. Big estates, like those belonging to Miwani, had weighbridges installed. After the checking and weighing, and ascertaining that all requirements were met, cane was received by the millers. It was at this point that the drivers and loaders were paid by the farmer, per ton of the cane load.[38] Previously, paying the driver and loaders was much easier since the farmers were also paid promptly. This was because before 1934, when the colonial state was not fully involved in the marketing of sugar, the Victoria Nyanza Sugar Company paid for cane it received from the farmers within a short period of time; the farmers were paid on the fifteenth of the month following delivery of cane.[39] After 1934 this changed, and the delay in payment was detrimental to cane growers since they had pressing financial obligations, such as payment for vehicles hired and for drivers and loaders who were to be paid immediately once the cane was accepted.

STRUGGLE AGAINST THE COLONIAL STATE: PAYMENT AND PRICING OF SUGARCANE

The payment for sugarcane received from growers changed by 1934 when the farmers started to be paid much later. This was because the millers too were not paid promptly by the marketing agency that received the sugar on behalf of the state ready for distribution. There was a waiting period of 150 days laid down by the colonial state before payment was made to millers who in turn paid the farmers. In most cases, the farmers had to wait longer. This was in accordance with Government

Notice No. 619 of 1934 Section 7, which was meant to regulate the sale and distribution of sugar.[40]

This was a long period to wait for proceeds since the farmers had to continue with cultivation of the ratoon crop, and it was frustrating.[41] The farmers advocated for a thirty-day wait instead since a majority of them had loans to pay for the land they had acquired. Taking into account the short duration of time involved in manufacturing sugar, they felt the 150-day waiting period was excessive. This new regulation generated a lot of tension between the farmers and the millers, though the millers did not control the payment of proceeds. Such regulations discouraged cane farmers and were perhaps seen by the Asian cane farmers as lack of appreciation of the economic potentiality of the sugar industry by the colonial state.

The state also failed to assist those Asian cane growers who dealt with the Victoria Nyanza Sugar Company in the early 1930s. In 1931 the cane farmers had put a lot of acreage under cane and hoped to get a price for it that was commensurate with the market price of sugar. Unfortunately most cane was damaged by locusts and the harvested cane did not pay well either because the cane farmers were left at the mercy of the factory, which offered prices not commensurate with market prices. The prices offered for cane were far below even the cost of cutting the cane, let alone other expenses such as labor and fertilizers. Consequently the farmers at Kibos found themselves confronted with disastrous financial losses partly due to manipulations by the Victoria Nyanza Sugar Company.[42] The situation was worsened because the colonial government was reluctant to solve the problem existing between the two parties by fixing the price of cane or ensuring that international market prices were adhered to. This left Asian cane farmers to struggle with the European sugar manufacturer, and the mistrust between the two parties was harmful to the growth and survival of the sugar industry.

The colonial state did not intervene on the part of the Asian farmers, partly due to the colonial government's policy that laid emphasis on European interests and partly because of its desire to control the pricing of commodities in colonial Kenya. This policy was resisted, and Asians in major urban centers of East Africa, especially Mombasa, Nairobi and Dar es Salaam, who had earlier organized themselves to form a more inclusive political organization, the East Africa Indian National Congress (EAINC), in March 1914 to fight such injustices, intervened.[43] The EAINC was modeled on the National Congress of India and was meant to serve as a forum to express opinions, make resolutions, and serve petitions to local government, the Colonial Office, and the government of India.[44]

The EAINC aimed for the political and economic betterment of Asians in East Africa generally and Kenya in particular. Therefore, in 1923 when the European settlers wanted to stifle Asian economic activity in Kenya, the congress called for a boycott and noncooperation as the only way to fight for their rights. This was when the Europeans formed an organization known as the European and African Traders Organization (EATO) whose main objective was to obstruct Asian commercial activities in the colony, including sugarcane growing. The organization's objective was clearly captured in its racial motto that read: "Every European and every African is an asset to Africa and everyone else is a liability."[45] Such objectives and motto expressed the threat that any Asian entrepreneurship was bound to face. These threats were portrayed in the European and African Trading Organization newsletter no.1, which stated that "to include the Asiatic in our economic situation is to include an element whose only ideas are the making and removing from the country whatever monetary gains it is able to accumulate and our destruction."[46]

Due to such agitation, the Asian cane farmers had to struggle for economic space. Therefore, the congress fought political and economic injustices in Kenya. For example, in 1928 the EAINC boycotted the executive and legislative councils whereby four of the five seats in the LegCo reserved for Asian elected members remained vacant.[47] The low representation in the LegCo meant that the Asian farmers and millers had almost no voice to push for their welfare. Hence, the EAINC was forced to intervene on behalf of the cane farmers who were suffering because during 1930 season the crops had failed due to locusts and drought. Thus on January 10, 1931, the EAINC asked the government to extend financial aid to the sugarcane farmers at Kibos, just as it had done to the European farmers, and demanded that the Asian cane farmers be given the full benefit of the protective duty that had been given to the European sugar manufacturers. Furthermore, the EAINC petitioned the government to fix the price of sugarcane so that farmers were not cheated and also to grant a subsidy to Asian cane farmers during the Depression as was done for the European settlers.[48] The colonial state did not help the cane growers during the Depression and dragged its feet. The cane farmers had to wait for a decade and a half before new cane prices were agreed upon in 1947, whereas European wheat and maize production seem to have been aided by an official rise in their price much earlier. It was difficult to justify the colonial government's failure to protect the sugarcane growers by similar means. The colonial state was indifferent and simply ignored the importance of the industry.

As the pricing of sugarcane continued to be controversial over the next decade, it caused friction between the cane growers and the manufacturers as the two parties could not easily reach an agreement. It would have been prudent for the colonial state to issue guidelines or fix the price of cane and save the farmers from anxiety and disagreement with the millers. For example, in 1946, when Mr. W. Wood was appointed the manager of the Victoria Nyanza Sugar Company, he tried to negotiate and offered a flat rate of Shs. 10 for hard cane and Shs. 11 for soft cane. The growers insisted on Shs. 12 for hard cane and Shs. 13 for soft cane.[49]

The situation stalled as no suitable agreement was arrived at. The farmers were adamant that the prices offered for cane were insufficient to cover the cost involved in growing sugarcane while the factory insisted that the low sugar prices made it impossible to comply with the farmers' requests.[50] Once again, the colonial state neglected the sugar industry and failed to give price guidelines that could have solved the impasse.

As the pricing of sugar continued to be a serious issue ignored by the colonial state, the new Asian management that took over the Miwani Sugar Mills in 1947 felt that the problem had to be addressed for the good of the industry. Therefore, Mr. Hindocha started to advocate for changes in the sugar industry in Kenya, drawing from his own successful ventures in Uganda. He not only revamped the factory, replacing the old machinery, but also insisted that the colonial state come up with guidelines on sugar prices. He reiterated that there was a dire need for the colonial state to fix the price of sugar, arguing that it was impossible to guarantee the price of sugarcane to the farmers until there was a guarantee of the price of sugar from the colonial government. Consequently during the same year, the whole sugar production position came under the review by the deputy PC and the agriculture department.[51]

Due to such initiatives and the stability in world prices, the governor in council fixed the prices of refined sugar on the recommendation of the East African Production and Supply Council. The price of sugar was increased to Shs. 17 per ton. Unfortunately, the governor did not fix the price of cane and as such the growers were left in a financial limbo again, risking exploitation by the manufacturers. The situation between the two groups worsened. The cane growers felt that they were given a raw deal by the manufacturers who still paid low prices for cane. As a result, many cane growers resorted to the only option left and in 1947 turned half of their produce into jaggery in protest.[52] This had repercussions; the factory started receiving less cane while Uganda was overproducing sugar, which led to the dumping of cheap sugar in the country. The situation made the Miwani to appeal to the government to fix the price for cane to correspond to that of sugar so as to protect the Kenyan manufacturers from their Ugandan counterparts.[53]

This pressure from Miwani made the colonial government realize that a solution had to be found after a decade of struggle by cane growers. For the first time a meeting was called between the representatives of the Production Board Local Committee, the sugarcane planters, and Miwani sugar mills in July 1947.[54] The attendees recognized that there was a need to revise the prices payable for sugarcane supplied to Miwani on a sliding scale. Table 4.1 shows examples of how the prices for cane would vary with the prices for sugar.

The prices shown in table 4.1 were for hard cane such as Uba; for soft cane, such as POJ varieties, 50 cents was added per ton. For instance, when Uba fetched Shs. 12 per ton, the price for the POJ would be Shs.

Table 4.1. Proposed Prices of Sugar and Cane, 1947

Price of Sugar (Shs. per ton)	Price of Cane (Sh. per ton)
300	8.50
320	9.20
340	9.90
360	10.60
380	11.30
400	12.00
420	12.70
440	13.40
460	14.10
480	14.80
500	15.50

Source: Extract from letter from D.K. Hindocha to Chairman, Production Board Local Committee, July 28, 1947, KNA: PC/NZA/3/2/1/8.

12.50 per ton. Those prices were to remain in force until 1950, despite the fact that Kenyan manufacturers wanted the prices revised due to cut-throat competition from Uganda.[55] Unfortunately, the colonial government's delay compelled the Kenyan manufacturers to pay low cane prices so as to remain in business. The colonial state could not protect them, revealing the contradictory nature of the colonial state whose decision could be said to have been blinded by racial bias grounded in protecting the European settler interest that led to poor legislation. This failure to protect the manufacturers inhibited the expansion of the sugar industry.

The pricing issues ultimately arose from failing to adhere to world market prices as guidelines for local pricing. Moreover, even when adopted it had to be used carefully because the world sugar prices were not always conducive, since the price of sugarcane paid to farmers varied from country to country and depended on variables such as the cost of cultivation of sugarcane, transport, and the quality of the cane. For example, irrigation of cane could be expensive; hence, the price of irrigated cane could not be the same as that of rain-fed cane. Furthermore, the distance to the factory was an important factor in setting the pricing of sugar. Comparatively, in India all sugarcane grown within a ten-mile radius of a factory was called "gate cane" and was always accepted at the factory delivery. Cane grown out of that zone was bought by the factory through contractors and its rates were lower than factory-delivered cane.[56] This was part of the conflict between the cane farmers and manufacturers. The Kenyan Asian growers felt strongly that the price of sugarcane could only be decided after a discussion with all the concerned parties, which, unfortunately, was only achieved in 1947.

The problem persisted even when Asian companies started manufacturing cane. Initially the Asian sugarcane farmers had complained about the European-owned Victoria Nyanza Sugar Company offering very poor prices for cane, and one might have thought that racial politics affected pricing in Kenya. However, even the Miwani Sugar Mills (an Asian firm) in 1948 was accused of being unreasonable and paying low prices.[57] It is clear that there were more than simply racial issues at work. Of equal importance was the capitalist mode of production that pitted Asian cane farmers against both the colonial state and the manufacturers.

PRODUCTION OF SUGAR: WITHIN THE FACTORY PROCESSING

Although pricing and payment of cane remained crucial to both the cane farmers and the manufacturers, the remainder of the process was equally important. Once cane was received, the production process began, a process that was difficult and delicate if the correct grades of sugar were to be attained. The process entailed proper coordination of different sections and was labor intensive, taking into account the machinery maintenance, fuel, labor, and finances for auxiliaries like bagging, which made the whole process expensive. It would have been better if such issues were ultimately to be considered and the opinion of both farmers and manufacturers recognized so as to harmonize the finances involved by both parties before the colonial state could determine the prices and marketing of sugar. Unfortunately this was not the case.

Once the cane had been received and weighed, as indicated previously, the factory then checked to ascertain the quality. The sampling took place outside of the factory gate before being taken inside. After that, a ground cane carrier put the cane into machines that cut the cane into small pieces. These pieces were then fed into the mill's crushers. The mills usually employed five rollers and four pans that served as a tank for receiving juice. The Miwani factory used heavy-duty crushers with a channel iron tram and fine teeth gears to make the running of the mill durable and noiseless. At the mills, the juice was extracted from the cane by passing it through a series of roller mills and the bagasse was removed and taken to the boiler to be used to produce steam that was needed for generating power to boil the juice. During the boiling process, lime and sulfur were added to make the juice clearer and create molecules that absorbed matter in the juice to make it settle down.[58]

The next step involved the contraction of the juice, whereby the boiled juice was channeled and passed to the pan station where it was boiled again until the grain was formed and developed into the required grade.

From here, the grain was taken to crystallizer machines where the crystal and molasses were separated. During the process, the sugar by-products, such as bagasse, molasses, and filter press mud, were separated to be used in different ways. The grain was then sieved by the grader to get the different grades and then forwarded to the bagging section from where it was ready for distribution. In cases where refined sugar failed to achieve the required standard, it was usually re-melted and processed again as other grades.

The process involved detailed machinery that was expensive and needed financial support as well as skilled labor. Unfortunately, more often than not, the millers experienced difficulties and delays in getting the required personnel and machinery, partly as a result of unnecessary bureaucracy created by the colonial government, despite the rapidly expanding local market.

TYPES AND GRADES OF SUGAR

Milling was a crucial part of the process that required utmost care to achieve the desired grades of sugar. Miwani Sugar Mills, with the biggest crushing capacity, produced most of the sugar among the five milling factories in colonial Kenya. By the start of 1960s, the sugar industry had progressed and manufactured different types of sugar, for example, brown sugar, cube sugar, mill white, castor sugar, icing sugar, and later refined sugar, which was a local sugar of quality.[59] There were also lower grades of mill white sugar such as grade II and grade III. By 1961, the colonial state started giving penalties to millers for the lower grades of sugar produced, while at the same time using incentives to improve the quality of cane. Such initiatives could have been implemented much earlier, not only to improve quality but also to show the state's involvement in the promising industry. The lower grades were partly due to the poor varieties of cane harvested, which could have been improved by the state. However, the colonial state claimed the scheme was expensive and abandoned it as soon as it was initiated.[60]

Despite the need to encourage the cane farmers and manufacturers of the potential of the sugar industry in colonial Kenya, the colonial government tightened its grip during this period. It introduced price controls on the icing sugar produced by Miwani Sugar Mills while importing icing sugar that had no restrictions. Furthermore, imported cube sugar was also free from control as per the legal notice number 284/61.[61] These double standards and contradictory policies were prevalent during the colonial period and hindered the expansion of the sugar industry. The colonial state seemed to have wasted money importing what could have been manufactured in the colony.

Be that as it may, the Miwani Sugar Mills was not constrained in its growth and started to produce refined sugar by late 1961; the price was Shs. 967.75 per long ton, exclusive of excise duty. Granulated white sugar was considered to be of refined quality too and was supposed to have moisture content not less than 0.04 percent. Despite the potential and capacity, the colonial state did not want Miwani to produce refined sugar. The colonial government preferred to import it from outside, from countries such as South Africa or Hong Kong. These measures were enforced by Mr. P. H. Jones, of the Ministry of Agriculture, Animal Husbandry and Water Resources, and Mr. G. D. Parkin, the deputy secretary for commerce, industry and communication who adamantly discouraged the production of any quality refined sugar by Miwani. Mr. Parkin had his own doubts about Hindocha and his proposed refined sugar. In July 1961, he asserted, "we must pin him down to selling refined sugar of the quality we desire, and not let him get away with mill white."[62]

Such a stance by a colonial bureaucrat did not encourage the manufacturers; it seems that the colonial state lacked the political will to support the industry. Despite the struggles that the sugar industry faced in colonial Kenya, the Miwani mill progressed steadily due to the resilience of the Asian cane farmers and manufacturers. The mill improved in production and technology and surpassed Uganda, the main producer in the East African colonies, and was ready to be the first to manufacture refined sugar in the region. The colonial government refused to acknowledge such an accomplishment and turned a blind eye to the economic potential of the sugar industry. The company had to contend with the stance taken by the colonial officials.

However, due to the determination and meeting the prerequisites set by the colonial administrators within a short period, Miwani was allowed to refine sugar on very strict conditions. It was first required that the degree of refinement be equivalent to British refined sugar or that imported from Hong Kong. Secondly, it was required that sampling had to be undertaken. The sampling was tedious and required the Miwani factory to supply a minimum number of twelve sampling containers and ensure that the samples were not tampered with until they arrived at the destination where tests were conducted. These tests, among many, required that one sample of five pounds of refined sugar be forwarded to the government chemist each week for analysis, which was very demanding. Miwani Sugar Mills complied, and the chemist then sent the results to the director of trade and supplies and a copy to Miwani.[63] Such were the strict conditions under which the millers were forced to operate.

As for the completed product, the factories packed the sugar after milling. It was recommended by the public health rules for food that the bagging materials used had to prevent both dirt and dust; hence, the use of multiple-wall paper bags or even the use of a polyethylene liner in the large jute sacks. Sometimes the factories also packed the sugar in open-weave sisal bags (gunny bags).[64] There were also cases where distributors and traders complained of insufficient weights. It seems that there was a problem in packaging; however, it was controlled through the efforts of local committees, and the district officers and the government ensured that the consumers were protected.[65] The state also introduced regulations that ensured the safety of the produce. However, according to the Public Health (Labeling and Advertisement of Food) Rules Amendment Act of 1958, sugar was exempted from some provisions. For example, it was not required to state the name of the food on the package, the name and address of the manufacturer, or the name and address of the packer, since sugar was mostly packed by the manufacturer.[66] However, Miwani Sugar Mills was required to comply with other rules within the act; for example, it was required to observe the limits on the amount of preservatives that could be added to sugar. The company used brown paper bags with polyethylene lining, which had a capacity to hold 112 pounds, or the gunny bags with a capacity of 224 lbs.[67] The quantity packed had to conform to the amount permitted by the weights and measures rules. This show how the Asian sugar industry had regulations to follow, which it complied with, and only needed the colonial government to play its part by being supportive to enhance sugar production. Nonetheless, Miwani's persistence and resilience made it possible for it to achieve such a goal and be the only East African factory that installed the refinery and produced refined sugar in the region by 1961.[68]

The implication of this was that the Asian sugar industry indeed had great economic potential. Had it been fully supported by the colonial state, it could have greatly reduced the overdependence on imported sugar and saved the funds that could have been channeled to other pressing needs such as infrastructure, education, or even health. This was an opportunity that the colonial state missed, that could have been exploited to attain the colony's policy of self-sufficiency particularly in the sugar industry. Despite the stringent measures Miwani Sugar Mills followed the regulations in the manufacture of sugar as was required by the colonial state. These included rules of manufacturing refined sugar, bagging of the finished product, and the weight of the product, which proved to be problematic throughout the colonial period. There were many weight irregularities that could have been solved through the colonial government's supervision and support.

STATE CONTROL OF MARKET IN COLONIAL NYANZA, 1930–1940

The marketing of sugar is the last part of the process of manufacturing, and a crucial component of that process. The colonial state's intervention in the marketing of sugar was part of the wider colonial policy geared toward control over the marketing of produce, especially maize, in colonial Nyanza and Kenya generally starting in the 1930s.[69] The colonial state imposed excessive regulations that inhibited the full potential of the sugar industry. Such measures affected the sale of both the cane and the sugar itself.

Both the cane growers and the manufacturers struggled for the market in a hostile colonial economy. The Asian cane farmers were limited in their production by the state, which did not allow them to import machinery but did allow the European manufacturers to do so. The Asian farmers in Kenya were further disadvantaged because the colonial state preferred the importation of Ugandan sugar over the expansion of Kenyan sugar production. The preference for Ugandan sugar led to exploitation of the farmers, consumers, manufacturers, and the Kenya government by Ugandan manufacturers. Despite these obstacles, the Asian cane growers and manufacturers continued to farm and petition the colonial state for help.

The farmers also faced challenges of marketing in colonial Nyanza. These challenges included the illegal importation of sugar and ineffective tariff systems, which led to flooding of Ugandan sugar into Kenya. Yet the colonial state remained unresponsive to market challenges. Such actions depict the contradiction imbedded in the colonial policies that inhibited sugarcane farming, a factor which contributed to Kenya's lack of self-sufficiency despite the potential of sugarcane farming in the area.

As Asian cane producers struggled for market, the rest of the country was not spared. The state's intervention in the marketing of sugar in CN was not an isolated issue. The colonial state began to control the market in the colony by early 1930s. It used the war period to intensify its control; these controls that began in early 1930s coincided with intensification of production and increased prices of maize that was brought along by severe food shortages due to famine in Kenya and the demands created by World War II.[70] These demands motivated the colonial state to develop a policy whereby the agricultural industry was boosted to meet the war needs; unfortunately, even during the war period controls were put in place to regulate the marketing of foodstuffs in Kenya. Likewise controls were also put in place to regulate the sugar industry. During this period, the colonial state focused on improving its budget position and so encouraged African agriculture; it especially sought to improve the quality of the produce, having realized that the colony could not rely on European farming alone.

The governor of Kenya, Sir Joseph Byrne (1931–1936), strongly believed that for improvement in the quality of produce to be effective, the colonial state had to control the market through regulations. This stance was shared also by the officials of the Colonial Office in London who believed that rural markets in Kenya were not efficient.[71] Therefore, many regulations came into effect to control the market. Unfortunately, Asian traders seem to have borne the accusation of poor marketing in

Nyanza as it was alleged that they did not care for quality of their cane, which led to poor prices.

However, due to the problems that the farmers faced in marketing, the market was eventually restructured. The colonial state posted agricultural officers to Nyanza to supervise the market in the sale of produce. Their work led to the reduction in the number of markets and petty traders in hopes of reducing competition and improving the maize prices.[72] In areas such as the Gusii highlands in South Kavirondo, market controls were intensified. For example, the AO, E. W. Gaddum, implemented a system of central inspection of maize, which not only improved the quality but also eliminated colored maize by 1935.[73] It was interesting that in sugar production the state was only keen on the quality of refined sugar and not the other grades. Also the colonial state toyed with the idea of giving the cane farmers an incentive to increase the quality, only to abandon it, citing financial constraints.

The colonial state was relentless in its control of the market. For instance, the colonial state introduced the Marketing of Native Produce Ordinance in 1935, aimed at increasing the quality of African produce meant for export. The desire to control the market culminated in the appointment of the marketing officer for African produce, C. O. Oates, in 1936. This appointment indicated the seriousness of the colonial state's desire to control the market. For example, by 1939 Oates opposed the small traders who wanted to join in the business of buying and selling of grains on the ground that it would lead to unhealthy competition, with some ending bankrupt.[74] Furthermore, in the case of maize quality, the agricultural officers stepped up and even introduced cleaning machines in Gusii market centers through which all produce had to pass.[75] Other control measures required traders to post on a board the price offered. To ensure success of the measures, the colonial state introduced a levy of 3 cents per bag of produce inspected in 1936 to finance the inspection system.[76] Though the system was intended to have a positive impact, many farmers found it problematic. Apart from the levy there were comprehensive market controls in colonial Nyanza by the end of the 1930s.[77] The colonial state had placed regulations on the distribution and marketing of sugar that began in the 1930s, intensified them during the war period, and continued them throughout the colonial period; these measures were generally counterproductive.

STRUGGLE FOR MARKET: THE COLONIAL STATE AND THE SUGAR INDUSTRY, 1930–1940

Having looked at the marketing of other produce in colonial Nyanza, we find that the sugar industry was not spared. Once the packing was done, sugar was ready to be released for marketing. The control of the sugar market involved the pricing and distribution of the produce. The price offered should have corresponded to the world sugar prices, but in most cases this was not so. During the early 1920s, the Asian farmers grew sugarcane mainly for jaggery since there was no mill. When the Victoria Nyanza Sugar Company started crushing cane in 1924, the Asians took their cane to the factory and some of them also doubled up and hawked sugar and jaggery in the African reserves since there were no regulations for marketing of sugar in place. The price of sugar in 1924 was good since it was sold at Shs. 17 per ton plus bonus, and the farmers at that time felt encouraged to grow more cane. During the Great Depression, the prices collapsed, however, and in 1930 the Victoria Nyanza Sugar Company paid farmers only Shs. 2.31 per ton of cane. This made the farmers to complain bitterly. During the same year, the deputy director of agriculture, H. Wolfe, reiterated that the Depression had impacted the prices of produce so much that a sound financial strategy was necessary to rescue the agricultural industry in Kenya as a whole. This sentiment culminated in the enactment of the Agricultural Advances Ordinance of May 1930.[78] Moreover, when the effect of the Depression became acute in 1932, the cane farmers suffered adversely. It was only after negotiations in 1933 with the manager, H. E. Moffit, that there was an increase in the price of sugarcane. Farmers were paid Shs. 7.50 per ton plus bonus.[79] But even then, only a few contracted farmers were paid this amount. A majority of farmers were ignored and not offered any agreements.

By the early 1930s, however, the colonial state started to be more involved in the marketing of sugar. For example, by January 1932 an excise duty was introduced in the colony on sugar and tea. This was as a recommendation of the Sir Wilfred Woods Report.[80] Accordingly, the duty of 2 cents a pound on sugar was increased to 4 cents a pound. This increment was justified on the grounds of generating more revenue in the colony.[81] This was followed up in 1933 when the colonial state started to introduce policies to control sugar production. According to Government Notice No. 46 of 1933, restrictions were put on importation of all milling and crushing machinery for processing sugar, which in essence curtailed the expansion of the young industry. However, given the fact that both the European sugar companies then had imported machinery, this possibly was meant to stop importation of all crushing and milling machinery for processing sugar by Asians, which in essence discouraged the Asian cane farmers. Furthermore, Government Notice No. 49 of 1933 prohibited the importation of distilling apparatus and machinery without a permit from the chief secretary. These two policies restricted the manufacturing of not only sugar but also jaggery.[82] This was done to protect the European manufacturers, namely the Victoria Nyanza Sugar Company at Miwani and Nottidge and Allen of Muhoroni. By denying the Asian farmers the possibility of erecting their own mill to produce sugar, the colonial state gave a monopoly to the two European companies, while the restriction on jaggery production also was disadvantageous as it was meant to ensure that the Asian cane farmers had no options but to supply cane to Victoria Nyanza Sugar Company. This ensured that the European-owned company benefited at the expense of the Asian cane growers. These policies reflected a bias against the Asians farmers on the part of the colonial state.[83]

Due to such monopolies, the European manufacturers were not eager to offer reasonable cane prices. The prices of cane and sugar continued to be a contentious issue that, despite the firm grip on marketing, the colonial state was not willing to solve. In theory, the Kenyan sugar price structure was based on the Commonwealth Sugar Agreement prices, which were fixed annually and were normally controlled by the price control ordinance. The retail price was calculated by equating the price paid to local producers, the cost of imported sugar from Uganda, and the cost of sugar imported from overseas, through the Sugar Equalization Fund, so as to reach an average cost per ton of all sugar supplies on the market. Added on top of the equated price were also the commissions for the main agents, commissions for the subagents, and a margin for retailers' profit. Prices were further adjusted to cover transport charges and other miscellaneous expenses, such as gunny bags.[84]

Unfortunately this was not the case for the sugar industry in Nyanza during the 1930s. The cane farmers realized they had to struggle to survive in an unfriendly economic environment perpetrated by the indifferent colonial government. Therefore, to improve marketing of their produce, the Asian farmers, many of whom belonged to the NIFA established in the early 1930s, resolved to form a group purposely charged with the responsibility of marketing their produce. The farmers formed the Nyanza Farmers' Cooperative Society (NFCS) in January 1936, composed almost entirely of cane farmers, which played a significant role in the struggle for market. The farmers were keen to sell their produce for good prices, especially when the Victoria Nyanza Sugar Company stopped taking cane from them in 1936.[85]

COLONIAL STATE CONTROL OVER THE SUGAR MARKET, 1940–1950

The formation of the NFCS was quite timely. It worked closely with the NIFA whose activities were influential in the next decade. For example, the Asian farmers through the NIFA asked the colonial state to reduce the railway freight in the transportation of cane since this affected the prices. Furthermore, sugar was overproduced in the world before World War II, which impacted the prices greatly; the prices were so low that restriction became necessary since an increase in cane growing in colonial Kenya was tied to the international agreement.[86] In addition, excise duty was also added to the basic price payable by manufacturers in late 1941. This discouraged more production of sugar due to the inflated costs of production.[87] This was contradictory to the earlier policy of May 1941 that had boosted the production of sugar when the local War Supplies Board issued a directive that all sugar machinery, not only in Kenya but in East Africa as whole, be put to full operation. This was due to the increase in demand in consumption and the needs brought by the war efforts. Such contradiction shows how the colonial policies during the war period were marked by confusion; sometimes the policies boosted the sugar industry and sometimes they discouraged local production.

Other policy changes were as a result of Sir Wilfred Woods' fiscal survey that was completed in January 1932. The survey was geared toward finding means of generating revenue for the three British East African territories. It advocated that excise duties be introduced on sugar, tea, cigarettes, and tobacco. Based on the report and wartime conditions, in 1943 a sugar consumption tax of one cent was imposed on all sugar consumed in Kenya with the specific objective of raising additional revenue so as to subsidize the sugar industry.[88] This tax was collected by the commissioner of customs and excise or his agent and was paid by the manufacturers or the importers who recovered it from the consumers.[89]

Other legislation was also put in place to control the production and marketing of sugar. For example, according to the control of sugar regulations of 1944, nobody was allowed to manufacture sugar without a permit issued by the controller. It also required that no manufacturer of sugar was allowed to sell, transfer, or dispose of sugar without a permit. Finally it required that anyone who sold sugar to furnish the Office of the Controller with information about his stock at hand and sales of sugar.[90]

Furthermore, price control was placed on castor and icing sugar during the same period. Unfortunately there was no documentation as to why the price control was imposed.[91] Most probably this was due to a shortage at the time since only a small amount of the two sugar brands was produced locally in the 1940s.[92] These taxes imposed during the 1941–1943 period affected the price of sugar and hurt the producers rather than providing them with the much-needed support. The harm caused to the industry was documented later when research conducted by the statistical accountant to the Board of Agriculture in 1950 indicated that the only way to increase sugarcane performance was to increase the price paid by the consumer and also to increase the subsidies to growers.[93]

Unfortunately, during this period the colonial state did not give the Asian farmers and producers of sugar leeway as they did to European settler farmers. Settlers were accorded different marketing arrangements to export their crops whereby centralized marketing boards were created to increase efficiency in marketing. Under the Sale of Wheat Ordinance of 1933, for example, all wheat was taken by the Kenya Farmers Association (KFA).[94] Furthermore, the European wheat growers in Kenya acted as local monopolists who not only sold wheat locally at an inflated price but also exported surpluses at lower than world market prices. Coupled with the tariff placed on imported wheat this provided advantages to European growers.[95] Comparatively the Asian sugar producers did not have such a monopoly; neither was the industry protected by tariffs on Ugandan sugar. Therefore sugar prices were not stable and were marred by the politics of control.

Needless to say, the pricing of sugar was key to the growth and expansion of the market.[96] Therefore, for the industry to survive there was a need for revision of policy and subsidy in the sugar industry. The colonial state waited for over a decade to make any changes. These took place in 1947, as discussed in the preceding chapter. During the same period, there were some irregularities in prices paid for sugarcane that were not impacted by the action of the colonial state. For example, in 1947 it was found that the Miwani Sugar Mills was cheating the farmers. They only offered farmers Shs. 12 per ton of cane, as stated earlier, and they also insisted that it took 15 tons of cane to manufacture 1 ton of sugar. This was not in line with practice in other countries; for example in India, 10 tons of cane made 1 ton of sugar, while in South Africa, 8 tons of cane made 1 ton of sugar. It seems that Miwani's rate was a complete rip-off for the farmers.[97] When Bresler was the accountant of Victoria Nyanza Sugar Company, the measurement was 10 tons of cane for 1 ton of sugar.[98] Sadly the Asian farmers were exploited by their fellow Asian millers just as much as they had been by the European millers.

Such irregularities should have been addressed by colonial state supervision of the marketing mechanism. But the colonial state was not very much concerned about the profitability of the Asian sugar manufacturers, especially when it came to pricing of sugar in colonial Nyanza. This was a short-sighted view that had a negative impact on sugar production in the colony. This negative effect on production was noted by W. Lyne Watt, the PAO, who reported in 1949 that compared with South Africa, Kenya's prices were 29 percent less, a situation that was blamed on insensitivity and poor sugar legislation. Many farmers and millers registered their dissatisfaction with how the government handled sugar prices. The Asian farmers in particular felt that due consideration had to be given to them in as far as the pricing of cane and white sugar was concerned, as was done with the coffee and maize farmers. This, they argued, would help them to counterbalance any rise in the cost of machinery and wages.[99]

In the meantime, the government kept on passing new regulations. For instance, in March 1948 the colonial state enacted the sugar consumption tax so as to fund the subsidy paid to the manufacturers for each ton of crystallized sugar manufactured in the country. This led to an increase in production amounting to 12,784 tons of sugar produced during the year for the whole colony.[100] Despite the fact that sugar fetched low prices at the time and also that the industry lacked the government's strong support, this was a marked improvement. The increase was also made possible by Hindocha's takeover of the factory. He not only revamped the operations by installing new machines but also accepted almost all the cane delivered, unlike his predecessors. This assurance of acceptance encouraged the farmers to plant more. There was marked increase in the production of sugar as more and more farmers took their cane to the mills, and this reduced the amount of jaggery produced. Table 4.2 shows how much cane, sugar and jaggery was produced in each of the nine Asian cane-producing areas in Nyanza during 1948.

The data in table 4.2 show that there was an increase in production generally, and the first four areas had a lot of mature cane. The performance was good compared with the only 5,760 tons of sugar that the farmers produced in 1946 and the 9,171 tons they produced in 1947.[101] The poor performance in 1946 was due to the dry weather and the prevalent labor shortage.[102] The substantial increase in 1948 was attributed to several factors, including an increase in acreage of planted cane, favorable weather conditions, and improvement in the procurement of available labor. Table 4.2 also suggests that few farmers turned their produce into jaggery. For example, out of 55,000 tons of mature cane in area 2 only 85 tons was converted into jaggery. The decrease in jaggery production in 1948 led to the increase in sugar production, resulting in 12,900 tons of white sugar in that year. The shift from jaggery production to delivering

Table 4.2. Production of Sugar and Jaggery in the Nine Areas, 1948

Sugarcane Areas	Approx. Total Acreage	Acreage of Mature Cane	Mature Cane 1947-48 (tons)	Annual Jaggery Output	Annual White Sugar Output
1	8,000	1,940	63,000	—	4,840
2	6,000	2,250	55,000	85	4,230
3	6,060	2,000	36,000	2,400	—
4	2,400	1,100	19,000	120	—
5	5,120	500	9,000	600	—
6	4,480	400	7,200	480	—
7	2,400	2,000	40,000	1,330	1,540
8	1,000	150	1,000	78	—
9	4,800	975	15,000	—	1,000
Total	40,260	11,315	245,200	5,093	11,610

Source: Description of Sugarcane Areas 1948, KNA: AK/11/50.

cane to the factory was because of the adaptation and success made by both the cane farmers and the millers as they struggled for market share.

THE COLONIAL STATE'S CONTROL OF SUGAR MARKET, 1951–1963

The previous section demonstrated how the sugar industry picked up in the late 1940s and the total acreage increased tremendously. The Asian manufacturers gave the industry a new lease on life despite the neglect by the colonial state. Even peasant commodity production flourished, despite the dominance of European settler production.[103] Once more the contradictory nature of the colonial state was witnessed as the Asian sugarcane production experienced both support and neglect during the colonial period.

By early 1950s, the colonial state was less supportive of its sugar industry than were Kenya's neighboring territories. This was seen in 1950 when the colonial state did not want to engage in international sugar discussions and adamantly refused to join the conference in London that was called to discuss export quotas. The colonial state, instead of attending to see how it could benefit the Asian farmers, dismissed the whole conference, arguing that Kenya had no prospects of exporting sugar in any foreseeable future, a position that was contrary to the potential of the Nyanza sugar belt.[104] The decision not to attend the conference was made partly because the colonial state saw that there was a constant demand for sugar due to the fast-growing population, a rise in purchasing power among Africans, and the new trend of adaptation to a varied diet. As such, the state adopted a policy geared toward self-sufficiency and not exports of sugar. Therefore, in 1951 the government set up prices for sugar; grade I was sold at Shs. 6.09 per pound while grade II fetched Shs. 4.85 per pound.[105] To attain their goal of self-sufficiency by 1953, the government further granted the sugar millers an increase in the price of Kenya's white sugar in 1953, which also led to an increase in the price of sugarcane given to the growers by Miwani Sugar Mills for various varieties as follows:

- Uba: Shs. 29.71 per ton
- CO: Shs. 30.11 per ton
- POJ: Shs. 30.71 per ton.[106]

Such prices were eagerly awaited since the cost of production had risen in terms of the cost of farm implements and wages for workers. Unfortunately, these prices were short-lived as there was a reduction in the price of sugar by 1954. For example, the wholesale price of sugar produced at Miwani, including excise duty, sold at Kibos for Shs. 100.22 per 224-pound bag, while sugar from other sources sold at Shs. 92.01 per 224-lb. bag. Sugar sold at a retail price of 57 cents per pound.[107] This meant that the sugar produced in Kenya was expensive and could not compete effectively with imports from other countries. The higher prices of Kenyan sugar encouraged wholesalers to buy sugar from outside, especially from neighboring Uganda. The presence of cheap imported sugar from Uganda during this period also led to some price irregularities. Kenyan sugarcane farmers suffered because their cane was not competitive due to higher prices. The situation was compounded by the fact that within a short period, in 1956, there was yet another increase according to Sugar Prices Order No. W.15 of March 20, 1956. Accordingly, the wholesale price at Kisumu railway station for imported sugar from overseas was Shs. 93.11 per bag, while that of East Africa, excluding Kenya, sold at Shs.102.11 for the 224-pound bag. In addition there was the subagent fee of Shs. 3.50 for transport from the railway station to the main agent's warehouse. This translated to a gross wholesale price of Shs. 114.43 per 224-pound bag for imported sugar and Shs. 114.52 for the 224-pound bag of the East African sugar.[108]

There was a further increase the following year in May 1957, when the wholesale price of sugar in the colony and imported sugar from overseas was increased again within a short span of time. This was in line with the commonwealth prices that went up worldwide by Shs. 5.93 per bag.[109] Consequently, the imported sugar sold at Shs. 121.22 per 224-lb. bag, while the sugar produced at Miwani sold at Shs. 121.17 per 224-lb. bag, and that of Uganda was sold at Shs. 102.11 per 224-lb. bag.[110] These prices made the Miwani sugar more expensive than that of Uganda, which affected marketing of the local sugar. The whole sugar industry was affected from growers to manufacturers and the colonial state was reluctant to help. Moreover, the subsidy given to both the growers and the manufacturers was reduced by half. For example, the growers' subsidy was reduced from Shs. 1.87 to Sh. 0.94 per ton of cane supplied, while the manufacturers' sugar subsidy was also reduced in 1957 from Shs. 4.48 to Shs. 2.24 per bag.[111] Furthermore, the minimal increase of sugar prices earlier on by the colonial state failed to adjust prices according to the increase of the commonwealth prices of sugar, and instead of addressing the farmer's needs, the state reduced the subsidy. Due to the minimal increase in the price of sugar, Miwani Sugar Mills was forced to revise the sugarcane prices based on the price stated by the colonial government, which in turn led to serious problems with the Asian cane farmers who were well aware that the prevailing international market prices in 1957 were good. The unreasonable control of the market through the reduction of the subsidy in the face of stiff competition from the Ugandan sugar made things worse for Asian cane farmers and manufacturers since they had to struggle to sell their expensive sugar in the local market, exposing the contradictory nature of state policy. It will be recalled that the state had embarked on rigorous experimentation to improve quality of cane, yet failed to protect the industry from unfair competition.

Moreover, the cane farmers suffered because the local price of sugar in Kenya did not correlate with the average world prices. Additionally, the sugarcane industry was disadvantaged by the cheap price of sugarcane grown under more favorable conditions in Uganda, whose industry was more established. To support the sugar industry in Kenya, even if it could not produce enough sugar to meet its domestic needs, as was the case, it would have been more expedient for the colonial state to impose an import duty on the Ugandan sugar so as to bring the price up and minimize the difference and also to only allow import of the amount of sugar needed to make up the Kenyan production deficit. Unfortunately this was not the case. The importation of sugar without duty continued unabated and hindered the progress of the industry in colonial Kenya.

In terms of consumer prices, on the other hand, the DC of CN, P. D. McEntee, authorized that sugar be regulated in the district in 1957. Hence, a new order was issued that detailed the retail price of sugar in each location in accordance to the regulation of the Defense Control prices of 1945. This brought a slight increase in the price of sugar; the retail price went up from 57 cents per pound paid in 1956, to 59–62 cents in 1957. Moreover, the new regulation controlled the prices in all markets and trading centers. Such initiatives were imposed by the colonial state in an effort to control the local market and protect the local consumers from the inflated prices by the local traders. The regulations affected the sale of sugar as the maximum prices were affected as they tended to fluctuate at different market and trading centers. By 1957, all the markets and trading centers found in the sixteen locations of CN had different prices. We find that the cheapest prices were found in Kano, Kisumu, and Kajulu. For example, nine markets and trading centers in Kisumu sold sugar at 59 cents per pound, while the most expensive price was found in Samia, where all the markets sold sugar at 62 cents a pound.[112] On the other hand, the markets near Kisumu Municipality, where the agent's store was based, sold cheaply compared with Samia, which was farther away at Port Victoria; distance and transport costs were major determinants in these prices. However, the detailed distribution of the prices at the location level meant that the traders could not go above the stated prices regardless of the difficulties they encountered, such as transportation costs. It also meant that the colonial state controlled the market even in the African reserves.

Even with the new prices, irregularities continued. For instance, in 1957, exploitative retailers were found at Rabour, Yala, and Nyangweso who sold sugar at Shs. 1 per pound instead of 60 cents. Such cases were rampant in almost all the locations; for instance, similar incidences were found among the retailers of Alego location, which culminated in a public outcry that the colonial chief Amoth Kowira was forced to order an investigation that revealed that some traders were charging 80 cents per pound. Such irregularities continued over the years and also involved magnates in the industry. In 1960, D. K. Hindocha was found to be charging Shs. 120.30 for a 224-lb. sack, which was the correct price for a subagent to charge as indicated in the Price Control (Sugar Amendment) of 1959 legal notice no. 427 of 1959. Hindocha was a main agent, and this was not the correct price for the main agent; hence it appears as though he was doubling also as subagent.[113] By 1961, the colonial government stopped purchasing sugar directly from the local factories; instead it was done through the main agents.[114]

At the end of the colonial period, the distribution of sugar was controlled by the colonial state, even in the African reserves. There was an expanding market that had to be satisfied by the Asian cane farmers. Therefore, many colonial chiefs constantly requested for more sugar for their respective areas. For instance, Chief Amoth Kowira of Alego, Chief Abok of West Alego, and Chief John Oloo of Yimbo wrote to the district officers in their division requesting for more allocation of sugar supplies.[115] There was a clear expanding local market that the colonial state could have taken advantage of and supported the Asian cane farmers so as to produce more for the home market and move toward self-sufficiency—or at least reduce the amount of monies used in importation of sugar. The state, however, placed more emphasis on the control of distribution of sugar, even in the African reserves, than in stimulating Asian sugar growers to expand production.

IMPORTED SUGAR

The controls placed on the sugar industry by the colonial state were partly due to lack of appreciation of the economic importance and potential of the Asian sugar industry that consequently led to greater deficiency than could have been the case, forcing the colony to over-rely on imports. Moreover, it led to exploitation through irregularities. The manufacturers were also not encouraged to mill other grades of sugar that was needed in the colony. The overall deficit in domestic production caused sugar to be imported into Kenya and distributed alongside the locally produced sugar. Small quantities of refined sugar were imported from Hong Kong, but the bulk of it came from the United Kingdom and was imported to the colony by the company Tate and Lyle Ltd. On the other hand, mill white sugar was imported from Uganda and Tanganyika.

There was a steady trend of importation of sugar into the colony especially from 1919 on. Initially the importation involved supplies for the Uganda Protectorate as well. By 1928, however, the focus was mainly on importation into Kenya, as is indicated in table 4.3.

The data in table 4.3 show the changing trends in imports due to the sugar consumption needs of the colony and the development of the sugar

Table 4.3. Imports of Sugar into Kenya, 1919 up to 1933

Year	Amount (tons)	Value (£)
1919	3,580	126,849
1920	2,400	167,554
1921	3,300	129,869
1922	3,270	82,288
1923	1,580	54,041
1924	1,370	42,904
1925	1,360	31,475
1926	978	20,265
1927	481	10,661
1928	4,010	68,903
1929	2,400	38,576
1930	1,730	23,510
1931	328	6,688
1932	221	4,182
1933	128	2,389

Source: *DAAR*, 1919–1933.

industry. Kenya imported a large amount of sugar between 1919 and 1923. This was because during that period Kenya had not started to manufacture her own sugar. The importation of sugar considerably decreased in 1926. This was because the Victoria Nyanza Sugar Company started crushing sugarcane in 1924. The opening of Victoria Nyanza Sugar Company caused the Asian farmers, who initially made jaggery, to acquire more land, and the cultivation of cane increased. This explains the low importation in 1926. The greatest decrease in 1927 was due to the increase in acreage of cane of the European cane farmers as Nottidge and Allen Company acquired land in Muhoroni and started to grow and produce sugar. The decrease in importation of sugar in 1929 was also partly due to increase in acreage as the number of Asian farmers increased from 40 with 3,693 acres of cane to 48 farmers in 1929 with 8,820 acres under cane.[116] This meant more sugar was produced, which led to a decrease in importation.

By 1931, the impact of the Great Depression was clearly felt in terms of decreased purchasing power among Kenya consumers; imports into Kenya from overseas continued to decline. For instance, there was a decrease from 1,730 tons in 1930 to only 328 in 1931.[117] Table 4.3 also shows a decline in imports in 1933; this was because in the face of depressed state of world sugar market, export was discouraged.[118] It seems possible that the idea to retain sugar in the colony led to the decrease in imports in an effort to secure the home market. Moreover, the Asian cane farmers by the early 1930s had increased their production of sugarcane. Despite the resilience shown by the Asian cane growers and the expanding local market demonstrated by the increase of African consumption needs, the colonial state failed to see the economic potential of the sugar industry in CN and did very little during this period to support the industry. During this period the data indicate the steady development of the industry in relation to viability of market and shows how the local market remained undersupplied by the local sugar.[119] This insufficiency of sugar and potential of the existing domestic market could have been used to encourage more farmers in the sugar-belt to produce more cane instead of relying on Ugandan sugar. Encouraging Asian production could have led to financial gains for both the farmers and the colonial state since the sugar industry was part of the wider colonial economy.

However, in 1937, the International Sugar Restriction Scheme was entered into by the United Kingdom on behalf of the crown colonies, and, as such, an export quota of 27,000 tons was allocated to East Africa as a whole. As a result, a meeting was convened in July 1937 to discuss the allocation of the quota among the three colonies. This, however, did not take place since the allocated quota was exceeded by the increase in consumption rate of the three colonies and so the quota was not necessary.[120] Consequently, a restriction scheme was implemented by September 1937 through an order issued under the Customs Management Ordinance of 1937 that prohibited the export of sugar outside the three territories except with the permission of the director of agriculture.[121]

In the postwar period, Uganda exports resumed. For example, in 1949, it was estimated that Uganda exported 17,000 tons of white sugar to Kenya, while during the same period Kenya produced 9,000 tons. However, the annual requirement was 36,000 tons; hence, there was need to import a further 10,000 tons to meet the domestic demand.[122] Therefore, Kenya had to import sugar from other countries such as Tanganyika. By 1951, furthermore, Kenya was compelled to import sugar from overseas, which cost even more than the local price payable to Miwani. Yet the colonial state was not ready to subsidize transport costs for the Kibos farmers so as to induce farmers to grow more. Ironically, the millers did not want to pay the farmers good prices for cane, yet the United Kingdom paid funds through a scheme known as Colonial Sugar Preference Certificates to manufacturers in Kenya for each financial year between 1946 and 1948 to help with the production costs.

During the period of 1946–1947, the Victoria Nyanza Sugar Company benefited to the amount of Shs. 6,035 while Ramisi at the coast got Shs. 14,394. The following season Miwani got Shs. 19,000 and Muhoroni Sugar Company got Shs. 2,600.[123] With this type of financial help, the millers could have also extended fair prices or given bonuses to the cane farmers.

The absence of such help meant that it was the sugarcane farmers who suffered most because such subsidies were not given to them as they were to the manufacturers. Furthermore, due to bare or nonexistent minimal tariffs, many countries were ready to sell sugar to Kenya. The colony spent a huge amount of money importing sugar, which could have been greatly reduced if the local industry was fully supported. In 1954 alone the cost of importation of sugar was £1,653,116.[124] This was a significant figure, bearing in mind that there was the possibility of increasing production of sugar in the colony to meet the local demands if the Asian cane farmers were accorded the preferential treatment given to settler wheat and coffee farmers during the same period.

In the meantime, Kenyan sugar prices continued to stabilize over time despite the competition from Uganda. Most cane farmers and manufacturers felt that there was a need to restrict the importation of Ugandan sugar, and they expressed this desire to the colonial state. However, their request was denied on grounds that the Asian production had not reached the required quality, and the farmers and manufacturers were urged to improve production.[125] The colonial state favored Ugandan sugar. For example, when the government in 1961 imposed the excise duty of 11 cents per pound, or 246.40 per ton, the prices were as shown in table 4.4.

The sugar prices shown in table 4.4 indicate that Ugandan sugar was bought by the government at high prices. Yet when the Kenyan manufacturers complained that the government was paying more for the Uganda sugar, which was affecting the local production, the government dismissed their claim.[126] In fact, a 1961 memo confirms that the two colonial governments had agreed to support Ugandan sugar at the expense of Kenyan sugar.[127] This was because the East African Production and Supply Council had established an inter-territorial agreement in 1954 covering staple foodstuffs, particularly wheat, maize, rice, and sugar. According to the agreement, Kenya had to ensure the security of markets for her agricultural produce, specifically the need to safeguard the local wheat industry (which was dominated by European settlers). Therefore there

Table 4.4. Comparative Prices of Imported Ugandan and Kenyan-Miwani Sugar, 1961

	Uganda Sugar (in Shs.)	Miwani Sugar (in Shs.)
Base price per ton	944.60	924.58
Excise duty per ton	246.40	246.40
Total price per ton	1,191 per ton	1,170.98 per ton

Source: Extract from a memorandum by G.D. Parkin, dated July 28, 1961, KNA: AE/3/916.

was an understanding that Kenya would buy sugar from Uganda and Tanganyika while Kenya, being the main grower of wheat, would export wheat products to her two neighbors.[128] Due to such arrangements, the Kenyan sugar manufacturers continued to suffer from the stiff competition of the Ugandan sugar.

Despite the hurdles faced by the Kenyan manufacturers, the Miwani Sugar Mills was able to make strides and became the only factory producing refined sugar in East Africa by 1961. Furthermore, we find that the refined sugar and even the special sugar produced in Kenya during this period sold very competitively despite earlier doubts by the government. For instance, while imported two pound sugar sold at Shs.1.05 that of Miwani sold at 80 cents only.[129] It was evident that by 1961, Miwani refined sugar was cheaper than other imported special sugars, despite the fact that prices were determined by the general rule of supply and demand. For example, the five-pound pack of Miwani refined was cheaper than the imported one, while the icing sugar and the cube sugar were cheaper by almost 25 or 30 cents. This was encouraging to the millers and growers alike, and if such a trend was to continue, sugarcane farming in Nyanza would have been more profitable.

As the population increased, moreover, the domestic demand for sugar was unrelenting, which forced the colonial state to import from other countries, such as the United Kingdom, Madagascar, Tanganyika, Mauritius, and South Africa. The high demand for importation was occasioned not only by population increases and the rise in purchasing power among the Africans, but also by new business ventures in the colony such as factories that processed cakes and sweets. For example, Vasani Products, based in Kisumu, started manufacturing candy in 1957 and used approximately 250 bags of refined sugar monthly for making sweets such as peppermints. Before that time, a lot of sweets were imported into Kenya from Madhvani Nagar Sugar Works in Jinja, Uganda. In the meantime, hotels, for instance the Pirani Tea Room, also needed a lot of sugar.[130] Other hotels, such as the Hindu Hotel owned by Himatlal K. Vyas, needed large amounts of sugar.

Businesses such as these were using sugar in large quantities, and Miwani, Muhoroni, and Ramisi could not meet the local demand. By 1957, Kenya became so desperate for sugar that it pressed Uganda to export more. The pressure caused Ugandan manufacturers to take advantage of the situation and start to dictate terms on how their sugar supplies were to be distributed, dictating even the amount to be paid for the Ugandan sugar.[131] Frustrated by the Ugandans' unreasonable demands before delivery, the Kenyan government was forced to turn to South Africa, which initially had only been supplying refined sugar in the country.

Uganda's behavior finally prompted Kenya to launch an investigation as to why the Ugandan sugar industry was performing better than that of

Kenya. The report indicated that it was because of the good ecological conditions in Uganda coupled with the fact that Ugandan manufacturers were paid better by the Kenyan government. Consequently, the government in 1957 revisited the whole pricing structure in the Kenya sugar industry and set the prices given to Kenyan millers equal to those it accorded Ugandan millers. In essence it removed the advantages that Ugandan millers had over Kenyan manufacturers. Moreover, Kenya started to have contacts with Ethiopia to learn more about the sugar industry there with an aim of emulating it.[132] It took the colonial government a longer period to realize the potential of the sugar industry; the colonial state could not listen to Asian sugarcane farmers and manufacturers and had to wait for Ugandan manufacturers to actually exploit the cane farmers, manufacturers, and the Kenyan consumers before it intervened to save the sugar industry.

Needless to say, the Kenyan manufacturers and traders had not been happy with the treatment accorded to Ugandan millers whose sugar sold cheaply in Kenya. Asian entrepreneurs felt that the government had hindered their effort in the import of special sugar into the country. Bhagwanji and Company complained during 1957 that the colonial government favored Europeans when issuing licenses for sugar importation.[133] By granting import licenses to Europeans, the colonial state deliberately discriminated against Asian traders who were struggling for economic space. The restriction of sugar imports to one race was an attempt to control the market and the expansion of the Asian sugar industry through restriction of sales.

ILLEGAL IMPORTATION OF SUGAR

The colonial government's neglect of the sugar industry through lack of supervision and ineffective tariff regulation affected the industry and encouraged the illegal importation of sugar. Some of this was illegal importation from Uganda because of the marked price differential between Kenyan and Uganda sugar.[134] Ugandan sugar sold cheaply in Kenya due to subsidized rail freight rates that the Ugandan cane producers enjoyed as well as Kenya's lack of effective tariff protection. The lack of supervision and the continued neglect by the colonial state made it easy for the illegal trade to take place.

Sugar was smuggled through the town of Kitale in western Kenya during the 1950s, and this affected the local production. For example, by 1956 the price of smuggled sugar was Shs. 10 lower per bag than the Kenyan price.[135] As a result, most local traders in the region preferred to buy the illegal sugar. During the same period, it was estimated that approximately 1,500 bags of sugar were smuggled per month.[136] The illegal sugar from Uganda got into the colony through road transportation and not the railway, which was the route for the officially imported sugar. Colonial administrators were informed of the activity but were slow in responding. It is evident that colonial state failed to take measures that could have stopped such importation, probably through inspection at the border towns. The failure of the state to take charge of the situation so as to protect the home industry contributed to the illegal movement of sugar along the Kericho-Kisii and Kisumu-Kisii roads. This movement caused immense setbacks for the Kenyan industry since it resulted in a lot of non-smuggled sugar in both the central and south Nyanza districts remaining uncollected.[137] Such unconventional means of trading undercut the profits of the agents and the Kenyan manufacturers and also disrupted the distribution of sugar, not only in Nyanza but in the colony at large.

Another illegal trade involved the manipulation of the use and or term of the word "sugar." Sugar, as defined in the Essential Commodities Control of Distribution No. 2 of 1943, included such products as "sugar water," "sugar honey," "artificial honey," and "golden syrup," which are all solutions of sugar in water and as such were covered under the term "sugar."[138] However, unscrupulous dealers misused the terms to their advantage so that it meant only crystallized sugar. They manipulated the situation and imported the sugar solutions into the country illegally. This trade thrived during the war, and the government needed to clarify what the term meant, as the importation of sugar water from Uganda in large quantities gained momentum. A firm in Nairobi known as the Nairobi Produce Store used to import 400 four-gallon tins from Jinja in Uganda every month. Each tin contained 40 pounds of sugar with the rest water. Consequently each consignment of 400 tins translated to 16,000 pounds or 70 bags of illegal sugar.[139] Such activities were rampant, especially during the war. What was interesting was that even when Miwani, the only factory with a refinery in the region, started to manufacture refined sugar whose main markets were Uganda and Tanganyika, there was no indication or any record that such a valued commodity was smuggled from Kenya into Uganda, as was the case with mill white sugar from the latter. The colonial state, having failed to protect the home market, made it possible for sugar to be easily smuggled into the colony and not out. The need to stop the illegal importation of sugar was important because it discouraged production of sugar by the Asian farmers and manufacturers in CN and undercut the profits meant for the local producers, which impacted the entire sugar industry in Kenya.

SUGAR IMPORT TARIFFS

Asian cane farmers and manufacturers struggled for the market as they had to compete with the illegal cheap sugar that was finding its way into the colony. This was because, prior to 1922, import duties were mostly designed for revenue collection and not to protect Kenya's agricultural industries. Later this was to change when the colonial state realized that the large-scale farmers, especially the European settlers, were being hurt by the cheap, illegal imports of various crops. Hence it was found that tariffs against the import of foodstuffs were essential to protect locally grown food crops.[140] This was the case with European wheat growers when tariffs on wheat imports were imposed to protect the local production, since the locally produced wheat exceeded the world prices by a wide margin.

Unfortunately, the measures that had been put in place for the European wheat farmers were not applied to the sugarcane farmers, despite passionate calls by the Kenyan manufacturers for tariffs to be placed on imported Ugandan sugar. The colonial state once again neglected the Asian sugar industry through its failure to protect it from outside competition by imposition of a tariff. This created a hostile economic environment that made it difficult for the industry to thrive.

This preferential treatment for European farmers in the levy of import duties began relatively early in Kenya's history. In 1922, under the influence of Lord Delamere, a series of import duties were imposed on specific items such as beer, cheese and butter, wheat, and manufactured goods. These duties remained in effect until 1930 but did not affect sugar. The state found it challenging to effect tariffs, and so a cost-of-living commission was set up in 1924 to review the tariff system and monitor prices. In 1929, the commission submitted its report, which was accepted in the LegCo.[141] In response a new tariff system was adopted, but it was geared toward protecting only certain local industries and was not inclusive. As with most colonial agricultural policies, it was tailored to favor the Europeans since it mostly reduced duties on items that were consumed by the Europeans. This was doubly so because the Europeans were the ones who formed the membership of the cost-of-living commission. For example, the commission saw to it that the import duty on refined sugar (mostly used by Europeans) was drastically reduced.[142] At the same time, the colonial government refused to place a tariff on the Ugandan manufactured mill white sugar that was flooding the Kenyan market.

However, it was problematic to compute an appropriate customs tariff that was fair and allowed flexibility as conditions changed. For example, between 1954 and 1955, the cost of imported sugar through the port of Mombasa varied from approximately £37 to £46 per long ton, a difference of 9 pounds per ton. Such variations occurred time and again and were also affected by the overall production of sugar in the world and the consumer demand. Moreover, it was compounded by the convertibility of the world currencies. As such the policy makers felt that it was difficult to fix an appropriate customs tariff that was equitable to producers and fair to consumers. When a tariff for wheat was imposed, the main concern was maximum profits for the European farmers, not affordability for the consumer.

The setting of tariffs was particularly challenging for sugar imports. For instance, it was generally believed that if tariffs were to be imposed, the tariff rate should be fixed for a period of six to twelve months to accommodate the time lag between the ordering and receipt of sugar; but this was not always the case.[143] Ironically, the colonial state rarely put tariffs on mill white sugar, and the sugar industry was not protected from adverse competition from outside sources, especially from Uganda. Perhaps if greater protective measures had been introduced for the Asian sugar industry through import tariffs, Ugandan sugar could not have been cheaper and could have reduced the competition. The lack of effective tariffs affected the marketing of sugar, which in turn affected the price offered to cane growers in colonial Nyanza.

EXPORT OF SUGAR

The exportation of sugar from Kenya shows again the contradictory policies of the colonial administration. Kenya had a deficit of sugar supply as indicated by imports from Uganda, yet the state found it prudent to allow exportation of sugar. Exportation could only have been viable if the colonial state supported the local industry to perform to its capacity. Kenya started to export sugar in 1924 when the Victoria Nyanza Sugar Company began crushing cane. The first export amounted to 149 tons valued at £6,268.[144] The following year there was an increase in the amount exported, which amounted to 277 tons valued at £9,614.[145] The exportation of sugar continued over the years concurrently with importations. For instance, Kenya exported sugar to Tanganyika and other areas. The export to Tanganyika amounted to 1,266 tons in 1927 valued at £37,052 and 597 tons valued at £19,258 in 1928.[146] By 1933, faced with the depressed state of world sugar market, the colonial state discouraged the exportation of sugar until prices stabilized.[147] According to the Sugar Control Bill of 1938, the director of agriculture authorized that sugar be exported to the Italian East African territory through the town of Moyale in the then Northern Frontier district. The exportation was to be in small quantities by traders and not manufacturers.[148] By the early 1940s, Kenya was also exporting its manufactured brown sugar to the United Kingdom where it was in great demand.[149] The sale was conducted through Dalgety and Company. Moreover, between 1924, after Victoria Nyanza Sugar Company started crushing sugar, and the year 1952, which marked the height of great experimentations undertaken by the sugar industry, Kenya's production had increased and the colony exported sugar as indicated in table 4.5.

The data in table 4.5 show that Kenya exported sugar despite the fact that the country was far from attaining self-sufficiency in providing for its own consumption needs. There was a substantial export between 1933

Table 4.5. Exports of Sugar from Kenya, 1924–1952

Year	Amount (tons)	Value (£)
1924	149	6,268
1925	277	9,614
1926	1,120	29,553
1927	1,270	37,052
1928	597	19,258
1929	782	25,503
1930	752	—
1931	1,060	24,930
1932	802	17,879
1933	4,130	66,963
1934	3,670	57,781
1935	4,300	52,976
1936	6,600	76,301
1937	4,640	51,341
1938	6,520	73,998
1939	4,070	50,308
1940	—	—
1941	2,370	32,586
1942	2,850	9,902
1943	1,610	27,320
1944	858	16,088
1945	1,380	27,521
1946	636	14,464
1947	1,450	—
1948	538	—
1949	185	4,508
1950	44	1,746
1951	51.2	2,412
1952	43	2,800

Source: *DAAR,* 1924–1952.

and 1939. However, export performance overall was erratic and kept fluctuating. There was a marked increase from 1948 to 1952, partly the result of an increase in the local consumption and purchasing power of the Africans. But exports fluctuated greatly over time. For instance, in 1928 export to Tanganyika dropped from 1,270 tons to 597 tons, a decrease that was occasioned by drought that led to reduction in sucrose content.[150] The decrease in tons for export in 1930–1931 was due to poor harvests as a result of the effects of locusts' visitation, which had destroyed much of the crop. The rise in 1933, on the other hand, was due to increases in acreage as both the Asian and European cane growers expanded the area under cane despite the depression of the world market. The increase was likely due also to the struggle to secure the home market that was threatened by Uganda imports.[151] The huge export witnessed in 1935 and 1936 was a result of increased demand created by the Italian-Ethiopian war.[152] Italy had invaded Ethiopia to rewrite the history of their humiliating defeat during the battle of Adowa of 1896. The presence of these soldiers created an expanded market for sugar. This continued until the outbreak of the Second World War. On the other hand, the decrease of sugar exported in 1946 was as a result of low production due to dry weather and the labor shortage experienced at the time.[153] Therefore, we find that there were many factors that affected production and led to fluctuations in export volume.

Asian sugarcane farmers and Asian capital contributed a lot to the growth and expansion of the industry right from its infancy when the first milling factory started crushing cane in 1924 all the way to 1952. The year of 1952 marked the period of intensification of experimentation supported by the colonial state, which revolutionized cane farming not only in Nyanza but in the whole colony, with the other four small factories also playing a small part. During this period, the immense support for cane experimentation led to the growth of the industry. Here we see the typical contradictory nature of colonial policy: On the one hand it supported the industry through experimentation while on the other it inhibited self-sufficiency through the sale of sugar via export, which curtailed the efforts of the farmers and manufacturers alike. On many occasions, the Asian cane farmers had to struggle to survive in a colonial state that was not always supportive.

Kenya also exported other crops such as wheat. However, both wheat and wheat flour had no ready expanding market in the colony, and so Kenya had no production deficiency relative to demand. Therefore, exportation of the wheat produce by the colonial state was welcomed, unlike that of sugar export which suffered from frequent insufficiency and underproduction. Exportation of sugar created shortages and defeated the state's policy of attaining self-sufficiency in sugar production. The exportation of sugar seen in table 4.5 can be compared with that of wheat and wheat flour, listed in table 4.6.

The data in table 4.6 show the performance that the export of wheat and wheat flour over a period of time and indicates profit value when combined. This performance can be compared with that of sugar export in table 4.5. Table 4.5 indicates that the value of sugar was better between 1924 and 1927. Moreover, the data show that the difference in export value of the two crops in 1932 was minimal while in 1939 the value of the two crops was almost the same. Although wheat did better between 1928

Table 4.6. Export of Wheat and Wheat Flour from Kenya, 1924–1946

Year	Wheat		Wheat Flour		Total Value (£) of Wheat and Wheat Flour Exports
	Tons Exported	Value (£) of Exports	Tons Exported	Value (£) of Exports	
1924	96	1,520	—	—	1,520
1925	31	482	94	1,588	2,070
1926	2	22	165	3,385	3,407
1927	14	200	275	6,038	6,238
1928	6,832	75,195	886	17,549	92,744
1929	3,480	69,281	1,601	31,996	101,247
1930	10,726	91,447	1,927	33,150	124,597
1931	3,182	15,313	1,404	20,263	35,576
1932	86	525	1,511	18,777	19,302
1933	6	50	2,054	25,158	25,208
1934	100	591	2,371	30,603	31,194
1935	51	479	3,878	49,730	50,209
1936	2,300	22,883	3,094	40,540	63,423
1937	390	4,746	3,470	45,264	50,010
1938	4	39	2,907	43,319	43,358
1939	49	393	1,194	50,655	51,048
1940	—	—	—	—	—
1941	—	—	—	—	—
1942	—	—	—	—	—
1943	45	794	5,194	117,750	118,544
1944	223	3,214	7,933	181,612	184,826
1945	10,059	116,319	6,001	153,950	270,269
1945	1,691	27,351	9,634	233.205	260,556

Source: *DAAR*, 1924–1946.

and 1931, it was overtaken by sugar again in 1933–1938. This shows that sugar was competitive.

Sugar exports competed effectively with wheat before 1943, and so there seems to have been no reason as to why wheat farming was supported more than sugarcane by the colonial state. The table indicates clearly that wheat and wheat flour exports only attained steady growth in value beginning 1943 when it became more profitable over sugar. From comparative analysis of these tables 4.5 and 4.6, it seems that sugar was profitable before 1943, which might have influenced the colonial state to export the little that was produced despite deficiency. On the other hand the export of wheat and wheat flour was occasioned by the fact that there was no ready expanding market, hence no deficiency, and moreover by 1940s wheat was more profitable.

MARKETING AND DISTRIBUTION OF SUGAR IN COLONIAL CN

Marketing remains a crucial stage in all agricultural production and includes many activities such as collecting the produce, acting as wholesalers, and even providing credit to the producers. In the sugar industry, once the sugar was processed it was conveyed into the hands of appointed agents and subagents who then distributed the sugar to retailers according to the instructions issued by the local board. From the 1940s on, the price was fixed by the government. However, during the period when the Victoria Nyanza Sugar Company first started milling, there was no clear distribution policy for sugar, which impacted its marketing, despite the fact that organized marketing of other produce had started by the 1920s.

By the early 1930s, therefore, the marketing of sugar was haphazard and required regulation. During this period, the produce was sold by Asian traders in major trading centers or by hawking it in the African reserves; as such these Asian traders acted as middlemen between the factories and consumers, especially the African buyers. Hence to improve marketing of their produce, the Asian farmers formed the NFCS, as earlier stated. The NFCS felt that the colonial government was not taking a keen interest in the Asian farmers' affairs. The colonial state supported the white wheat and maize growers in many ways (e.g., low railway rates) but provided no such subsidies to the sugarcane farmers.[154] Such biases led to the formation of the NFCS, and the society made an appreciable difference in the sale of jaggery. Unfortunately, their efforts failed to make much of an impact on the sale of sugar due to lack of supervision and regulations; moreover, unorganized marketing of produce often led to poor returns as opposed to what would be expected in a structured marketing system.[155]

State intervention would be required if the sugar industry was to improve its performance. Hence to solve the problem, the colonial government appointed agents to distribute sugar and started to grant licenses to subagents and retail traders for the purchase of sugar. This started during a period of rationing and control of sugar that was carried out under the defense regulations during the World War II period. By 1943, distribution had been streamlined. Accordingly, the internal distribution of sugar was placed squarely on the hands of the deputy sugar controller, and this worked well until 1949 when there was an acute shortage of sugar in the world. The colonial state then allowed factories to nominate main agents to distribute sugar.[156] However, the effects of the restriction on the sale of sugar soon began to be felt. The idea of seeking permission to sell sugar restricted the movement and sale in local markets and exposed the dilemma of the colonial state since restricting the movement of sugar curtailed the growth of the market and the expansion of the sugar industry. Although Miwani Sugar Mills asked the government to relax the restriction order for sugar and its by-products, nothing much was achieved. The granting of licenses seems to have been used as a means of control, since colonial state policy limited the quantity of market outlets in order to curtail rural capital accumulation and protect European farmers from competition.[157]

Such controls were witnessed when it was decided that as of June 1949, the government of Kenya would purchase all sugar allocated to Kenya. Sugar produced in Kenya and in Ugandan factories and that imported from outside East Africa was controlled by the deputy sugar controller, as mentioned earlier. The controller then issued instructions to various mills on how to dispose of their sugar output and the imported sugar at different points in the country under the control of local boards as covered by section 3 paragraph 1 of the sugar regulations.[158] Sugar was then passed to appointed wholesalers who in turn distributed it to traders. Accordingly, factories were required, on receipt of a disposal order, to immediately send by rail the quantities stipulated in the order, which after full payment for the consignment required them to distribute sugar to the agents. Hence it became the prerogative of the controller and Kenyan government to arrange for the distribution of sugar.

However, section 3 paragraph 2 of the regulations was contradictory and problematic in implementation. For example, it required that the deputy sugar controller ensure that no person or wholesaler could deliver sugar to retailers except by his directions, which specified monthly allocations for different districts in the province. This proved to be cumbersome since it meant that every month the controller had to write to every wholesaler, approximately 300 in number. Miwani Sugar Mills, for example, was supposed to supply sugar only to those parties identified by the controller in the quantities specified, which was time-consuming.[159] Furthermore, some of the distributors who had close ties with

Ugandan sugar factories complicated matters. Due to restrictions enforced by the colonial state, they wanted to get sugar from both sources. For example, the Anandji Group and R. K. Bhanyani wanted to get sugar from both Miwani and Uganda.[160]

Despite such stringent policies and complaints from sugarcane farmers, manufacturers, and traders, other regulations were put in place. For instance, the CN district, including the Kisumu municipality, became a prohibited area in 1953 under the provision of section 14 of the sugar ordinance, Government Notice nos. 936 and 937 of 1946. When a district was declared a prohibited area, it meant that no person was allowed to import sugar into the prohibited area without a permit from the controller. Furthermore, no person was to possess or sell any sugar in excess of ten pounds in weight without a permit in writing from the controller. This overregulation of the market forced traders to seek permission to sell more than ten pounds. Yet there were areas where sugar was highly consumed, for instance, in areas where there were many government departments such as the police, agriculture, and veterinary and mission schools and hospitals, such as the Yala area. This meant that the distributor in Yala had to constantly seek permission to sell the required amount.[161] Such controls were not cost-effective. Moreover, in the absence of sufficient economic incentives and lack of consultation of the cane growers and manufacturers in marketing, the colonial state curtailed the distribution and stifled the market for sugar in CN.

Due to such problems and complaints from the sugar manufacturers, the colonial government in 1955 instituted the Control of Import and Export Order, which was followed closely by the Control of Essential Supplies (Sugar) Order of 1958, by which the government took control of distributing sugar from the office of the deputy controller and appointed subagents instead.[162] D. K. Hindocha was appointed as the authorized agent for sugar distribution for both the municipality of Kisumu and the whole of CN district. In addition, various retailers were appointed by the colonial state through the office of the DC to distribute sugar in their respective areas, as indicated in table 4.7.

Table 4.7 indicates that there were numerous retail distributors in CN, all appointed by colonial officials. The data show that out of 22 retailers, 18 were Asians. The data also show that not all applicants were given sugar allocations. For example, the R. K. Bhayani and Anandji Stores applications were rejected since both traders were found to be receiving sugar from Uganda. Furthermore, traders that were appointed by 1954–1955 received very few bags of sugar. For instance, Sgt. Major Mukudi of Ukwala got only one bag per month despite the high population of the area.[163] In a span of only four years the allocations increased. For example, R. P. Rana, operating in the same area as Mukudi in 1959, was allocated twenty bags.[164] This could be explained by an increase in population and African purchasing power over the period. For instance, Uk-

Table 4.7. Sugar Retailers and Merchants in CN by 1959

Sugar Distributors	Area	Year Appointed	Quantity of 224-lb. Bags per Month
Hasham Brothers	Kisumu	1954	5
Dhamecha Brothers	Kisumu	1954	12
Sidi Ahmed	Kisumu	1954	5
Jeremiah Ochieng	Kisumu	1955	10
Sgt. Major Were Mukudi	Ukwala	1955	1
Blake Muswa	South Gem	1956	4
Damodar Jamnadas & Co.	Kisumu	1957	10
Hassanali Brothers	Kisumu	1957	20
Shivji Karim Surani	Kisumu	1957	20
Ramzans Diamond Stores	Kisumu	1957	10
Madatali Fancy Stores	Kisumu	1957	6
K.R. Soratha	Kisumu	1957	20
Messrs Thakarshi	—	—	—
Ranchhod & Sons	Yala	1959	10
R.K. Bhayani	Kisumu	1959	Nil
L.P. Malde & Co.	Kisumu	1959	20
Jamnadas & Sons, Ltd.	Kisumu	1959	20
Anandji Stores, Ltd.	Kisumu	1959	Nil
Nyanza Produce Stores	Kisumu	1959	31
Saif Hashid	Yala	1959	30
R.P. Rana	Ukwala	1959	20
D.K. Hindocha, Ltd.*	Kisumu Municipality		Limitless

Source: Sugar, Newcomers Quotas 1959, KNA: DC/KSM/1/3/57.
*D.K. Hindocha was both a manufacturer and merchant.

wala division had the highest population in the district by 1957. This fact could also explain why the allocations by 1959 were high compared with the earlier years. For example, Saif Hashid of Yala was allocated thirty bags as the major distributor to the Arab population.

Both the agents and retailers in CN received letters of appointment from the director of trade and supplies that detailed their responsibilities. The subagents were not allowed to distribute sugar beyond the limits of their existing areas of distribution, and they were also not allowed to sell sugar to retailers after the 19th of each month.[165] This meant that if they had any surplus, then disposing of it would not be easy. They could not sell in the neighboring locations even if they had a surplus. Such stringent measures by the colonial state were counterproductive and curtailed the progress of the industry by restricting the movement of sugar and controlling the market. The colonial state could have corrected the situation by removing the legal and policy barriers that created inefficiencies and monopolies, so as to allow proper functioning of the market.[166] To improve the market, the colonial state needed to reduce the overregulation, but both the Asian cane farmers and the manufacturers, as always, had minimal say in the development of regulations. Regulations were designed to allow the state to exercise control over all aspects of marketing,[167] and they were imposed by those in control to stifle competition.[168] In the struggle for the market between the state and the Asians, the latter were the losers and the Kenyan sugar industry suffered.

As the colonial state tried to control the sugar market, the annual consumption needs for sugar increased. As already suggested, this was occasioned by population increase and an increase in the Africans' purchasing power.[169] As the demand for sugar increased, more traders asked for permits. By 1959 the government issued letters of appointment to new distributors of sugar. The new allocations, like the previous ones, were on a monthly basis. These new traders were expected to purchase and hold sugar in reserve stock for a period of one and a half months. This was difficult as millers and traders wanted to sell sugar as long as there was demand.[170] Moreover, most stores were too small to hold stock because besides sugar, other goods were also sold. These rules showed the firm grip that the colonial state had on the market. These regulations made the traders, Asian cane farmers, and the millers unhappy. For example, while the millers detested the sugar regulations, claiming they were drafted in haste and enacted without consultation, the cane growers despised them because the once lucrative market for jaggery had become chaotic. This was because jaggery got into hands of businessmen who paid the minimum price for the produce while much of it ended in the black market.[171] Hence they wanted to be consulted before any rules were passed.

The colonial state continued with such controls over the years, despite the opposition of the Asian cane producers. Their requests for less regulation were ignored, and it was only in 1960 that the colonial state claimed that the rules were "temporary" measures to streamline the sugar industry.[172] This was clearly not the case, as the overregulation had done nothing to streamline the market, and the excuse was intended only to calm the discontented sugar producers. This was necessary at that particular time due to the Kenyan political scene. As Africans pushed for independence, the colonial state did not want to antagonize the Asian entrepreneurs, including those in the sugar industry, for fear of possible Asian alignment with the African "agitators."

The Asian cane farmers and producers had developed a market consciousness, and their persistence finally yielded fruit when the colonial state for the first time involved them. This was seen when the state established the Sugar Industry Board and, although farmers were not represented, the initial membership included two representatives of the sugar millers. Ironically, the function of the board still stifled the sugarcane industry. The board's functions, among many, included regulation of cane planting, which had never been regulated before. The board started to issue planting licenses in conjunction with the minister of agriculture. This was detrimental to the sugar industry, and questions emerged as to why the Asians now needed licenses to plant cane when the European settlers in the same district did not require a license to plant wheat that was not even competitive. The board also determined the price at which millers bought sugarcane from farmers and in addition recommended a new levy on every ton of sugar manufactured.[173] This was not only ridiculous but also contradictory to the colonial state's policy of invigorating agriculture as stipulated in the Swynnerton Plan of 1954.

The continued control of the marketing of sugar was completely contrary to the colonial state's relaxed grip on European and African farmers, especially regarding the sale of maize. For example, in some areas of colonial Kenya such as the Bukusu of North Kavirondo the local native councils were active and facilitated the collection and sale of the farmers' produce in bulk.[174] Such opportunities were not accorded to the Asian farmers, despite the fact that they were also organized in groups such as the Nyanza Indian Farmers' Association and other production committees that constantly worked to improve their lot. The colonial state failed to acknowledge the Asian sugarcane farmers' efforts and give them an opportunity to be involved in the sale of their produce as they did to European wheat and maize growers in the colony. The Asian farmers and manufacturers could not effectively bargain since pricing and marketing was controlled totally by the colonial government. As a result the marketing of sugar in CN performed poorly.

These stringent regulations seem to have been put in place due to the fact that between 1958 and 1960, the output of sugarcane had increased in CN, and the colonial state imposed the regulations to restrict production. This made little sense since population growth and increased demand provided a market for the increased output. Consumption needs in the colony had to be met, as indicated in table 4.8.

Table 4.8 shows the steady increase in cane production and the even greater increase in consumption capacity. This was significant since during the colonial period, Nyanza was the major producer of sugar; the other mills at the coast and near Nairobi had very small crushing capacities. As such Nyanza, being the main producer, faced the challenge of trying to meet the local demand. The colony also continuously increased the quantity of sugar imported to meet the local needs over the years. The

Table 4.8. Production, Consumption and Import of Sugar in 1000 Metric Tons, 1955–1963

Year	Domestic Sugar Produced	Consumption	Imports
1955	17	55	28
1956	20	65	45
1957	20	69	49
1958	28	71	43
1959	28	78	43
1960	30	88	58
1961	33	93	60
1962	33	100	67
1963	38	98	60

Source: M. O. Odhiambo, "Sugarcane Procurement Problems in Kenya: The Case of the Nyanza Sugar Belt," *East African Agricultural and Forestry Journal* 53 (1987): 30.

table shows that by 1963, the country produced 38 metric tons during the season, but the consumption was 98 metric tons; hence a deficit of 60 metric tons had to be imported to meet the domestic demand.

With such demands, the need to regulate cultivation of sugarcane and marketing and distribution of sugar as required by the colonial state was ill-conceived. The regulations created more problems. For example, in 1961 the Miwani Sugar Mills refused to enter into a contract with the government that required the mill to give advance notice of three months to supply a definite monthly quota of refined sugar in bulk.[175] This was a difficult condition given the fact that the amount of sugar available could be affected by natural calamities such as drought and locust invasions, as had occurred in the early 1930s. In such instances, the manufacturers could not be in a position to ascertain the amount to be received from the cane growers. Despite the mill's opposition, the government was unrelentingly in its demand that Miwani Sugar Mills comply with the distributional Bulk Control Order regarding bulk supplies to subagents.[176]

The arrangement was not conducive to the success of the industry. It seems like the Asian sugarcane farmers and manufacturers in colonial Nyanza had their hands tied and were reduced to mere spectators as far as marketing of their produce were concerned. They rarely succeeded in changing the overregulation of the sugar industry. However, such uncalled-for restrictions did not discourage the Asian cane farmers who continued to produce more cane and needed more outlets for their produce. The increased production generated fresh calls by 1961 for another sugar factory to be built since the Muhoroni factory had ceased to operate and Miwani Sugar Mills could not cope with the output. By 1961, when Miwani started producing refined sugar, the factory's annual output of mill white sugar was greatly reduced.

However, by 1962, the colonial state changed its policies toward the distribution system, and it moved ahead and decontrolled sugar distribution. This was meant to get rid of all the agents and subagents and hand over distribution to the manufacturers, which could have been done much earlier to benefit the millers and the farmers. Unfortunately, in early 1963, before any benefits had accrued, the colonial government came up with a new rule and introduced the stockholding policy that required Miwani to hold stock of 3,000 tons of sugar at the beginning of each month.[177] The main reason for such a policy was to avoid any sugar shortages that could be caused by strikes and other incidences as the clamor for independence gained momentum. Within a short period, the colonial state actually increased the stock that Miwani had to maintain to 4,500 tons and prohibited locally manufactured sugar from leaving the area.[178] Miwani rejected the stockholding policy since it prohibited the company from selling all its sugar when there was demand for it. The company preferred to sell its sugar as soon as it was produced so they could pay the sugarcane farmers in a timely manner.[179]

The company petitioned the government to allow it to sell sugar as before, but the colonial government refused, arguing that it had a right to compel Miwani to follow the policy since it guaranteed the purchase of its sugar.[180] In addition, it began to intimidate the owners of the Miwani mill, accusing them of repatriating large sums of money to their sugar firms back in India.[181] The intimidations were unfortunate, making the economic and political environment even more hostile to the cane farmers and manufacturers. The regulations provided no competition or incentives to efficiency. And although they provided a measure of protection to the consumer in relation to the prices, they negatively impacted the amount of sugar available to consumers and traders. Consequently, state intervention in the market affected the growers who needed a more functional market structure. It appears that throughout the colonial period the colonial state was not keen on the sugar industry, which was, after 1948, solidly in the hands of Asian capital as the colonial state prepared to relinquish its power and grant Kenya self-government in June 1963.

CONCLUSION

The process of getting cane to the mills, having it be accepted and purchased by the mills, and marketing it for retail purchase provided an endless series of pitfalls for the Asian cane farmers.

The transport of cane to the mills was particularly difficult. The colonial state subsidized rail transport for the European farmers but not the

Asian farmers. The Asian farmers had to rely heavily on truck transport, which was often unreliable because of excessive rainfall and impassable roads. The reluctance of colonial state to take the lead left the Victoria Nyanza Sugar Company to take the initiative to install the tramway on its own; but the colonial state failed to take advantage of the railway that lay adjacent to the cane farmers, a move that could have boosted the industry. Time and again, the colonial state failed to capitalize on the potential offered by the sugar industry.

Once the cane arrived at the mills, the farmers could not always count on the cane being accepted for processing. The mill owners sometimes refused to accept cane that they had previously contracted for if the mills were overstretched in capacity. In these cases, the farmers still had to pay transport costs for getting the cane to the mill but had no means with which to pay for that transport.

This was compounded by the vagaries of pricing in the industry. There were no clear regulations and guidelines on the pricing of sugarcane. The mills set their own prices, which were often well below world market value. In addition, the colonial state regulated the terms of payment to farmers and eventually enacted a law that allowed mills to pay farmers at 150 days rather than 30 days, which again made it difficult for farmers to pay their transport costs, much less reinvest in their farms. The sugarcane farmers attempted to fight these practices by joining forces with the only Asian political party in Kenya; the East African Indian National Congress on issues of concern to them, but with only limited success.

The colonial state's policies continued to waver between neglect and unwelcome control. While neglecting the Asian farmers in terms of support for transportation and adequate price assurances, the state's excessive control of other aspects of production was detrimental to the farmers' interests. The state opposed all attempts by the Asian community to start their own mill or purchase heavy equipment that would increase the farms' productivity. In the midcentury period the colonial state set up arbitrary quotas for production, distribution, and sometimes pricing that operated without input from the Asian community and without regard to variations in costs and output from year to year. Laws such as the Sugar Act of 1944, which controlled the issue of permit and disposal of sugar, were unreasonable and problematic.

Worst of all, the state did nothing to protect the Kenyan industry from the dumping of Ugandan sugar into the Kenyan market. The tight state controls inhibited production, despite the fact that Kenya did not produce enough sugar to meet local demand. Instead, the state encouraged the import of sugar from Uganda and Tanganyika. The colonial state refused to put tariffs on imported sugar because they wanted to keep tariffs low on wheat exports to those same countries. Wheat was a crop produced mostly by European farmers, and so the colonial officials; nearly all European themselves, set policies to protect their own production. The irregular importation and the ineffective tariff system all contributed to denying the industry a substantial profit and the opportunity to reinvest and grow.

NOTES

1. KNA: AG/25/70, Solicitors Allen and Hamilton to Colonial Secretary, June 27, 1922.
2. Martin Okeyo (O.I.), Kibos out-growers' Farm Manager, June 24, 2009.
3. Mohinder Singh (O.I.), large-scale farmer Kibos, June 22, 2009.
4. Joseph Otieno (O.I.), truck driver, Chemelil, June 26, 2009.
5. KNA: DC/CN/1/4/3, K-LDAR 1931.
6. KNA: DC/CN/1/4/3, K-LDAR 1931 and 1932.
7. KNA: DC/CN/1/4/4, K-LDAR 1933.
8. KNA: BV/1/249, Executive Engineer to Commissioner for Local Government, Lands and Settlement, July 13, 1932.
9. Peter Wanyande, "Management Politics in Kenya's Sugar Industry: Towards an Effective Framework," *African Journal of Political Science* 6, no. 1 (2001): 123–140.
10. KNA: BV/6/605, Manager, East Africa Sugar Industries, Muhoroni, to Director of Agriculture, October 7, 1942.
11. KNA: PC/NZA/3/2/34, Petition by NIFA to Colonial Secretary, October 6, 1937.
12. KNA: PC/NZA/3/2/34, Colonial Secretary to NIFA, December 3, 1937.
13. Kanogo, "Kenya and the Depression," 112.
14. KNA: BV/111/119, Cane Transport, 1960.
15. KNA: AG/25/70, Governor to Duke of Devonshire, Secretary of State for the Colonies, January 3, 1924.
16. KNA: AG/25/70, Victoria Nyanza Company's Tramway Ordinance, 1923. Coryndon, to Colonial Office, January 3, 1924.
17. KNA: AG/25/70, Ag. Director of Public Works to Colonial Secretary, September 26, 1923.
18. KNA: AG/25/70, General Manager, Uganda Railway to Attorney General, May 19, 1923.
19. KNA: AG/25/70, Agreement made between Uganda Railway and Victoria Nyanza Sugar Company, 1923,
20. KNA: AG/25/70, Agreement made between Uganda Railway and Victoria Nyanza.
21. KNA: AK/11/30, Miwani Sugar Mills to Lalsingh Sandhu, April 22, 1952.
22. KNA: PC/NZA/3/2/34, DC K-L to PC Nyanza, October 1, 1937.
23. KNA: AK/11/50, Moon to Director of Agriculture, October 2, 1944.
24. KNA: PC/NZA/3/22/1/8, Regional Officer, Regional Distribution Board, Kisumu, to PC Nyanza, November 4, 1948.
25. KNA: PC/NZA/3/22/1/8, Regional Officer, Regional Distribution Board, to PC.
26. KNA: PC/NZA/3/22/1/8, Regional Officer, Regional Distributional Board, to PC.
27. KNA: PC/NZA/3/22/1/8, Report by Magor of the visit to the Asian Farms, October 6–10, 1947.
28. KNA: AK/11/30, District Agriculture Officer (DAO) CN to PAO Nyanza, May14, 1952.
29. KNA: AK/11/30, PAO Nyanza to Member for Agriculture and Natural Resources, June 2, 1952.
30. KNA: AK/11/30, Milwant Singh Sandhu to Chairman, Agriculture Production Committee, April 22, 1952.
31. Kanogo, "Kenya and the Depression," 112.
32. Smith, "An Overview of Agricultural Development Policy," 119.

33. Bruce Berman and J. Lonsdale, "Coping with the Contradictions: The Development of the Colonial State in Kenya, 1895–1914," in *Unhappy Valley: Conflict in Kenya and Africa Book One: State and Class* (London: James Currey, 1992), 79.
34. KNA: AK/11/51, Miwani Sugar Mills to PAO Nyanza, October 12, 1953.
35. KNA: PC/NZA/3/88/1/8, Petition presented to the PC, Nyanza by a deputation from Kibos-Manpower Committee on August 5, 1944.
36. Rama Mohan (O.I.), Chief Chemist, Kibos Sugar and Allied Industries, June 22, 2009.
37. Joseph Otieno (O.I.), Chemelil, truck driver, June 26, 2009.
38. Joseph Otieno (O.I).
39. KNA: PC/NZA/3/2/34, DC, K-L to Ag. PC Nyanza, October 3, 1934.
40. KNA: PC/NZA/3/2/34.
41. KNA: PC/NZA/3/2/34, NIFA to DC K-L, October11, 1934.
42. KNA: NZA/3/2/34, Resolutions passed at the EAINC meeting on January10, 1931.
43. Gregory, *India and East Africa*, 93.
44. Gregory, *India and East Africa*, 94.
45. KNA: PUB.1/1/1/6, EAINC papers, Part II, 37.
46. KNA: PUB.1/1/1/6, European and African Trading Organization newsletter no.1, as quoted in EAINC Papers, Part II, 38.
47. Gregory, *India and East Africa*, 369.
48. KNA: PC/NZA/3/2/34, PC Nyanza to Colonial Secretary, February 24, 1931.
49. KNA: AK/11/50, NIFA to Chairman, Agricultural Production Board, August 6, 1946.
50. KNA: AK/11/50 NIFA to Chairman.
51. KNA: PC/NZA/3/22/1/8, Regional Distribution Officer to the Executive Officer, Central Distribution Board, August 26, 1947.
52. KNA: DC/CN/1/3/1, CKDAR, 1947.
53. Gilbert Muyumbu, "A Historical Study of State Policies in Kenya Sugar Industry," M.A. Thesis, Egerton University, 2006, 51.
54. KNA: PC/NZA/3/22/1/8, D. K. Hindocha to Chairman, Production Board Local Committee, July 28, 1947.
55. KNA: PC/NZA/3/22/1/8, Hindocha to Chairman, Production Board.
56. KNA: AK/11/51, AAO (Sugar) S. Singh Gill to PAO Nyanza, September 24, 1953.
57. KNA: AK/11/173, M. J. Patel to Chairman Supply Board, November 27, 1948.
58. Engineer J. S. Cheema Singh (O.I.), Factory Manager at Kibos Sugar and Allied Industries, June 23, 2009.
59. KNA: AE/3/916, Miwani Sugar Mills to Director of Economic Coordination, October 9, 1961.
60. KNA: AK/11/173, PAO Nyanza to DC CN, April 11, 1950.
61. KNA: AE/3/916, J. P. Starling to Director of Trade and Supplies, September 16, 1961.
62. KNA: AE/3/916, Parkin to Jones, confidential, July 10, 1961.
63. KNA: AE/3/916, Proposed Analytical Standards for Refined and Mill white sugar, July 4, 1961.
64. KNA: BV/111/119, Confidential Report on Kenya Sugar Industry, 1967.
65. KNA: DC/KSM/1/3/58, Chairman, Gem Location Committee to DC CN, July 5, 1954.
66. KNA: AE/3/916, Parkin for Permanent Secretary (PS) Commerce, Industry and Communication to Hindocha, Director, Miwani Sugar Mills, August 26, 1961.
67. KNA: AE/3/916, Refined Sugar.
68. KNA: AE/3/916, Managing Director, Henckell du Buisson (E.A) to Mr. Starling, Department of Trade and Supplies, July 10, 1961.
69. Maxon, *Going Their Separate Ways*, 66.
70. Maxon, *Going Their Separate Ways*, 61.
71. Maxon , *Going Their Separate Ways*, 63.

72. Maxon, *Going Their Separate Ways*, 63. Maxon also demonstrates how such measures were also affected in Vihiga in North Kavirondo District, where the petty traders were eliminated in various market places.
73. Maxon, *Going Their Separate Ways*, 66.
74. Maxon, Going *Their Separate Ways*,116.
75. Maxon, *Going Their Separate Ways*, 67.
76. Maxon, *Going Their Separate Ways*, 68.
77. Gavin Kitching, *Class and Economic Change in Kenya: The Making of an African Petit Bourgeoisie 1905-1970* (New Haven, CT: Yale University Press, 1980), 9.
78. Colony and Protectorate of Kenya, *DAAR 1930* (Nairobi: Government Printer, 1931), 5–7.
79. KNA: AK/11/50, Dhanwant Singh on a brief History of cane growers presented at a meeting between the Asian cane growers and Manager of Victoria Nyanza Sugar Company, June 22, 1946. During the meeting the new manager Mr. Wood was briefed on what had been agreed upon by Moffit.
80. KNA: PC/NZA/3/22/1/8, Extract from Sir Wilfred Woods' Report on Fiscal Survey, 1932.
81. KNA: PC/NZA/3/22/1/8, Extract from Sir Wilfred Wood.
82. Muyumbu, "A Historical Study of State Policies," 34–35.
83. Probably the state did not want any involvement in importation of machinery during the Depression so as to preserve the foreign exchange.
84. KNA: AE/3/916, Hindocha to PS Ministry of Agriculture, Animal Husbandry and Water Resources, July 25, 1961.
85. KNA: PC/NZA/3/2/34, Department of Agriculture, Nyanza to H .G. Pilling, February3, 1936.
86. KNA: AK/11/50, Director of Agriculture to Ag. SAO Nyanza, December 17, 1945.
87. Muyumbu," A Historical Study of State Policies," 41.
88. KNA: PC/NZA/3/22/1/8, Extract from Sir Wilfred Woods Report on a Fiscal Survey, 1932.
89. KNA: PC/NZA/3/22/1/8, Sugar Excise Duty.
90. KNA: AG/16/186, Government Notice Number 320, Defense Control of Sugar Regulations, 1944.
91. KNA: AE/3/916, A. C. Kinner to P. J. Riley, Price Controller, September 28, 1961.
92. KNA: AE/3/916, Kinner to Riley.
93. KNA: AK/11/173, PAO to DC, CN, April 11, 1950.
94. Winter-Nelson, "A History of Agricultural Policy in Kenya," 34.
95. Winter-Nelson, "A History of Agricultural Policy," 34.
96. J. C. de Wilde, "Price Incentives and Agricultural Development," in R. H. Bates and M .F. Lofchie, eds., *Agricultural Development in Africa: Issues of Public Policy* (New York: Praeger Publishers, 1980), 46. See also Martin Ravallion who argues that prices offered by the market have a strong effect on the economic decisions and choices made by farmers. Martin Ravallion, *Markets and Famines* (Oxford: Clarendon Press, 1987), 20.
97. KNA: PC/NZA/3/22/1/8, Sugarcane Cultivation in Miwani-Muhoroni area, October 1947.
98. KNA: PC/NZA/3/22/1/8, Sugarcane.
99. KNA: AK/11/173, W. M. Lyne Watt to Chair, Central Commodity Distribution Board, March 2, 1949.
100. Colony and Protectorate of Kenya, *DAAR 1949* (Nairobi: Government Printer, 1950), 12.
101. *DAAR 1947, 9.*
102. Colony and Protectorate of Kenya, *DAAR 1946* (Nairobi: Government Printer, 1947), 10.
103. Berman and Lonsdale, "Coping with the Contradictions," *Unhappy Valley Book One*, 84.

104. KNA: AE/3/296, Secretary, Commerce and Industry to the Economic Secretary, East Africa High Commission, September 12, 1949.
105. KNA: AK/11/30, PC, Nyanza to Member of Agriculture, June 21, 1951.
106. KNA: AK/11/51, Miwani Sugar Mills to PAO Nyanza, January 5, 1953.
107. KNA: DC/KSM/1/3/58, DC CN to Chief John Muganda, North Ugenya Location, June 21, 1954.
108. KNA: DC/KSM/1/3/58, Sugar Prices, Order No.W.15, March 20, 1956.
109. KNA: AK/11/52, Miwani Sugar Mills to PAO Nyanza, December12, 1957.
110. KNA: KSM/1/3/58, DC CN to all District Officers, CN, March 20, 1957.
111. KNA: AK/11/52, Miwani Sugar Mills to PAO Nyanza, December 12, 1957.
112. KNA: DC/KSM/1/3/58, Control of Prices Order, 1957.
113. KNA: DC/KSM/1/3/57, Inspector of Weights and Measures, to Director of Trade and Supplies, May 13, 1960.
114. KNA: AE/3/916, G. D. Parkin to P. H. Jones, confidential, July 10, 1961.
115. KNA: DC/KSM/1/3/58, Sugar Allocation-African Reserves, 1958.
116. *DAAR 1929*, 37.
117. *DAAR 1931*, 33.
118. *DAAR 1933*, 6.
119. *DAAR 1933*, 6.
120. Colony and Protectorate of Kenya, *DAAR 1937* (Nairobi: Government Printer, 1938), 17.
121. *DAAR 1937*.
122. KNA: AG/16/186, H .L. Adams, Chairman, Kenya Supply Board to Accountant General, Produce Controller and to Chairman, Central Commodity Distribution Board, June 7, 1949.
123. KNA: DC/CN/1/3/1, DC CN to W. Lyne Watt, December 2, 1947.
124. KNA: BV/6/1704, T. M. Loudon to Major, Hon. F. W. Cavendish-Bentinck, Minister for Agriculture, Animal Husbandry and Water Resources, confidential, April 25, 1955.
125. KNA: AK/11/50, Sugar Repot by PAO Nyanza, August 1947.
126. KNA: AE/3/916, Miwani Sugar Mills to PS Commerce and Industry, July 25, 1961.
127. KNA: AE/3/916, G. D .Parkin, Office of PS for Commerce, Industry and Communication to PS Commerce and Industry, Uganda, confidential, July 26, 1961.
128. KNA: BV/6/1704, Secret Minutes of Meeting to Discuss the Problems of East African Self-Sufficiency in Staple Foodstuffs, September 1954 (sic).
129. KNA: AE/3/916, Miwani Sugar Mills, Retail Price of Specialty Packs in Sugar.
130. KNA: DC/KSM/1/3/57, R. G. Waller, Executive Officer, C C D B to G. M. Vasani, Kisumu, March 1, 1957.
131. KNA: AE/11/30, Director of Trade and Supplies, Kenya to Director of Trade, Uganda, May 2, 1957.
132. Muyumbu, "A Historical Study of State Policies," 75.
133. KNA: AE/1/30, Director, Bhagwanyi and Co. to Secretary of Commerce, January 15, 1957.
134. KNA: DC/KSM/1/3/57, DC CN to Executive Officer, C C D B, March 6, 1956.
135. KNA:DC/KSM/1.3/58, Executive Officer C C D B, to DCs Kisumu, Kitale, Bungoma and Kakamega, February 28, 1956.
136. KNA:DC/KSM/13/58, Executive Officer.
137. KNA: DC/KSM/1/3/58, DC South Nyanza to DC CN, January 24, 1958.
138. KNA: AG/16/186, Deputy Sugar Controller to Attorney General's Office, Legal Department, September 17, 1943.
139. KNA: AG/16/186, Deputy Sugar Controller to Attorney General.
140. Swainson, *The Development of Corporate Capitalism*, 41.
141. Swainson, *The Development of Corporate Capitalism*, 41.
142. Swainson, *The Development of Corporate Capitalism*, 42.

143. KNA: BV/6/1704, Confidential Report by Committee of Economic Co-operation on the Future of the East African Sugar Industry, August 29, 1955.
144. *DAAR 1924*, 6a.
145. *DAAR 1929*, 651.
146. *DAAR 1928*, 17.
147. *DAAR 1933*, 6.
148. KNA: BV/6/613, DC, Moyale to Director of Agriculture, August 3, 1939.
149. KNA: BV/6/613, Dalgety and Company to Director of Agriculture, January 8, 1941.
150. *DAAR 1928*, 17.
151. *DAAR 1933*, 6.
152. *DAAR 1935*, 4.
153. *DAAR 1946*, 10.
154. KNA: AK/11/173, NFCS to Director of Agriculture, January 17, 1941.
155. Irvin D. Talbot, "Agricultural Innovation and Policy Changes in Kenya in the 1930s," Ph.D. diss., West Virginia University, 1976, 150.
156. KNA: AG/16/186, Produce Disposal-Sugar Control, 1949.
157. N. Makana, "Changing Patterns of Indigenous Economic Systems: Agrarian Change and Rural Transformation in Bugoma District 1930– 1960," Ph.D. diss., West Virginia University, 2005, 259.
158. KNA: AG/16/186, Produce Disposal-Sugar Control, 1949.
159. KNA: PC/NZA/3/15/142, Miwani Sugar Mills to DC CN, November 1, 1949.
160. KNA: DC/KSM/1/3/57, J. S. Courtenay-Bishop, Officer of Trade and Supplies, to PC Nyanza, July 8, 1959.
161. KNA: DC/KSM/1/3/57, Notice, Sugar Ordinance Cap.194 –Laws of Kenya.
162. KNA: AE/1/60, Director of Trade and Supplies Board to Secretary of Commerce, August 2, 1958.
163. KNA: DC/KSM/1/3/57, Sugar, Newcomers Quotas.
164. KNA: DC/KSM/1/3/58, DC CN to R. P. Rana, June 18, 1959.
165. KNA: DC/ KSM/1/3/5/57, The Defense (Control of Distribution and Rationing of Foodstuff) Regulations, 1944, Sugar Distribution.
166. Amitabha Sadangi, "Integrating the Poor into Market System," in B. Debroy and A. U. Khan, eds., *Integrating the Rural Poor into Markets* (New Delhi: Academic Foundation Press, 2004), 21.
167. Peter O. Ndege, "Struggle for Market: The Political Economy of Commodity Production and Trade in Western Kenya 1929–1939," Ph.D. diss., West Virginia University, 1993, 254. Ndege discusses in depth the role played by the colonial state in the control of the market forces in his work whereby he illustrates how everything was controlled, for example the market buildings and specification of the days and hours when markets were to be held.
168. Robert H. Bates, *Beyond the Miracle of the Market: The Political Economy of Agrarian Development in Kenya* (Cambridge: Cambridge University Press), 10–11.
169. Heyer, *Agricultural Development*, 334.
170. Muyumbu, "A Historical Study of State Policies," 89.
171. KNA: PC/NZA/3/22/1/8, Report on a visit to the Asian farms in the area of Kibos-Miwani-Kibigori-Chemelil- Muhoroni by Mr. Magor, October 6-10, 1947.
172. KNA: BV/125/10, PS Ministry of Agriculture to Chairman, Nyanza Sugar Planters Association, November 2, 1960.
173. Muyumbu, "A Historical Study of State Policies," 81.
174. Makana, "Changing Patterns," 112–113.
175. KNA: AE/3/916, P. H. Jones to Miwani Sugar Mills, August 26, 1961.
176. KNA: AE/3/916, Sugar Supplies from Miwani, 1961.
177. KNA: AE/22/630, PS Commerce and Industry, to Minister for Commerce and Industry, March 23, 1963.
178. KNA: AE/22/630, PS Commerce and Industry, to Minister for Commerce, March 23, 1963.

179. Muyumbu, "A Historical Study of State Policies," 89.
180. Muyumbu, "A Historical Study of State Politics," 89.
181. Muyumbu, "A Historical Study of State Politics," 89.

FIVE

Labor in the Nyanza Sugar Industry

Asian farmers suffered from the colonial state's neglect and also from its misguided control of the industry. The previous chapter looked at farmers' relationship with the mills and general market. This chapter examines the question of labor in the sugar industry, especially as it relates to the African workforce.

The chapter examines policies that were put in place to ensure the availability of labor, especially to the European settler farmers, and the contradictions that emerged because of them. Both Asian and European farmers struggled to find workers, and, as in the marketing practices discussed in the earlier chapter, the colonial government discriminated against the Asian cane growers and millers. The Asian farmers had to compete for labor with the settlers and the colonial government to get their farmhands and the few who worked in the factories.

The chapter also shows how labor recruitment efforts affected the sugar industry and how labor conditions were impacted by the changes in the colonial economy marked by the two world wars and the Great Depression. The chapter describes working conditions including wages and the tasks undertaken by laborers, such as planting, cutting of cane, and factory work. Housing of the workers, wages, and food rations were also significant factors impacting the recruitment and retention of labor in the sugar industry.

The colonial state did attempt to stabilize the labor supply, especially for the European farmers. But these attempts were problematic in many ways. The chapter ends with a consideration of the state's attempts at control and the increasing protests on the part of African workers.

EARLY LABOR RECRUITMENT

The introduction of wage labor during the colonial period served many purposes. Not only was it meant for the colony's sustainability through the introduction of taxation, but more so it was meant to get Africans out to work. This was meant to meet the needs of European settlers and the colonial state, and it also benefited the Asian farmers and other potential employers. Initially, the settlers did not give attractive salaries since most of them, just like the Asians, were poor during the pre-World War I period. During this period, most of the Africans did not engage in wage labor since they were economically satisfied in their indigenous economies, as cash had little value and the payment of taxes was often in kind; moreover, taxes had not been increased exorbitantly.[1]

Needless to say, the pioneer Asian cane farmers experienced problems in getting labor, first, because the Africans did not want to go out and work, and secondly because they had to compete with the European-owned estates and European farmers in the region. Thirdly, they had to compete for labor with the colonial government, which required workers for public projects such as construction of roads, bridges and administrative centers.[2] This problem was compounded by the fact that the colonial state and settlers needed to retain the laborers and ensure that they worked satisfactorily. However, during the pre-World War I period, most cane farmers, especially the small-scale ones, relied on casual, short-term laborers from the surrounding locations of Kajulu, Karateng, and Kano. These were paid on a daily basis. Most of these workers preferred to commute from their homes due to other family obligations such as attending to their *shambas* (farms) and looking after their cattle.[3]

In order to create a larger and more stable workforce, the colonial state came up with policies that forced the Africans into wage labor. First, the colonial state introduced taxation to induce the Africans to seek wage labor. This resulted in contradictions; taxes served two purposes in that they encouraged wage employment but also encouraged peasant commodity production.[4] Many Africans produced more so that they could have a surplus to trade and meet the tax obligation without necessarily going out to seek wage employment. Such was the contradiction of the colonial policies.[5] For instance, instead of a decline in agriculture, it was found that peasant commodity production increased, particularly in areas that were considered labor-exporting zones such as Nyanza. That being the case, it was not easy to get labor since the tax obligation could easily be met by proceeds from the sale of commodities. As such coercion was used by chiefs and headmen to recruit labor for construction of roads and even for settler farms. Furthermore, the working conditions in most places were inhumane and so they could neither attract nor retain workers, especially from the areas where African commodity production excelled—hence the use of coercion. This method was used widely during the early years of colonial rule because both the state and settlers were undercapitalized.[6] In essence taxation was used to make the Africans to go out to seek wage labor, as attested by Governor Belfield who asserted that "It was the duty of the native to contribute something either by means of labour . . . We consider that taxation is the only possible method of compelling the native to leave his reserve for the purpose of seeking work . . . only in this way can the cost of living be increased for the native."[7]

Consequently, the state introduced a hut tax of one rupee per dwelling place annually in 1901.[8] Although the hut tax was approved in 1901, in Nyanza collection began in 1902. The tax was levied on dwellings which a man had; thus the more houses one had in his compound the more taxes one paid. The hut tax was eventually increased in 1903 to 3 rupees per annum; this increased the burden on Africans who found it necessary to seek wage labor since failure to pay meant harsh punishment such as burning of huts and granaries. The colonial chiefs used coercion to force out labor. Some were ruthless and were therefore seen as agents of oppression. However, there were some who were respected by the local people; for example, Chief Oriri Kamidigo, who was loved by the people of Kajulu, encouraged them to seek wage labor to meet the tax obligation but did not force them.

Before 1908, the colonial government allowed chiefs and headmen to recruit labor, and the greater the number of workers one recruited; the more he was seen as effective. As a result, some chiefs abused their power and used coercion to bring out more laborers. This generated a lot of problems that forced the colonial state to rescind this policy and adopt a policy of "encouragement" where chiefs and headmen were deterred from direct recruitment of labor, and the work was taken over by professional and private labor recruiters.[9] The recruiters were given a license for twelve months by the DC to conduct their business. Because such agents still had to be guided by the chiefs who knew their areas well, coupled with the fact that some chiefs were paid according to numbers of laborers they recruited, there were no major changes.[10] Therefore, complaints continued since there was not much difference so long as the chiefs and headmen were still consulted. Moreover, even the agents were seen by the Africans as part of the oppressive colonial government.

Meanwhile, the colonial government introduced another tax in 1908; this was a poll tax of 3 rupees per annum paid by all unmarried males ages sixteen years and older (if a man could prove he had paid the hut tax, he was not required to pay the poll tax).[11] This was meant to increase the financial burden on Africans so as to force them to seek wage labor. However, during 1912–1913, many Africans were able to meet their tax obligation from income generated from household production.[12] But as the taxes continued to be increased, as was witnessed when the tax was raised to 5 rupees in 1915 and to 8 rupees by 1920, these increases made taxation burdensome and forced many to seek wage labor. Many people in Kisumu district were forced to go and seek employment in the Asian sugarcane farms, especially between the months of September and March when taxes were collected.

The issue of accessibility to land for peasant production was also a factor. It was found that the continued access to land discouraged Africans from joining the wage labor force.[13] Therefore, to maintain the flow of the laborers, the colonial government introduced a land alienation policy and the creation of African reserves as a way of shrinking land so as to make it difficult to sustain a family, hence pushing the Africans to wage labor. These colonial policies deprived Africans of their basic means of production and forced some people to seek labor as pressure increased on land. Reserves acted as labor reservoirs as expressed through the views of a European settler who asserted that "the ideal reserve is a recruiting ground for labor, a place from which the able bodied go out to work, returning occasionally to rest and beget the next generation of laborers."[14]

However, this had its own paradox in the sense that most young men who were found to have voluntarily joined the labor force came from areas that witnessed aggressive peasant commodity production. Despite the government regulations on land alienation and insistence that taxes could no longer be paid in kind in the form of grain, goats, sheep, or chickens as initially was the case, African commodity production increased. This helped to spur an interest in wage labor as some people saw the opportunity to enter wage labor as means to get money to buy European hoes *(jembes)* to improve peasant farming. Some people joined wage labor specifically to intensify their own production. For instance, in 1913 and 1914 approximately 30,000 hoes were sold in the district.[15] These were found to be more effective than the traditional wooden tools.

This situation produced another contradiction, for while men in Nyanza sought wage labor for enhancement of household production; such commodity production meant that African laborers were not easily available throughout the year. This was because most workers preferred to work on their own farms during the planting, weeding, and harvesting periods. Consequently, instead of the peasant commodity production being destroyed as a result of preference to settler production, it expanded.[16]

These were not the only reasons for seeking wage labor; the need to buy consumer goods, such as cloths, bicycles, and other obligations such as payment of dowry, also brought men out to work. Nevertheless, the turnout was not sufficient due to some of the reasons mentioned earlier. Nonetheless, laborers were made available due to the cooperative role played by the local chiefs toward the labor recruitment effort.[17] This was because Africans were reluctant to enter wage labor since their indigenous economies were sufficient to sustain them during the early colonial period. Many were only drawn into it due to the various colonial policies that were put in place. The unwillingness of the Africans to engage in wage labor created periodic shortages of labor. Although both the Asian cane farmers and the European settlers experienced problems, it was much more difficult for the Asian cane farmers to get labor, a situation that became worse with the onset of World War I.

WORLD WAR I AND POSTWAR LABOR CONDITIONS

World War I had a direct impact on labor issues in colonial Nyanza. The war put more pressure on the need for labor as more Africans were required to serve as porters in the carrier corps and also as *askari* (soldiers). The Africans were not keen to offer their services toward the war efforts because of the poor working conditions of the *askari* and carrier corps. Many more died due to the poor conditions than actual combat.[18] This was one of the factors that also forced Luos to work on the Asian farms. Those afraid to serve toward the war effort opted to seek wage labor on Asian and European farms. Meanwhile, the settlers continued to demand more labor and the Asian farmers had to compete to get laborers not only with the settlers but also the colonial state. As shortages continued to be acute, the colonial state resorted, after 1916, to armed raids to replenish the diminishing numbers of the carrier corps. This unleashed a period of forced labor since that was the only reliable method for the government to secure labor. During this period, the colonial state needed a large labor force for public works.[19] There was a need for the construction of roads and bridges to facilitate the movement of the troops. The demand for labor during the war period fueled the Asian cane farmers' competition with the state in acquisition of labor.

The war also made it possible for the settlers to bargain and benefit from the available labor. This was seen in 1916 when the War Council was formed. Once the European farmers achieved membership in the War Council, they used it to benefit their own labor recruitment and retention. For example, they were able to change the Master and Servant Ordinance in 1916 so that desertion was now treated as crime liable to arrest without any warrant by the police.[20] This was meant to stop laborers from leaving employment regardless of the conditions. Moreover, the War Council was able to influence the colonial state to reduce the wages of the carrier corps from 10–15 rupees per month to 5–6 rupees per month.[21] Wages were reduced to be on par with what the settler farmers were offering. This was to minimize the competition for labor between the government and the settlers, since it no longer made sense to risk life serving in the carrier corps when one could make the same amount of money in settler farms; the settlers capitalized on the war to get what they wanted. These policies favored the settlers while the Asian cane farmers were left in a precarious situation as they struggled with the settlers over labor. It was not easy for Asian sugarcane farmers to get labor as settlers were favored by the War Council, which had no Asian members and therefore had no voice to articulate their problems.

As the war progressed, a severe famine hit CN, which also influenced some people to join wage labor, particularly in the sugarcane farms. During 1917–1918, a severe drought known locally as *Kanga* caused famine in the district.[22] The famine continued into 1919 and the areas most affected were Kano and Nyakach locations. The people of these locations were forced to move to other areas, such as Nandi and Kericho, to look for food, while others went to work on the sugarcane farms. This improved slightly the labor situation on the farms. Unfortunately, many of those who got to the farms were found to be malnourished and too weak to work.[23] Many also flocked to Kisumu to look for food and work.[24] For instance, it is estimated that about 300 families from the Ramogi clan in Nyakach location moved to Muhoroni and entered wage labor.[25]

However, this did not make things easier for the Asian sugarcane farmers as they had to compete for labor with the colonial government, which was involved in construction work such as building of roads and extension of the railway and created a rise in demand of labor after the war. For example, the recruitment of labor for the construction of the Uasin Gishu railway complicated issues since it offered higher wages: instead of the normal Shs. 12, the laborers were offered Shs. 16 per month.[26] By 1922 the currency in Kenya changed from rupee to shillings.[27] Be that as it may, many laborers deserted the sugarcane farms to seek employment where better wages were offered. This meant that most laborers would be attracted to seek work on the railway rather than on the sugarcane farms, which made it difficult for both the Asian and European cane farmers to get labor, especially the small-scale Asian cane farmers who were undercapitalized and so could not offer attractive wages. Therefore the problems of labor continued.

STRUGGLE FOR LABOR BY FARMERS AND INDUSTRY

There was a struggle for labor in colonial CN as people in Nyanza were hesitant to work during the war and in the postwar period as a result of the forced methods used for recruitment of the carrier corps. They only engaged in labor as last resort, and Asian cane farmers continued to experience scarcity. Attempts to solve the labor problems in CN saw the colonial government's intervention when, during the period 1917–1922, professional labor recruiters were barred from Nyakach and Kano locations. These two locations were set aside by the colonial government to provide laborers to the Asian farms due to the stiff competition that existed. Thus most wage seekers from these locations were absorbed by the Kibos-Muhoroni sugarcane plantations.[28]

Meanwhile in other parts of CN, especially in European farms, professional labor recruiters were engaged to recruit labor. These professional labor recruiters were found to flout many requirements such as ascertaining if a laborer was fit medically. Cases were reported where they sent sick laborers to employers. The recruiters would persuade a man to leave his home, sign his registration certificate, and provide transport to the workstation. These recruiters were Europeans, Asians, and privately employed Africans who used any methods to obtain labor. Recruitment was a business venture during this period, and many of those involved in it were aggressive. Some recruiters expanded their business. For example, recruiters like John Riddoch, Andrew Boustead, and Miss Mclean also employed the use of agents.[29] Furthermore, people believed that the recruiting agents were government officials, who were detested. There were instances when the labor agents used dubious means to get labor and at times cruelty was involved; this indicated their desire to get more workers. The presence of the labor agents and their actions exposed the struggle that the Asian cane farmers faced to get labor as a result of competition for labor with African households, the colonial government, and European farmers in the area, some of whom also were cane growers. Due to such competition, Asian small-scale farmers found it increasingly difficult to obtain labor, a fact that was compounded by their undercapitalized situation that made the Asian cane growers unable to offer higher wages. The situation was made worse by the cruelty employed by professional recruiters and their associates like the chiefs and headmen. This discouraged workers, and the colonial government faced a lot of criticism. The professional recruiters were condemned and their activities were opposed by the Native Labor Commission of 1912–1913. Both the Economic and Finance Committee in 1925 and the Labor Commission in 1927 recommended that the professional agents be abolished.[30] Therefore, district officers were given the responsibility of getting labor by encouraging the Africans to enter wage labor.

The struggle for labor with Europeans continued, especially after World War I. The settlers came from the war more assertive in their demands, which included the increased availability of cheap labor, and the implementation of the soldier-settlement scheme at the end of the war made the labor issue crucial. This was because the establishment of the scheme saw more than fifteen European ex-soldiers settled in the Kibigori-Chemelil area, which was already settled by some Asian farmers. This coincided with the expansion of the Asian farming when 8,000 acres were put under cane. Meanwhile, other Europeans in the area and Koru were also cultivating sugarcane, coffee, and sisal.[31] All these groups needed labor. This in essence meant that the Asian cane farmers continued to struggle for labor as the competition increased. This was compounded by the fact that the shortage had started to take a toll as many people shied away from wage labor; the war experiences, forced recruitment, and diseases contributed to the fall in numbers of wage seekers after the end of the war.[32]

The European settlers, sensing the situation, called upon the government to employ the same means used for recruiting the carrier corps during the war that had worked so well to solve the labor problem then. In 1919, Governor Sir Edward Northey, eager to please the settlers whose support he needed in his administration, thought that he could achieve his aim by instituting forced recruitment of labor, as has been done in World War I. He therefore issued the 1919 Labor Circular, which mandated that the local administration be directly involved in recruitment of labor. Northey was bent on helping the settler farmers at the expense of the Asian cane farmers. Here again, the colonial state was very receptive to the needs of the settlers, while on the other hand it continued to ignore the needs of the sugar industry. The Northey circular was significant in that it made it easier for professional labor recruiters to get labor, especially for settlers and the large-scale sugarcane farmers and estates, at the expense of the small-scale farmers who were left struggling to get labor as competition stiffened.

Needless to say, the circular created a lot of tension since it was tantamount to forced labor and many called for its withdrawal. Those demanding its removal included missionaries in the whole of East Africa, and not Kenya alone. Greatest criticism and persistent condemnation came from Archdeacon W. E. Owen of Nyanza Province. The Northey circular showed the contradictions that beset the colony by highlighting the different perspectives of administrators, settlers, and the missionaries on the issue of labor. What was appalling about the circular was that it recommended that women and children be included in wage labor, especially those whose reserve was near the settler farms. Moreover, the coercive methods used for recruitment were inhumane, as people were forced out for labor against their will. Critics of the labor laws like Tiyambe Zeleza argued that "Labor coercion just changed its form and acquired a new name; and was baptized 'encouragement,' though the underlying motive was the same as compulsory labor and was legalized."[33] Despite the protestation, Northey was adamant due to his desire to please settlers whose agriculture could not have survived without cheap labor; indeed, "the Kenya settlers could not have operated within or indeed survived a market in which demand for free labor exercised an upward pressure on real wages. They had to rely on extra-economic coercion to produce forms of semi-servile labor extracted from the peasant sphere at a price below its costs of production."[34]

Northey was bothered by the plight of the settlers who were the protectorate's main exporters, but not that of Asian cane growers.[35] Therefore, during the interwar period, the labor phenomenon continued to affect the Asian cane growers. The problem was exacerbated due to the fact that during that period, the Asian sugarcane plantations witnessed expansion, despite periods of economic depression in the 1920s and 1930s, in that more acreage was cultivated. For instance, approximately 8,000 acres cultivated during these decades required more laborers. The huge increase was made possible when in 1923 the Victoria Nyanza Sugar Company was established at Miwani in Kisumu District. The company consequently purchased cane from the Asian farmers at Kibos, which led to increased production. Furthermore, the Victoria Nyanza Sugar Company also had acquired a large sugarcane estate, which also required labor. Consequently, the demand for labor increased in the 1920s. Getting labor remained a problem, especially in the sugar belt region.

Needless to say, the demand for labor became more acute in Nyanza when the soldier-settlement scheme was implemented during the interwar period, as more settlers came to Kenya. The new arrivals increased the demand for labor. The demand caused the colonial government to intervene to ensure that there was sufficient labor through the introduction of the *kipande* system (pass), which was made possible by the Registration of Natives Ordinance of 1915, which was implemented in 1920.[36] Under the ordinance, every adult African male over the age of sixteen years was to obtain the *kipande* document. It contained personal information of the bearer, such as name, ethnic group, district, location, and even village. Furthermore, the employer also recorded the type of work the laborer performed, duration of employment, locality, and wages.

The introduction of the *kipande* did not immediately end labor shortages. The situation became even worse with the economic recession of 1921. It led to a fall in the prices of the agricultural produce and to high prices of manufactured goods. This led to reduction of wages of African laborers by almost one-third.[37] Such a situation discouraged many laborers who found it more beneficial to work on their farms. As a result there was acute labor shortage in the farms in the Kisumu district. In an effort to get the Africans back to work, and also in an attempt to contain the financial crisis that had befallen the colony as a result of the war and postwar recession, the colonial state decided to exploit the Africans further and raised the hut and poll taxes to a total of Shs. 16 per person annually.[38]

This led to an outcry from Africans who were becoming more politically conscious at the time. For example, the taxation issue was voiced by the Young Kavirondo Association (YKA), which was formed in 1921 at a meeting held in Lundha in CN by mission-educated men.[39] The association was composed of educated Luo and Luhya men, led by Jonathan Okwiri and Simeon Nyende among others, who advanced the African grievances and furthermore were concerned with ameliorating the working conditions of the African laborers. The association was able to secure a meeting with Governor Northey at Nyahera near Kisumu in July 1922 and presented their grievances. Northey agreed to abolish labor camps and the reduction of taxes from Shs. 16 to Shs. 12 per annum. However, Northey refused to abolish the *kipande* system, asserting that it was introduced for identification purposes, which were to benefit the Africans.[40]

The Young Kavirondo Association also protested against the forced recruitment of women, children, and old men, especially when it came to the odious compulsory labor such as working on road construction. The YKA also demanded an increase in wages. Northey did not concede to the changes on labor camps because he had "seen the light" that forced labor was wrong but rather because the association was a threat. The colonial state could not risk dealing with a situation similar to the protests of Harry Thuku's East African Association. The protests of early 1920s by Africans in Nyanza were significant because they posed a threat to settler dominance; and so possibly the need to retain the status quo influenced Northey.[41]

Due to the support from missionaries and pressure from the 1920s protests, things began to change. The British secretary of state for the colonies, Winston Churchill, sent a dispatch to the colonial government stating that forced labor could only be used for essential services. Furthermore, government officials were prohibited from taking part in the recruitment of labor for private employment. This in essence meant that forced labor was not totally abolished, as stated in part of the circular: "when unemployed young men are found in the Reserves enquiries should be made as to whether they have paid their Poll tax . . . no actual force can be employed to compel a man to go out to work, he can however be made to pay his tax."[42]

Such content indicates that though the Northey circular was amended, it still contained loopholes that could be manipulated to force people into wage labor.[43] The burden of taxation and low prices induced some people to enter wage labor so that by 1923, approximately 1,500 Africans from the nearby reserves took up employment in the sugarcane farms.[44]

Nevertheless, Asian farmers struggled with the colonial state to get labor as the government diverted African labor through its policy that encouraged the African farmers to grow cotton in Nyanza, especially CN. This was seen when in 1920 the state distributed 1,500 seeds in Kano and Kajulu locations of CN district. Unfortunately, the initiative registered little success because the crop was labor intensive and the price offered was not lucrative. For instance, by 1922 the price offered ranged from 25 to 30 cents per pound, and this kept on declining. The cotton scheme failed, despite engaging labor that could otherwise be used in the Asian sugarcane plantations.[45]

The mid-1920s also brought further labor demands as the new factory at Muhoroni, the Nottidge and Allen Sugar Company, acquired land and started milling sugar. The company was able to employ a workforce of

400 by 1927.[46] This increased the competition for labor as more Europeans joined the sugar industry. However, the procurement of labor was not easy during the post-World War I period due to new difficulties, notably the depressed financial state of the colony.

THE DEPRESSION PERIOD AND LABOR SHORTAGES ON SUGARCANE FARMS

During the Great Depression period starting in 1929 and including the early 1930s, cane farmers, both small-scale and large estates, found it difficult to remain solvent due to the slump caused by the depressed market, fluctuations of prices, and locust visitations.[47] The Depression impacted the large estates and by extension the small-scale farmers; both experienced hardships, which meant that the small-scale farmers faced a cut in profits that forced them to reduce the number of their laborers.[48] The sugarcane farmers did not have enough laborers as the prices of primary commodities collapsed in the world market and continued to fall until 1935.[49] Labor was affected as wages dropped significantly as a result of the depressed market.[50] Farmers found it difficult to retain laborers as wages were expensive, especially for the small-scale Asian cane farmers. This was necessitated by the Minimum Wage Ordinance that was passed in 1932, following the ratification of the International Labor Organization Convention on minimum wages by Britain in 1928. As such, all British colonies had to reevaluate wages given to workers.[51] To make wages affordable, the colonial state also intervened but only to keep the settler farmers solvent. For example, in 1931 the rail freight for wheat, maize, and barley was reduced.[52] Such help was not extended to Asian cane growers, who also faced hard economic times. Moreover, this assistance meant that the settlers could technically afford to pay for laborers while the Asian farmers had to struggle to get laborers, as they were forced to offer lower wages.

The situation changed by 1933 as there was improvement in the colonial economy. The Asian cane farmers had adequate labor because of economic pressure exerted in the reserves by the tax collectors, a situation that was partly made possible by the presence of W. P. Shields, a labor officer who resided in Kisumu during that particular year.[53] The state was keen to get revenue from taxes during the Depression. It was ironic that despite the difficult economic situation the colonial state did not find it prudent to adjust the tax obligation of Africans; instead taxation was pursued vigorously and Africans contributed approximately 37 percent of the colony's revenue.[54] This impacted Africans who resorted to wage labor and by 1933 it was reported that labor supply exceeded the demand in Kisumu-Londiani district.[55] For example, during 1933–1934, the Victoria Nyanza Sugar Company employed about 1,400 laborers, while Muhoroni also increased its laborers to about 500.

However, the shortage of labor in the Nyanza sugar industry started again by 1934 and persisted for a long time. The shortage was caused by the discovery of gold in Kakamega in North Kavirondo District in 1931.[56] The discovery sparked an influx of European gold prospectors into Kakamega. As gold mining continued, labor became scarce in the sugarcane farms as large numbers of Africans as well as some Asians flocked to Kakamega to seek employment and try their luck in the gold fields.[57] Most of the European farmers who rushed to Kakamega during the gold rush were those who had been hard hit by the depression and had become bankrupt.[58] In the following year, the effects of labor shortages were minimal. For example, in 1935 as there was adequate labor; the Victoria Nyanza Sugar Company employed approximately 1,400 workers while both the Muhoroni Sugar Mills and the small-scale Asian farmers employed each about 500 laborers.[59] Unfortunately, the situation changed in 1936 and labor shortages once again were felt. There were incessant complaints by European cane and coffee farmers, especially those from Songhor and Koru, who complained of a lack of discipline and prolonged unauthorized leaves on the part of workers.[60] Although the Victoria Nyanza Sugar Company also suffered shortages, the problem persisted on European farms and the labor officer in the district was informed of the situation.[61]

Be that as it may, labor shortages continued as peasant production proved to be more rewarding; many Africans preferred to work on their farms. For example, in 1937 there was marked improvement in peasant production in the whole province as agricultural produce fetched higher prices. Due to the increase, many people were encouraged to work on their own farms. For example, the maize prices that had started rising by end of 1936 remained high even in 1937, partly as a result of speculative buying by Asian traders. This motivated the peasant grain growers in the province as a whole to produce more. During the same period, the Gusii planted and sold more maize than before.[62] The increase in the price of produce encouraged farmers, and many men wanted to work their farms. This contributed to a situation where either laborers overstayed their leave or simply deserted. To control the situation, the colonial state introduced the "leave/pass" legislation, which greatly reduced the number of deserters.[63]

Despite the legislation, there were many complaints, especially from the farmers at Songhor and Koru, who reported that the situation did not improve much as laborers either still overstayed their leave or performed poor quality of work. Therefore the farmers requested the assistance of the DC to curb this tendency. Consequently, a system was adopted whereby the DC would send a notice to the chiefs in the reserve stating that the laborer had overstayed his home leave; the chiefs were given the task of ensuring that the workers went back to work.[64] This example suggests that the increase in peasant production impacted labor in Asian sugar farms as African labor was tied down in the reserves. Therefore, the Asian cane growers faced another challenge as they had to compete for labor with the African commodity producers.

The labor situation improved in 1938 as commodity prices fell and labor became plentiful in the sugarcane plantations. The increase in labor was made by the introduction of a bonus payment for work done, which proved to be a good incentive.[65] Generally, during the interwar period labor supply remained erratic and was far from stable. It was further affected by the outbreak of World War II in 1939, as the labor situation worsened.

WORLD WAR II AND POSTWAR LABOR ISSUES

The World War II period was also marked by challenges of getting labor as the colonial state recruited many Africans for the war initiatives against Italy in Ethiopia and Somaliland.[66] Thus World War II created labor shortages through military efforts as many Kenyans were recruited to join the King's African Rifles. Moreover, the war created a demand for more food production to meet the war needs; hence agricultural production increased.[67] Changes were undertaken by the colonial state that provided the settlers with fertilizers, capital, and technical efficiency; hence mechanized settler farming now became possible.[68] Furthermore, large numbers were conscripted to produce much-needed crops such as sisal and rubber. As such, both the settlers and peasant farmers were engaged in commodity production, which made it more difficult to get labor, especially for the small-scale Asian cane farmers. Despite the demand created by the war during the period, the Asian sugarcane farmers were not accorded the same support as the settlers and were still struggling to acquire machinery, especially for breaking the land.

Furthermore, the war conditions, just as in World War I, created an opportunity for the settlers once again to demand state support in getting labor and other services. The colonial state cooperated with the settlers to avoid antagonism during the war. The state accorded settlers the help they needed by allocating to them money that was used to build roads and help settlers with the fencing in of their land.[69] The state also used coercion to acquire labor, which was made possible through the Defense (Native Personnel) Regulations of 1940, which empowered the PC to produce labor quotas for military and essential services.[70] The essential services included working for private employers; many people were captured by chiefs on DC's order. The support seems to have deliberately benefited the settlers as it prevented effective competition from African peasant commodity production.

Unfortunately, contradictions emerged instead because despite the extensive support that was accorded to the settlers, the increase in yields of wheat and barley were minimal. The colonial state was once again forced to change its policy and supported peasant production by the mid-1950s.[71] The change in policy tied down the African labor in the reserves, and so the Asian cane farmers continued to struggle to get labor. The colonial state was not very supportive to the sugarcane farmers. It neither acknowledged that the sugar industry needed attention just like the settler and African produce production, nor did it recognize that the sugar industry was an integral part of the wider colonial economy.

The situation was compounded by the fact that, apart from the military services and labor for farms, many Africans were recruited for government projects such as the construction of airfields, roads, and military camps.[72] This forced the cane growers to compete with the state for labor. By 1940, most of the laborers on the Asian farms were either casual or on long-term contracts. Labor recruitment for Miwani was done by African clerks and headmen. Most of these workers were unskilled and some came from far away. For example, the mills transported casual laborers from as far as Nyakatch, Ahero, and Awasi daily.[73]

Another critical factor during World War II was the expansion experienced by the sugar industry. This called for more labor. For instance, Miwani Sugar Mills expanded its nucleus estate and employed about 60 percent of the total labor force in the industry.[74] Asian farmers were also able to acquire more land from the profits of their businesses; this was seen when European farmers who had settled at Chemelil were eventually bought out by the Asian farmers in 1942.[75] In the same year, an Asian entrepreneur also bought the 4,800-acre Muhoroni sugar plantation and factory. However, the expansion of Asian sugar industry was not commensurate with the number of workers available. This was because a large number of African males were recruited toward the war efforts. The first priority was given to the colonial state because according to the colonial government's defense regulations of 1940, the PC had the power to enlist people for military and essential purposes; and so when the provincial administration used these powers later in the war, the Asian farmers and the European-owned Victoria Nyanza Sugar Company faced labor shortages.[76]

The ceaseless search for labor made the colonial government look for ways of drawing Africans into the labor market to solve the crisis; those found unfit for the military were sent to work at the European and Asian farms. Unfortunately, there was still a shortage of labor due to the number of men enlisted in the army; the provincial administrators in such a situation encouraged women to seek wage labor in the farms because of the pressure of the wartime needs. Before this time, women were usually left at home to work on the *shambas*.[77]

Ironically, after the war, in spite of the large number of eligible male laborers in reserves surrounding the Victoria Nyanza Sugar Company, it was not easy to get sufficient labor since the labor market was made free with the removal of the conscript labor regulation in September, 1946.[78] Furthermore, the ex-soldiers still had some funds to live on and so many did not enter wage labor after the experiences of army life. Therefore, shortages of labor continued to be felt. Miwani Sugar Mills, which had just taken over from the Victoria Nyanza Sugar Company, decided to minimize its problem. When Hindocha moved to Miwani from his sugar estate at Kakira in Uganda, he brought along approximately 350–400 of his former Ugandan workers.[79] Most of these workers were semi-skilled and skilled such as drivers and artisans; many were Acholi. Apart from Miwani, some employers, like the large-scale farmers, preferred migrant laborers from outside the colony since this group normally signed up for long contracts of three years. Furthermore, they were more reliable since in most cases they brought along their families and were provided with food rations, accommodation, and train tickets or travel warrants back to their homes.[80]

Labor continued to be a problem after the war because more white farmers were given land in the white highlands through the Agricultural Settlement Board, which was established in 1946.[81] This increased the demand for labor as more white settlers arrived, which put more pressure on cane growers. The board also arranged for special agricultural courses at Egerton College that taught about local agricultural conditions. The colonial state did not include the African peasants and the Asian cane farmers in the program at Egerton; the colonial state wrongly envisaged that it was only the white settlers who were meant to benefit from the agricultural courses taught at Egerton College, despite the fact that the Asian cane growers and the African peasant commodity producers formed part of the bigger colonial economy and had demonstrated that their productive activities were as viable as the settlers'. Needless to say, the labor shortages during the war and in the postwar period clearly indicate the contradictions found within the state's policies. For instance, when the shortages persisted, the state reverted again to the use of coercion and recruiters. Despite such challenges to the Asian sugarcane growers and millers, by late 1940s the industry was solidly financed by Asian capital, which was expanding and required more labor.

Moreover, toward the end of 1940s the sugar industry was revamped and a lot of land was brought under cane; even Muhoroni Mills had started to receive cane from the farmers. This increased production and more labor was needed. Yet by 1952, approximately 62 percent of men in CN district were already contracted as laborers, so for the expanding sugar industry to meet its workforce, mills such as Miwani (former Victoria Nyanza) had to engage recruiters and open centers for such purposes. For instance, Miwani established a recruiting office at Kisii and

Migori where the Gusii and the Kuria were recruited. But the undercapitalized small-scale cane growers did not engage recruiters. Despite such initiatives, employers still faced shortages, as demonstrated by the frequent problems with desertion that affected sugarcane producers.

DESERTION AND LABOR SHORTAGES

Desertion and a lack of adequate labor were related problems, experienced by both the sugarcane farmers and the millers. These actions were a form of protest against poor wages and working conditions. Desertion entailed leaving the workplace before the expiration of a worker's contract. Several factors contributed to such a phenomenon, ranging from ill-treatment to excessive workloads to poor living conditions. Many workers deserted after a few days' work and others even before arrival; the latter were arrested. Some of those who left were found to have justification, such as inadequate food rations. Also there were cases where desertion was caused by employers who granted large advances on engagement, sometimes up to six months' pay.[82] John Ainsworth, the PC of Nyanza in 1907–1908, believed that the desertion and labor shortage in the plantations within the province was due to poor working conditions, poor rations, cruelty, poor housing, long working hours, and the "advance system" method of payment. The advance method was particularly detrimental since once the money was spent, the laborer was not motivated to work. Although the practice was common and was used to attract more laborers, it was found to be self-defeating on the part of the employer.[83]

Desertion, on the other hand, became a crime in early colonial Kenya, and when caught the offender could be fined Shs. 100 or sentenced to jail for two months.[84] To enforce such measures, the colonial state required every male over sixteen years of age to be registered.[85] The registration required the Africans to carry a certificate of identity that had his personal details. This served the best interest of the employer. This certificate or *kipande* was meant to curb desertion, as it allowed the laborers to be easily traced. The *kipande,* though beneficial to the employers, mostly settlers, was injurious and hated by the African workers, since it not only hindered the bargaining power of the laborers for better wages but also restricted their movements to the locality of employment.[86] Moreover, the government refused to let employers contract any African laborer without a *kipande.*

Even with the enforcement of the *kipande* to contain desertion, shortages were still experienced. Retention of labor was a huge problem, especially in settler farms as opposed to the Asian farmers who were found to retain labor as they treated their workers fairly.[87] Although the Asian small-scale farmers faced labor shortages, they were less affected by desertion as the European farmers since most Asian small-scale farmers employed casual workers who were paid on a daily basis. Moreover, most small-scale farmers did not use professional recruiters, as most of their laborers came on their own accord due to reasons discussed. Furthermore, the small-scale farms were found to have lax regulations and there was an interpersonal relationship with the employer that was lacking in the large estates.[88] As indicated earlier, there were many fewer cases of deserters in small-scale Asian sugarcane farms than in the large sugar estates and settler farms. Those who deserted from small-scale farms mainly did so due to low wages, poor rations, and delayed payment caused by the insufficient finances of the small-scale farmers. On the other hand, majority of deserters were from large estates and settler farms and did so due to ill treatment.

The ill treatment on large-scale sugarcane farms could partly be attributed to neglect of the industry by the colonial state. For example, the 1928 inspections carried out by the labor office indicated that while the sugar industry was inspected only three times, those of coffee and sisal were inspected 172 and 65 times, respectively.[89] Such negligence by the labor office also contributed to lack of retention of laborers, as mistreated workers had no one to help them and many deserted. Consequently, labor continued to be scarce in the early 1930s, though desertion was not the only factor responsible. Since the interwar years, the Asian cane growers had to compete for labor with the Kericho tea plantations that also got most of its laborers from CN. This was because the neighboring Kipsigis community where the tea estates were located was not keen on wage labor and was found to be unreliable. By 1933 the tea estates had employed about 8,000 workers, most coming from Central and South Nyanza.[90] The demand created by the tea plantations gave African workers more options, and this affected the cane farmers, as competition for wage labor increased. These options increased the occurrence of desertions witnessed over the years.

Desertion continued to be a significant labor issue for large-scale cane growers. By the 1950s, desertion was so high at Miwani Sugar Mills that *askaris* (policemen) were stationed at the Miwani railway station to ensure that the deserters did not board the train. The majority of the deserters were Gusii, Kuria, and Luos of South Nyanza.[91] The laborers were basically employed as cane cutters and loaders. Many Kuria did not anticipate the difficulty of the work that awaited them, and many deserted after working for only a few days at Miwani. The problem was so severe that it required the attention of the colonial state. For example, in 1951 alone, out of 1,412 recruited labor force, 499 deserted.[92] Desertion remained a crucial issue that had an impact on the Nyanza sugar industry. The recruitment and retention of labor remained a major problem for the farmers. However, there were other reasons that made workers uninterested in wage labor during the colonial period, most notably working conditions on the farms and in the factories.

WORKING CONDITIONS IN SUGARCANE PLANTATIONS AND FACTORIES

In the sugar industry, just like any agricultural industry, the issue of working conditions remained a major factor in the success of the industry. Working conditions involved issues as diverse as terms of service, wages, housing, the types of tasks performed, and labor unrest. All these issues affected sugar production, as most of them were major determinants of the engagement and retention of labor in the industry.

TERMS OF SERVICE FOR LABORERS IN THE SUGAR INDUSTRY

Labor in the sugarcane farms was composed mostly of African workers, although there were Asians who were employed too. The latter were mainly the new arrivals, mostly young and poor who were ready to do any work in order to save enough money to venture into entrepreneurship. Some were called upon from home by relatives to help them in the sugar plantations, especially those who had agricultural experience. The African workers comprised men, women, and juveniles (as casuals); while some men were on long contract, others worked for a shorter period.

There were different types of labor contracts for these African workers: the verbal contract, which entailed thirty working days; piece work, whereby laborers were paid wages according to the task performed; and the long-term written contract, which involved several months' service. Most Africans in the sugarcane farms preferred a thirty-day working contract.[93] This was convenient for the laborers who apart from family obligations were also involved in peasant production. The number employed on a farm varied according to the size of the farm. For most farms, between 15 and 300 people were employed.

Other terms of service included the supply of provisions; both the small-scale cane farmers and the large-scale provided rations as means of encouragement and retention due to the odious and busy schedule involved in cane production. For example, once the laborers were employed they were provided with rations and other basic necessities. The large estates also provided their laborers with tools. The Victoria Nyanza Sugar Company may be used as an example. Upon employment, the employer provided the laborers on contract with a blanket, basin, eating bowl, plate, sleeping mat, and working tools such as a cutlass (*panga*) or digging hoe.[94] The cutlass and the hoe were returned by the workers at the end of the contract, while the other provisions, like the blanket and plates, were kept by the laborer. Just like Muhoroni, Victoria Nyanza Sugar Company also provided rations as follows:

- 2 pounds *posho* (maize flour) per day
- 1 pound of sugar and salt per week
- Beans and cooking oil[95]

By 1952, the rations at Miwani consisted of a slightly different mixture:

- 16 oz. of *posho* (maize flour) per day
- 4 oz. of beans per day
- ½ lb. of sugar per week
- 6 oz. of vegetable oil per week
- 4 oz. of salt per week
- 90 cents per month for meat[96]

This ration was found not to be adequate; as a result, the laborers brought foodstuffs such as vegetables from home and also grew their own vegetable gardens in the estates.[97] The inadequacies found in the ration indicated that the diet in the African reserves was much better than in the labor camps because it was more varied and nutritious.[98] The rations offered in Asian labor camps were also slightly different from those found in settler farms. For example, the workers in the coffee farms were given two pounds of *posho* per day along with a ration of beans weekly and salt.[99]

These rations remained unchanged over the years. When compared with what the Victoria Nyanza Sugar Company offered during the same period, the ration in settler farms was found to be much less sufficient; the workers were offered neither sugar nor cooking oil. The reason given for the small ration on settler farms was that as opposed to the Asian farms, there was enough room for laborers to grow their own food.[100] Perhaps this was made possible because the settlers had more land. Comparatively, Magadi Soda Company by 1923 offered the best rations to its laborers; the rations included the basic two pounds of *posho* daily, a pound of beans weekly, a pound of meat weekly, salt weekly and one lemon weekly.[101] This made Magadi Soda Company popular among the workers and therefore a strong competitor for labor. Not only did it provide better wages, it also provided a more diverse ration. These factors made it difficult for the Asian small-scale farmers to acquire labor.

TYPES OF TASKS IN THE SUGAR INDUSTRY

Another important element in retention was the type of tasks that workers were recruited for. Labor was hierarchically organized in the sugarcane plantations. In terms of tasks there was racial division of labor; Africans did more strenuous, menial, and urgent tasks than the Asian workers. In terms of hierarchy, the Asian or European farmer stood at the top; then there was the supervisor. In Asian farms, the supervisor or *nyapara* was in most cases Asian. These supervisors were mostly cruel and handled the workers badly. For example, insults were very common and as colonized people in a three-tier racial hierarchy, the African laborers were often subjected to abuse, but withstood the cruelty due to their need for money to pay taxes. On the other hand, the Asian workers were treated well because the Asian farm owners normally invited friends and relatives from India to come and work as a *nyapara* on their farms. Most of these overseers were trained; a few had studied agriculture at Punjab University. However, they had to apply for immigration permits for entry into Kenya, which could be turned down by the colonial state.

The *nyapara* was responsible for making sure work was performed and production efficient. Some had exceptional abilities and knew much more about sugarcane farming than the farm owners.[102] Other Asians worked at the weighbridges, drove tractors, and worked as farm assistants. Once in the country, especially the skilled Asians had to apply for a temporary employment pass in order to take up employment. For example, Rattan Singh had to apply to work on the farm of Shamji Harji and Bros. as a specialist in machine maintenance.[103]

In the factory, a similar hierarchy applied; all the senior posts were occupied by Asians and Europeans. Africans basically did the manual work, though there were also African skilled workers, like artisans, who despite their skills were paid less for the same qualifications and same tasks as the Asians. Many recognized the discrimination involved in this system. In 1955, the DC of CN, A. D. Shirreff, objected strongly to this system and complained that it was prejudicial to local people.[104] Nevertheless, Asian workers also faced challenges of unfair policies. The colonial state's discriminatory immigration policies denied Asian farmers and manufacturers much-needed skilled labor, such as supervisors. This was important to the cane farmers since some of the cane farmers had no knowledge of or expertise in sugarcane growing, and so seeking skilled manpower was important. Unfortunately, the colonial state failed to see the urgency of supervision and made it difficult for skilled workers to come from India.

There were specific tasks needed in the cane farms. Sugarcane farming was demanding and the farmers needed labor to perform different tasks throughout the year. The peak period was between March and November. This was the time for planting, weeding, and harvesting. Miwani Sugar Mills, having the largest plantation, needed laborers throughout the year, both on the farm and at the mill.[105] This required a more organized system to ensure that there was constant supply of laborers. For instance, they had to employ workers during field preparations in tasks such as tree-felling and de-rooting, which were done manually due to lack of machines, which by then were very expensive. Labor was also needed in the factory.

There was a clear division of labor; work was allocated based on gender and also age. There were a few instances when women were engaged for work. However, most workers were men. They did odious jobs like de-rooting, clearing bushes, breaking land using the plough, hoeing (which had to be ready before the onset of rainfall), cane cutting, distributing manure, and loading and offloading cane. Furthermore, it was only men who worked in the factories and jaggery plants. During the colonial period the laborers in Asian cane farms worked minimal hours; the laborers not only had a two-hour lunch break but also left the field at 4:00 p.m., which seems to have been a very generous terms.[106] Today some of the Asian cane farmers such as Mohinder Singh, a large-scale farmer, believe the colonial laborers were better than the current ones, since they used to work even in the afternoon hours, compared with today's workers who work only till midday.[107]

Be that as it may, during the colonial period, most of the work on the farm began at 6:00 a.m., and the laborers worked for an average of six hours daily due to the odious nature of the work such as weeding and cane cutting and the fact that they worked in hot weather. During weeding, the laborers were supposed to manually remove weeds that could not be completely cleared by the mechanical weeder. The workers normally worked from Monday to Saturday except for emergency cases such as when the farmers needed to load already harvested cane to the factory; during those times they would also work on Sunday.[108]

Cane cutting was the most difficult and risky task. It was done in rows. During manual cutting, total concentration was needed to avoid cutting oneself.[109] Occasionally, laborers had accidents of flying pieces of cane striking them. Secondly, when the firing of cane was stopped, there were cases where some people were bitten on the hands by spiders. The presence of snakes also made the work risky. Cane knives were sterilized during the cutting season. The blades could be left overnight in a sterilizer to be used the next day. This was done to prevent the spread of ratoon stunting disease.[110] Another important task in the production process involved transport. Some men worked as lorry drivers, and they were in great demand during the harvesting period. They basically transported cane either to the trolley lines or the sugar mills. By 1948 approximately eighty African drivers were employed in the Asian sugar farms.[111]

The factory workers, on the other hand, had a different working schedule from farm hands. They worked on three eight-hour shifts, from 8:00 a.m. to 4:00 p.m., 4:00 p.m. to midnight, and midnight to 8:00 a.m., alternating weekly. This was because during the rolling season, the furnace was kept burning for twenty-four hours. The factory had many workers, skilled and unskilled, who not only worked different shifts but also in different places. The minimum manpower in a factory with capacity of producing 1,000 tons of sugar per annum was approximately fifty-five.[112] The manpower included technicians; in the factory unlike the farms, more skilled workers were needed to man the various important sections such as the evaporating plant, vacuum pans, condensing and cooling ponds as well as in the crystallizers section which was one of the most sensitive areas as it ensured that the proper grade of sugar was attained. This labor force excluded those who worked as cane inspectors, accountants, and general administrators. It was not uncommon for some personnel to double up duties. For example, the factory chemist could also be the cane inspector.

Nonetheless, in 1943 the colonial government introduced the task system in agricultural work on sugarcane farms. This was meant to protect both the worker and employer against possible exploitation. The system required a laborer to perform a given task per day. It was hoped that the task system would increase production. In most instances, work was measured. For example, three men would be expected to cut enough cane to fill two trucks per day.[113] Organized daily tasks worked as follows: "Each man is set a daily task, when he has completed it he may go, even if it has only taken two hours to complete it. Each completed task is recorded on a ticket with which he is issued at the commencement of the contract. This ticket states his daily (task), rate of pay, and the contract is either for a month or forty-two days. The monthly contract has twenty-six working days and the forty-two day contract has thirty working days. Each worker is paid at the expiry of his contract."[114]

The task system was productive since it rewarded the worker for his extra effort and diligence by setting him free once the task was done. This system was also used in Uganda's sugarcane plantations. This system was useful to both the farmer and the worker since the work was done on time.[115] Consequently, working conditions/terms were among the major determinant for return to wage employment. Other determinants were also important since many of laborers were migrant workers, conditions such as adequate food and housing became important for retention.[116]

WAGES IN SUGARCANE FARMS AND FACTORIES

The issue of wages was always a significant one in the Nyanza sugarcane farms and estates. There was no standardized wage at any time. The wages paid differed due to many reasons, not only in areas of employment and the employer (Asian or European) but also based on gender and age. For instance, the few women employed were paid less than men. There was also a difference in the wages paid for farmhands and factory workers.

Asians paid low rates compared with most European farmers. This was because the Asian cane farmers, especially the small-scale farmers, were undercapitalized and had little support from the colonial state. For instance, by 1910 when many Asians paid 3–5 rupees, European farmers paid 6–7 rupees.[117] By 1912, the African laborers in Kibos-Muhoroni numbered approximately 700 and were paid between 5–7 rupees.[118] Although Asian employers paid low wages compared with European farmers, they extended credit facilities as an inducement to the laborers.

Needless to say, the 5–7 rupees paid was not enough to pay for the tax and also meet the families' basic needs. Therefore many workers only sought wage labor jobs during the off-peak season. Furthermore, European employers at this time were not willing to pay good wages either, arguing that an increase in wages did not improve their condition but rather led the African laborers to engage in bad habits.[119] Consequently, farmers generally continued to have labor shortages. Regardless of the low rates, many people were recruited. By 1915, approximately 36 percent of the people of Kisumu area entered wage labor force due to two factors: the proximity of the Asian plantations and the cooperative role played by the local chiefs toward the labor recruitment effort. As such, wage employment was not entirely the result of the attraction of high wages.[120]

In addition, the level of wages fluctuated over the years. By 1928 wages had increased to Shs. 12–14 per thirty-day work ticket plus rations with the price of cane at Shs. 12 per ton.[121] However, the Depression impacted wages so that by 1933 the average wage for farmworkers had dropped to Shs. 8–12 for a thirty-day working contract while factory workers were paid Shs. 12–14 under the same contract.[122] The drop discouraged many Africans from seeking wage labor. The Depression led to an increase in those seeking wage labors. For instance in 1934 there were about 700 workers on the small-scale Asian farms, up from 500 of the previous year.[123] As the effects of the Depression took a toll, there was reduction in wages due to poor cane prices. The price of cane fell from Shs. 12 a ton to Shs. 6 a ton while that of jaggery slumped from Shs. 5 a frasila to Shs. 2 a frasila. Hence wages decreased to Shs. 7–9.[124] The drop in prices and wages shows how the Asian cane growers struggled to remain in business during the hard times. The hardships created by the falling price of sugar meant that the small-scale farmers faced a cut in profits and so had to lower wages. With low wages, the struggle to obtain and retain labor became difficult.

By 1935, the situation had not improved and labor supply in Muhoroni, Songhor, and Koru farms were found to be far below what was expected, possibly because wages remained low. The wages had stagnated for a long time and ranged between Shs. 6 and Shs. 8 per month plus *posho* (maize meal).[125] These were the rates offered in the Asian farms for farm hands. However, Victoria Nyanza Sugar Company offered higher wages for the same workers at the rate of Shs. 8–12 a ticket for thirty days in 1933.[126] By 1935 the wages in the Asian small-scale farms still remained at the rate of Shs. 6–8 a ticket, while the large estates, in particular the Victoria Nyanza Sugar Company, paid its workers higher wages. For example, the average wage in the company during the same period was Shs. 13.50 for a thirty-day ticket.[127] This slightly increased and in 1938 the wages in the large estates were Shs. 12–14.[128]

Wages also differed depending on the task. For example, cutting, which was strenuous and dangerous, attracted few laborers. To solve the problem, the farmers started a system whereby laborers were paid 75 cents per day for cutting and loading cane on a truck. Therefore, if one worked for thirty days at that rate, the wages would be Shs. 22.50.[129] However, sometimes wages offered to laborers differed in various employments. For example, by 1941 the Victoria Nyanza Sugar Company increased wages and paid laborers Shs. 9–15 for a thirty-day working contract while the Asian cane farmers paid Shs. 6–10.[130] The wages offered by the Asian cane farmers were far below those offered by Victoria Nyanza Sugar Company in 1941. This was caused by many factors such as lack of supervision during the war and also the fact that many Africans preferred to work on the farms rather than enlist toward the war efforts.

By 1950 unskilled laborers were paid Shs. 18–25 per thirty days' work on Asian farms.[131] Furthermore, wages differed according to the worker's skills and as well as type of employer. For example, by 1955, Miwani Sugar Mills paid unskilled laborers between Shs. 20–30 per monthly ticket, while other Asian farms paid Shs.18–20 for the same ticket.[132] However, wages continued to go up slowly. For example, by 1963, due to the activities of the trade union, discussed in the next section, farmworkers and unskilled factory laborers were paid Shs. 85 per month.[133]

Many of the cane farm laborers did not earn much. These wages were impacted by the fact that the colonial state did not put in place good policies for the workers, as they were more concerned about pleasing the settler farmers, as indicated in the Master and Servant Ordinance. The policies were full of contradictions and failed to have clear guidelines on plantation workers' wages, and so laborers were easily exploited as wages varied on different farms growing different crops. For instance, wages offered by the sisal and maize farms were different from those offered in the sugarcane plantations. For example, by 1923 the sisal estates paid Shs. 24 without food for 1,000 leaves cut or Shs. 12 with food for 1,500 leaves.[134] During the same period workers on maize plantations in Nyanza were paid Shs. 10 per month with rations.[135] The wages for maize laborers were lower than other agricultural sectors apart from the sugar industry, and the reason given for that by employers was due to the "picking" (stealing maize) during harvest.[136] These wages were higher than those paid by small-scale Asian farmers in 1935; it was apparent that cane farmers faced stiff competition in labor procurement, which possibly contributed to the poor wages in the Asian cane farms as farmers were heavily undercapitalized and could not afford to pay their laborers good wages. The Asian cane growers had to struggle not only to get laborers but also to retain them. Yet despite the fluctuating rates, wages were a key variable in the attraction and retention of labor, more important than housing.

HOUSING IN THE SUGARCANE PLANTATIONS

Housing was also one of the most significant conditions that attracted labor, and it had a huge impact on the retention or return of workers after a short period of visitation to families. Housing on the Asian farms and estate plantations was not very adequate. Until the 1930s, Asian farmers used traditional huts for housing their workers, as did the Victoria Nyanza Sugar Company. These huts were made of mud and wattle or mud and grass-thatched roofing, but they were not maintained as well as the ones found in the reserves. Moreover, they had only one single room that served as kitchen, dining room, and bedroom. From the 1930s there was marked improvement, and sanitary conditions were good in all the camps on the estates belonging to Victoria Nyanza.[137] This was probably because it was easy to maintain the outside compound, which was constantly swept and the trash burned, but the maintenance of individual huts was challenging since as many as three people shared a structure. On those farms where housing was found to be inadequate, some laborers were granted permission to build their own houses, especially on the large estates such as Miwani.

Housing was a sensitive issue since labor laws stipulated that employers were expected to provide housing. In most cases housing was overcrowded and posed great health risks since the structures were built haphazardly. Unfortunately, the large estates such as Victoria Nyanza Sugar Company were reluctant to build permanent dwellings for their workers, citing financial constraints. This forced the chief labor inspector, Dr. Fisher, to intervene in the 1920s, when the Victoria Nyanza Sugar Company refused to improve the housing conditions of its workers.[138] The Victoria Nyanza Sugar Company's head office in Sydney, Australia, was also adamant and refused to cooperate with the labor office in Kenya to improve the poor housing and health conditions in their labor camps. Yet according to the Master and Servant Ordinance of 1910, servants were to be properly housed, fed, and given an adequate supply of medicine. Furthermore, it gave authority to medical officers and labor inspectors to enforce laws in regard to workers' health.[139] The ordinance stressed healthy working conditions and remained important, which led to the amendment of a law in 1920 that set basic regulations on the welfare of the laborers.[140]

Unfortunately, due to the unwillingness of the company to rectify the situation, Dr. Fisher pressed on and managed to get a copy of the *Sydney Bulletin,* which published the profits that the company had made from its sugar investments in Kenya. The newspaper revealed that from 1923, when the company started crushing cane, to 1931, the Victoria Nyanza Sugar Company had extracted £331,246 in profits, of which £265,000 had been exported to Sydney.[141] Such information made it possible for Dr. Fisher to push ahead and engage the company to improve the housing conditions, despite the difficulties he faced. Unfortunately, very minimal measures were taken to improve housing; the company was less concerned about the welfare of its workers than about the profits it made.

The labor officer's report in 1933 decried the poor living standards and reported that the shortage of housing in sugarcane plantations was so rampant that two to three people shared a hut while juveniles shared housing with adult laborers, mostly women. The houses were so poor that in most cases they were unable to adequately shield the occupants from cold, wind, and even rain. The situation was so bad that another report of the labor officer, also in 1933, recommended gradual erection of permanent houses.[142] Sensing that the colonial government was adamant about stopping the poor living conditions, the company embarked on an immediate solution and within the same year of the release of the report, there was improvement of the living arrangements. A satisfactory sanitary system was put in place and the company installed piped drinking water. There was also improvement in the outward general cleanliness of the labor camps. Victoria Nyanza Sugar Company promised, despite the Depression, to set aside 5 percent of its profits for the construction of permanent houses.[143] Unfortunately, this was not carried out. The new houses were temporary structures that were only a little better than the old ones, and it was only later in early 1960s that the successor company, Miwani Sugar Mills, embarked on building permanent houses for its labor force.

Housing on the larger sugar companies in Nyanza such as Miwani and Muhoroni were divided into estates. For example, by 1954 Miwani had five housing estates. These were one-room huts measuring 15 by 17 feet. The housing was congested, often leading to outbreak of diseases such as typhoid and dysentery. Most of the living conditions of the workers were pathetic since there were no adequate disposal systems and the inappropriate use of bucket latrines posed a great health danger.[144] Although the company policy on housing stated that the living conditions of the laborers should be good with adequate accommodation, good latrines, and piped water, this was not the reality.[145] Many workers lived in congested dwellings. For example, the living arrangements at the Miwani Sugar Mills housing estates in 1954 were divided into five zones namely; Factory Camp, Section 1, Section 2, Section 3 and the Ox-Camp. A total of 2,155 laborers had to share 774 rooms, excluding possible infants brought along by the female workers.[146] This shows the degree of congestion in the housing estates. Most of these housing estates had piped water supplies, with the exception of the Section 1 estate. However, the situation remained inadequate and the houses were not improved to permanent status. The housing estates also had no hospital despite recommendations from the medical department. Former workers alleged that the general manager refused to spend money on improvements even though the situation was bad and cases of laborers dying of dysentery were reported.[147] Congestion in the labor camps worsened as the huts continued to be depleted over the years. For example, the factory camp had the worst housing and sanitation, and that led to serious outbreaks of bubonic plague in 1947 and 1952. The same disease broke out in Section 3 of the camps and claimed seven lives.[148] Such poor living conditions as well as low wages and poor working conditions caused labor unrest to emerge among the cane workers in order to draw the attention of their employers to their plight.

LABOR UNREST IN THE SUGAR INDUSTRY

Labor unrest was not unique to the sugar industry. There was general unrest in the whole country as early as the 1930s that culminated in widespread strikes. These early strikes marked a change in the type of protest seen and initially were mostly found in the urban areas and were organized by highly skilled manual workers.[149] For example, in 1934 a strike took place in Mombasa when the port workers of the Kenya Landing and Shipping Company demanded a stoppage of the proposed wage reduction, as wages offered were not even commensurate to the increasing prices of commodities and rent. The colonial government's findings agreed with the workers that there was a need to increase wages.[150]

However, rural workers were also discontent, as demonstrated by the successful strike by employees of Asian fish merchants at Kisumu in 1935, which secured an increase of Shs. 2 on their wages.[151] This set a precedence and was soon followed by another successful strike by the boat men at Asembo Bay in 1936.[152] In the meantime, labor unrests continued in the urban areas; for instance, stone masons in the quarries near Nairobi went on strike in May 1937. Thus the year 1937 was marked by labor unrests all over the country as African workers demanded better wages and terms of service. There were strikes even at the sisal farms at Mwatate at the coast, the Shell Oil Company installation site at Mombasa, and labor unrest at the railway marine workshops in Kisumu and at the Victoria Nyanza Sugar Company's plantation estate at Miwani.[153] Although these early strikes met with minimal success, they were significant in the sense that they marked the growth of consciousness among workers in Kenya. Many workers were sensitized to their own needs, and there was general labor unrest in the whole country by the end of the 1930s.

After World War II there was general labor unrest in the whole country, and the unrest in Nyanza, particularly in the sugar industry, was no different from that found in postwar Kenya. This was caused by the increased prosperity that led to higher living costs and a demand for better wages.[154] However, the 1940s strikes were also found to be more organized as trade unions that had emerged by the mid-1930s had become more effective. Among the earliest was the Labour Trade Union formed on April 18, 1935, with Makhan Singh as the secretary. This union called the sixty-two days-strike in the motor service industry and won a wage increase in 1937.[155]

The sugar industry was not spared the unrest that swept across the country. Like other workers, the cane workers too had grievances. Labor unrest in the sugarcane plantations and factories were related to poor or delayed wages, sometimes even uncalled-for deductions. These issues, coupled with poor rations, terms of service, and poor housing in some of the sugarcane farms, led to labor unrest. Work stoppages proved to be expensive, especially in the factory during the rolling season. The Asian farmers faced difficulties and suffered due to their undercapitalized status, which made it hard for them to offer their laborers sufficient wages.

However, during the post-World War II period, a marked difference was found in the attitude of laborers to work stoppages. This was partly because many workers were more knowledgeable about their rights; moreover, the joining of the ex-soldiers into the labor force and the level of agitation in country during the postwar period impacted the workers. The strikes, in most cases, were spontaneous and lacked proper organization. The initial ones involved only a few laborers. For example, in 1942, about fifty laborers at an Asian sugarcane farm in the Miwani area put down their tools for three days, demanding an increase in wages.[156] Unfortunately, the strike did not yield fruit since it was contrary to the Defense Regulations that forbade any strikes by essential services workers during the war. Their demands were rejected and the laborers forced to resume work unconditionally.

Nevertheless, as discontent grew, workers in the sugarcane plantations for the first time were able to organize themselves into a bargaining position. In 1946, for example, workers at the Victoria Nyanza Sugar Company's Chemelil plantation went on strike for two days over deductions from their wages and the amount of work demanded. Many laborers were discontented and did not do a good job. Investigation by the labor office found that the deductions were not justified. Within a short span of time, the Victoria Nyanza Sugar Company suffered another strike from the factory workers who demanded higher wages and did not resume work until an agreement was arrived at based on a more-work, more-pay policy.[157] A lack of interest at work created more problems. By 1947 many field hands were coming out of the field as early as 10:00 a.m., having worked for only two or three hours.[158] Many workers were dissatisfied; there were two other strikes reported at the Miwani Sugar Estate in 1947, after the takeover by the Asian manufacturer. The strikers protested against the new regulations put in place by the new management. For example, laborers were previously not restricted from chewing cane during working hours in the farms.[159] The workers were against the rule that forbid them from chewing cane when at work in the farm, as they were accustomed to.[160]

As for the issue of wages, those who were in the best position to bargain effectively were skilled artisans produced by mission schools. Missions in Kenya produced artisans such as carpenters, masons, and plumbers who easily found jobs in the sugarcane plantations and factories. Moreover, at the end of World War II, some ex-servicemen also came back with skills such as driving; hence there was an increase in skilled laborers who obtained employment in Asian farms. Such opportunities were occasioned by the fact that some Asian farmers had acquired ex-military lorries that were reconditioned and used for transporting sugarcane. For example, by 1948, about eighty Africans, mostly ex-servicemen, were employed as drivers in the Asian sugarcane farms.[161] The entrance of the ex-soldiers into the labor force sensitized the workers, as the ex-servicemen were aggressive and started to demand better wages and better treatment. Due to exposure to ex-soldiers and those ex-soldiers' organizational skills, most strikes in the sugar industry occurred after the war. The ex-soldiers' entrance into wage labor was significant because it ushered in a change in type of protest. The workers started to face the employers and bargain for better working conditions and wages instead of simply deserting the workplace as before.

This group of workers seemed to have been more persistent and aggressive in their demands, especially the ex-soldiers who were militant. Therefore, within three years, another strike occurred at Miwani Sugar Mill in 1950 that involved many workers. Approximately 570 laborers went on strike demanding higher wages and better working conditions, but their demands were rejected.[162] Despite the frustrations, the laborers continued to press for better wages. In 1952, another strike took place; the casual laborers refused to work and demanded an increase of their daily pay from Shs. 3 to Shs. 5. But even this was not successful. Nevertheless, the farmers felt the drain as the lorries sent out to pick up the casual laborers came back almost empty.[163]

The most effective strike in the postwar period took place in December 1959. During this particular strike, the numbers involved were high since the stoppage involved laborers from Miwani Sugar Mills and the other Asian farms. Unlike other protests, the 1959 strike brought cane production to a halt for almost a week; this strike was specifically called to show solidarity with the railway workers who went on strike for better wages.[164] It was better organized than earlier strikes but ultimately was unsuccessful: the workers had no union and there were many unemployed workers in the area who could be used as strike breakers. This made the Asian cane employers unresponsive to demands.[165]

Despite the initial setbacks, the strikes were significant in the sense that the cane laborers started to organize and saw collective bargaining as a better response to solve their grievances. This led to change the following year as the workers formed their own trade union, the Kenya Union of Sugar Plantation Workers. (KUSPW)

The formation of KUSPW in 1960 coincided with many labor unrests due to the political fever that was sweeping across the country as Africans' demand for independence increased. The political fever gave rise to a new spirit of militancy among the workers.[166] Workers started to demand better conditions. There was change of tactics as the strikes were now called by the workers' trade unions. For example, in 1960, the strike by workers of the coffee industry was successful and was only brought to an end with an agreement on the picking rates between the union (the Coffee Plantation Workers Union) and the Kenya Coffee Growers Association.[167] In Nyanza the situation was similar and workers no longer tolerated labor injustice. For example, a bitter strike took place in the Macalder mines in South Nyanza during 1961 where workers were exposed to racial hatred on a daily basis; they only agreed to go back to work after some of the most hated supervisors were removed.[168] Similarly in 1962 and 1963, the cane workers, through the KUSPW, also organized, and a strike was averted only after the cane growers and manufacturers agreed to increase wages. Trade unionism led to new ways of bargaining for the laborer, as will be discussed.

These strikes were significant because they revealed a nascent form of labor consciousness among the workers as they bargained for wages despite minimal success. Furthermore, there were other advantages, as the laborers in sugarcane plantations became sensitized to their rights and cooperated with other laborers in other fields in the district. This was the case when they associated with the railway workers. Such solidarity meant that they now had learned the art of collective bargaining, and therefore the Asian cane farmers and manufacturers were bound to reassess their wages as competition stiffened. Unfortunately, the strike also did not achieve much because the workers were not properly organized and the colonial government was not as responsive to the sugar industry workers as they had been to the dock and rail workers. For instance, during the strike by the dock workers at the port of Mombasa, there was government intervention when it sent Eliud Mathu to Mombasa to talk to the workers; this was not the case with strikes in the sugar industry where the colonial state looked the other way. Perhaps the strikes in the sugarcane plantations were as a result of the neglect of the sugar industry by the colonial government since there was little supervision; neither do we see any intervention to help the Asian farmers and manufacturers in the sugar industry as the workers became more militant and more organized in their trade union.

TRADE UNION IN THE SUGAR INDUSTRY

In Kenya the formation of trade unions to articulate the interests of workers began in the mid-1930s. During this period, the Labour Trade Union was formed under Makhan Singh.[169] This early union, though meant to be multiracial, was largely composed of Asian workers.[170] Other trade unions that emerged included the Nairobi African Taxi Drivers Union formed in 1943 and the Kenya Houseboys' Association formed in 1946, just to mention but a few.[171] The emergence of a trade union in the sugarcane plantations of CN was not an isolated issue but rather part of an ongoing phenomenon in the whole country. The need to form a trade union started to emerge among the cane workers around 1953. It was a slow process because it was bedeviled by suspicion from the Asian employers who resented such a move. On the other hand, the workers were hesitant in joining for fear of repercussions from the employers. Workers also were reluctant to pay the dues of monthly subscription of Shs. 2 for fear that the money would be pocketed by union officials.[172]

However, KUSPW formed in early 1960, as earlier indicated, was composed of both manual and skilled workers of the sugar industry.[173] The main objective of the union was to improve the terms of service of the cane workers, better the relationship between workers and employers, and engage in settlement of disputes between the two.[174] Unfortunately, although the officials were hard-working and some workers joined, most workers responded slowly, such that by 1960 the members were only approximately 2,125 out of a workforce of about 12,000 workers.[175] It was difficult to get the workers to join the union due to intimidation by the cane farmers. This was because different Asian cane grower groups—such as the Nyanza Indian Farmers' Association, Kibos Planters' Association, and the various production subcommittees—unified to vehemently oppose the emergence of trade unions among the workers.[176] The fear and resentment of KUSPW by the Asian cane farmers and factory owners was witnessed when the secretary general of the union, J. Ongundha, was accused of trespassing when he visited Miwani Sugar Mills and was arrested and sent to prison for two years in 1961.[177]

The arrest of the secretary general did not hinder the increased recruitment of members. Instead, the arrest of Ongundha motivated many to join the trade union and demand his immediate release.[178] Despite the frustrations and slow take-off, once the workers had built up trust, it was easy to organize because the workers stayed in the same camps; so reaching out to them and dissemination of information were fast. However, organization was tricky since the employers could regulate those visiting their premises. As the political situation in the whole country became increasingly charged as the quest for independence grew, people became more conscious of their rights and more were willing to join the trade union. All over the country trade unions were active.

In Nyanza, the sugar industry workers became more involved, especially after the rallies held at Miwani by the nationalist politicians Oginga Odinga and Tom Mboya. It was alleged that after the rallies the membership of KUSPW rose from about 2,125 members in 1961 to 4,500 by 1964.[179] Meanwhile, the struggle for better terms of service continued. During the 1962 period the union achieved one of its major milestones when it was able to negotiate a pay increment for the workers through collective bargaining because more workers were willing to put down their tools. This achievement made the union strong and popular, and many members joined the union.[180] The following year the union was able to engage the Asian employers once again, which led to better working conditions for the workers and better wages. For example, the minimum wage for the unskilled worker was fixed at Shs. 85 per month.[181] The union continued to articulate the workers' grievances and this led to improved housing. For example, pressure from the union made Miwani start the construction of permanent housing. In the meantime, through the relentless efforts of KUSPW officials, mistreatments such as beatings by the *nyapara* (supervisors) were brought to an end through constant prosecution.[182] These were no mean achievements, and the union became so popular that by 1963 other branches were opened along the sugar-belt at Kibos, Miwani, Muhoroni, Chemelil, and even Ramisi at the coast with the national headquarters at Kisumu.[183]

The success of KUSPW was due to the fact that it acted in concert with other organizations that supported it. For example, the Kenya Federation of Labour (KFL), which today is known as Central Organization of Trade Unions (COTU), and the Plantation Workers International Federation (PWIF). These organizations extended funding that helped the union to build its offices in Kisumu and other areas and also offered scholarships to officials to enable them to undertake labor courses at the Labor College in Kampala.[184]

The rise of trade unionism in the sugar industry led to new methods of addressing grievances. In most cases it involved dialogue with the cane farmers and employers, and such negotiations improved wages and working conditions, which in turn led to retention of labor since desertion was no longer needed as a response to labor disputes.

CONCLUSION

An adequate labor supply was one of the many challenges faced by the Asian farming community. For much of the colonial period, many Africans rejected wage labor. Africans had their own plots to tend, and wage labor was not appealing: wages were poor, working conditions were hard, and travel to and from farms was often difficult or impossible, especially in bad weather. The unpredictability of the weather in the

Kibos region made recruiting even more difficult. Farmers would suddenly need a great number of workers, or conversely could suddenly have no demand for workers.

During World War I, Africans were recruited to serve as carrier corps and some as combatants who joined KAR. The European farmers took advantage of this situation and advocated for Master Servant Ordinance of 1916 to be changed to favor them: not only could they get more laborers but also the wages of carrier corps were reduced to be commensurate with what the settlers offered on their farms. The colonial state reduced the incentive to work anywhere other than on the European farms.

The Asian farmers tried to make up the labor shortage by inviting skilled personnel from India, but even these initiatives were frustrated by the colonial state's immigration regulations. The Asian farmers could not offer wages that were on par with the Europeans' wages because the European settlers received government financial support. In response, the Asian farmers tried to improve working and living conditions to make their farms more attractive to workers. In times of particular shortage, the colonial state enacted cash taxes that forced Africans to enter wage labor to pay their tax bills. But these measures caused even more suffering and discontent.

The tight labor market was even more challenging during wartime, when the colonial government needed additional labor for public works. For example, during World War II, the government enacted the Defense Native Regulations of 1940 that enabled the PC to recruit Africans for essential services. During these times, competition for workers became even more intense.

These labor problems persisted throughout the first half of the twentieth century and worsened after World War II when workers began to organize and protest. Strikes and desertion were common means of protest, and workers began to form trade unions. The workers had legitimate grievances about wages, working conditions, and living conditions, but the undercapitalization of Asian farmers left them unable to respond with sufficient improvements. In addition, the state responded to labor unrest that affected European industries but left the Asian farming community to fend for itself. Such discriminative labor policies prevented the Asian cane producers from adequately getting labor in colonial Nyanza. As such, the sugar industry struggled but managed to survive despite the biased policies on the part of the colonial state.

NOTES

1. Fearn, *An African Economy*, 51.
2. Paul P. Zeleza, "The Establishment of Colonial Rule *1905-1920*," in *A Modern History of Kenya*, ed. William Ochieng' (Nairobi: Evans Brothers Kenya Ltd, 1989), 50.
3. Joseph Ouma (O.I.), June 12, 2009.

4. Zeleza, "The Establishment of Colonial Rule," 50.
5. Zeleza, "Establishment of Colonial Rule," 50.
6. Zeleza, "Establishment of Colonial Rule," 50.
7. Anthony Clayton and Donald Savage, *Government and Labour in Kenya 1895–1963* (London: Frank Cass and Company, Ltd., 1974), 41. See also Wolff, *The Economics of Colonialism*, 99, for more discussion of the governor's speech.
8. Clayton and Savage, *Government and Labour*, 28.
9. Sharon Stichter, *Migrant Labour in Kenya: Capitalism and African Response 1895–1975* (Harlow: Longman Publishers, 1982), 38.
10. Stichter, *Migrant Labour*, 38.
11. Sticher, *Migrant Labour*, 42. See also Maxon, *East Africa*, 2nd edition, 161.
12. Berman and Lonsdale, "Coping with the Contradictions," 91.
13. David Siddle and Kenneth Swindell, *Rural Change in Tropical Africa from Colonies to Nation States* (Oxford: Basil Blackwell Ltd, 1990), 27.
14. Wrigley, "Kenya The Patterns of Economic Life," 246.
15. KNA: DC/CN/1/51, Agriculture.
16. Berman and Lonsdale, "Coping with the Contradictions," 84.
17. Stichter, *Migrant Labour*, 39.
18. Zeleza, "The Establishment of Colonial Rule," 50.
19. Zeleza, "The Establishment of Colonial Rule," 50.
20. Zeleza, "The Establishment of Colonial Rule," 50.
21. Zeleza, "The Establishment of Colonial Rule," 50.
22. Osamba, "Colonial African Labour," 58.
23. Chanan Singh (O.I.), July 2, 2009.
24. Roger Van Zwanenberg, *Colonial Capitalism and Labor in Kenya 1919–1939* (Nairobi: East African Literature Bureau, 1975), 108.
25. Osamba, "Colonial African Labour," 58.
26. Van Zwanenberg, *Colonial Capitalism*, 50.
27. Ross, *Kenya from Within*, 215–216.
28. KNA: PC/NZA/3/20/4/4, Inquiry on Free Flow of Labour from Nyanza.
29. Van Zwanenberg, *Colonial Capitalism*, 172.
30. Van Zwanenberg, *Colonial Capitalism*, 176.
31. Osamba, "Colonial African Labour," 65.
32. Maxon, *Struggle for Kenya*, 146.
33. Zeleza, "The Establishment of Colonial Rule," 51.
34. Berman, *Control and Crisis*, 143.
35. Maxon, *Struggle for Kenya*, 182.
36. Kanogo, "Kenya and the Depression," 113–114.
37. Ross, *Kenya from Within*, 234.
38. Ross, *Kenya from Within*, 234.
39. Maxon, "The Years of Revolutionary Advance," 81.
40. Ross, *Kenya from Within*, 19.
41. Maxon, "The Years of Revolutionary Advance," 82.
42. Osamba, "Colonial African Labour," 71.
43. Clayton and Savage, *Government and Labour*, 134–136.
44. Colony and Protectorate of Kenya, *Native Affairs Department Annual Report 1923* (*NADAR*) (Nairobi: Government Printer, 1924), 28.
45. Osamba, "Colonial African Labour," 82. For how the colonial state implemented the cotton scheme in Nyanza see the discussion by Fearn, *An African Economy*, 68–78.
46. Colony and Protectorate of Kenya, *NADAR 1927* (Nairobi: Government Printer, 1928), 86.
47. Colony and Protectorate of Kenya, *NADAR 1931* (Nairobi: Government Printer, 1932), 129.
48. Colony and Protectorate of Kenya, *NADAR 1933* (Nairobi: Government Printer, 1934), 129.
49. Kanogo, "Kenya and the Depression," 115.

50. Kanogo, "Kenya and the Depression," 116.
51. Van Zwanenberg, *Colonial Capitalism*, 56.
52. Van Zwanenberg, *Colonial Capitalism*, 118.
53. *NADAR 1933*, 122.
54. Kanogo, "Kenya and the Depression," 116.
55. KNA: DC/CN/1/4/4, K-LDAR 1933.
56. Priscilla M. Shilaro, *A Failed Eldorado: Colonial Capitalism, Rural Industrialization, African Land Rights in Kenya and the Kakamega Gold Rush, 1930–1952* (Lanham, MD: University Press of America, 2008), 14.
57. *NADAR 1933*, 122. See also KNA: DC/CN/1/4/5, K-LDAR 1936, which indicates how the sugarcane farmers faced problems, as shortages continued to be felt. The farmers had now to compete also with miners with the discovery of gold in Kakamega.
58. Kanogo, "Kenya and the Depression," 116.
59. Colony and Protectorate of Kenya, *NADAR 1935* (Nairobi: Government Printer, 1936), 184.
60. KNA: DC/CN/1/4/5, K-LDAR 1936.
61. KNA: DC/CN/1/4/5, K-LDAR 1939.
62. Maxon, *Going Their Separate Ways*, 61.
63. KNA: PC/NZA/4/4/88, Labour in Kisumu 1938.
64. KNA: DC/CN/1/4/5, K-LDAR 1936.
65. Colony and Protectorate of Kenya, *NADAR 1938* (Nairobi: Government Printer, 1939), 107.
66. Maxon, *East Africa*, 213.
67. Maxon, *East Africa*, 214.
68. Van Zwanenberg with King, *An Economic History*, 44.
69. Van Zwanenber with King, *An Economic History*, 46.
70. Tiyambe Zeleza, "Kenya and the Second World War, 1939–1950," in W. Ochieng', *A Modern History of Kenya*, 147.
71. Van Zwanenberg with King, *An Economic History* 46.
72. Zeleza, "Kenya and the Second World War," 148.
73. Osamba, "Colonial African Labour," 105.
74. Osamba, "Colonial African Labour," 95.
75. KNA: DC/KSM/1/3/122, Kibos Planters Association.
76. KNA: DC/KSM/1/1798, Regulations regarding African labor for essential work, 1945.
77. KNA: DC/KSM/1/17/22, Recruitment of women labour, 1944.
78. KNA: DC/CN/1/3/2, K-LDAR 1946.
79. Mohinder Singh (O.I.), June 22, 2009.
80. Mohinder Singh (O.I).
81. Van Zwanenberg with King, *An Economic History*, 45.
82. KNA: DC/CN/1/4/1, K-LDAR 1922.
83. KNA: DC/CN/1/4/1, K-LDAR 1922.
84. Norman Leys, *Kenya* (London: Frank Cass, 1973), 13.
85. David Anderson, "Master and Servant in Colonial Kenya 1895–1939," *Journal of African History*, 41, no. 3 (2000): 459–485.
86. Anderson, "Master and Servant," 464.
87. *NADAR 1934*, 174.
88. Colony and Protectorate of Kenya, *NADAR 1936* (Nairobi: Government Printer, 1937), 178.
89. Colony and Protectorate of Kenya, *NADAR 1928* (Nairobi: Government Printer, 1929), 109.
90. *NADAR 1933*, 127.
91. KNA: Lab/2/1895, Labour Inspection Report 1951.
92. KNA: Lab/2/1895, Labour.
93. *NADAR 1923*, 29–30.

94. KNA: Lab 9/2132, Miwani.
95. *NADAR 1933*, 129.
96. KNA: Lab 9/256, LIR, Miwani, November 29, 1952.
97. *NADAR 1935*, 184.
98. Van Zwanenberg, *Colonial Capitalism*, 39.
99. *NADAR 1933*,128.
100. *NADAR 1923*, 28.
101. *NADAR,1923*, 27. According to the report, it was noted that the Magadi Soda workers were in physically good condition during the period of employment.
102. Vitalis Ogolla (O.I.), June 6, 2009, Agriculture Manager, Kibos Large scale Farms.
103. KNA: DC/KSM/1/19/270, DC CN to Principal Immigration Officer, October 7, 1958.
104. KNA: Lab 9/256, Labour in Sugarcane Farms, 1955.
105. Osamba, "Colonial African Labour," 95.
106. *NADAR 1934*, 174.
107. Mohinder Singh (O.I.), Kibos large scale farmer, June 22, 2009.
108. Absalom Apondi (O.I.), Cane Cutter, Muhoroni, July 26, 2009.
109. Chanan Singh (O.I.), July 26, 2009.
110. KNA: AK/11/52, Report on a visit to Potential Sugarcane areas in Kenya by P.O. Wiehe, December 1, 1956, 20.
111. Mohinder Singh (O. I.), July 26, 2009.
112. KNA: AK/11/52, Extract from P.O.Wiehe to L.H.Brown, Department of Agriculture, Nairobi, May 29, 1957.
113. Mohinder Singh (O.I).
114. W. Elkan, *Migrants and Proletarians* (London: Oxford University Press, 1960), 113–114.
115. Mohinder Singh (O.I.).
116. Van Zwanenberg, *Colonial Capitalism*, 60.
117. Osamba, "Colonial African Labour," 56.
118. KNA: DC/CN/1/5/1, Kisumu District Quarterly Report (KDQR), June 30, 1912.
119. Van Zwanenberg, *Colonial Capitalism*, 41.
120. Stichter, *Migrant Labour*, 47, 56.
121. *NADAR 1928*, 99.
122. *NADAR 1933*, 129.
123. *NADAR 1934*, 174.
124. *NADAR 1934*, 174.
125. KNA: PC/NZA/3/1/465, Intelligence Report K-LD, September 1935.
126. *NADAR 1933*, 129.
127. *NADAR,1933*,129.
128. *NADAR 1938*, 107.
129. *NADAR 1935*, 184.
130. KNA: DC/KSM/1/7/95, Miwani, 1941.
131. KNA: Lab 9/ 2132, LIR, Miwani, April 15, 1950.
132. Chanan Singh (O. I.), July 26, 2009.
133. Michael Juma ((O.I), July6, 2009
134. *NADAR 1923*, 27.
135. *NADAR 1923*, 32.
136. *NADAR 1923*, 32.
137. *NADAR 1934*, 173.
138. Van Zwanenberg, *Colonial Capitalism*, 59.
139. Van Zwanenberg., *Colonial Capitalism*, 60.
140. Anderson, "Master and Servant," 459–485.
141. Van Zwanenberg, *Colonial Capitalism*, 59.
142. KNA: PC/NZA/3/13/8, Labour Officer to Provincial Labour Inspector, Nairobi, May 6, 1933.

143. *NADAR 1933*, 129.
144. Osamba, "Colonial African Labour," 117–118.
145. *NADAR 1934*, 173.
146. KNA: Lab 9/256, LIR, May 13, 1954.
147. Peter Abwogo (O.I.), Miwani, June 21, 2009.
148. Joseph Okongo (O.I.), Miwani, June 21, 2009.
149. Stichter, *Migrant Labour*, 122.
150. Sticher, *Migrant Labour*, 166.
151. Clayton and Savage, *Government and Labour*, 214.
152. Clayton and Savage, *Government and Labour*, 214.
153. Clayton and Savage, *Government and Labour*, 215.
154. Clayton and Savage, *Government and Labour*, 215.
155. Clayton and Savage, *Government and Labour*, 212.
156. KNA: Lab 9/1895, LIR Miwani, July 22, 1942.
157. KNA: DC/CN/1/3/2, CKDAR 1946.
158. KNA: PC/NZA/3/22/1/8, Lyne Watt to Labour Office, November 11, 1947.
159. *NADAR 1933*, 129.
160. KNA: DC/CN/1/3/1, CKDAR 1947.
161. KNA: DC/KSM/1/17/34, Skilled labour.
162. Osamba, "Colonial African Labour," 129.
163. Zakaria Olang (O.I.), Miwani, July 26, 2009.
164. Osamba, "Colonial African Labour," 130.
165. Osamba, "Colonial African Labour," 130.
166. Alice Hoffenberg Amsden, *International Firms and Labour in Kenya 1945–1970* (London: Frank Cass and Company Ltd, 1971), 75.
167. Amsden, *International Firms*, 75.
168. Amsden, *International Firms*, 79.
169. Clayton and Savage, *Government and Labour*, 211.
170. Clayton and Savage, *Government and Labour*, 211.
171. Clayton and Savage, *Government and Labour*, 326.
172. Zablon Wakiaba (O.I), June 20, 2009
173. Office of the Attorney General, *Kenya Registrar-General Annual Report* (Nairobi: Government Printer, 1960), 377.
174. Martin Otieno (O.I.), June 20, 2009.
175. Osamba, "Colonial African Labour," 183. Osamba gives a detailed account of the strong opposition that the trade union officials faced that led to the sugar employers also organizing and forming their union, the Kenya Sugar Employers Union (KSEU), in April 1960 for better bargaining power.
176. Osamba, "Colonial African Labour," 170.
177. Osamba, "Colonial African Labour," 183.
178. Martin Otieno (O.I.).
179. Michael Juma (O.I.), July 6, 2009.
180. Osamba, "Colonial African Labour," 183.
181. Osamba, "Colonial African Labour," 186.
182. Michael Juma (O. I.), July 6, 2009.
183. GOK, *Register-General Annual Report*, 1963, 26.
184. Osamba, "Colonial African Labour," 190.

SIX

Challenges of Sugarcane Growing in Colonial Nyanza

Asian cane farmers had a plethora of challenges to face: labor shortages, discriminatory policies, land access, and more. This chapter discusses some of the remaining problems that they encountered during the colonial period. Many of these are ecological problems: poor soil, erosion, irrigation, droughts, heavy rainfall and waterlogging, locusts, and disease. The farmers' ability to address these problems was hampered by the chronic undercapitalization of the Asian cane farmers. Their inability to secure loans from the government or European banks and the lack of support from the state meant that they rarely had the funds needed to jumpstart the industry after the devastating effects of drought and locusts.

The chapter also examines in more detail some issues that have been touched upon in earlier chapters, such as how the Asian cane farmers had to deal with biased and contradictory colonial state policies on land and issues such as the dumping of sugar from other countries, such as Uganda, that affected the growth of the sugar industry. The farmers in addition had to deal with manufacturers who more often than not were uncooperative, while at the same time dealing with the minor problems that included outbreaks of fires, theft of cane, and immigration difficulties.

Finally, the chapter discusses, in greater detail than earlier, the efforts made by the Asians to solve the problems they faced in the sugar industry.

COLONIAL STATE POLICIES AND THEIR IMPACT ON CANE FARMING

Colonial policies were very important in determining the success of the sugar industry in colonial Nyanza. These policies affected a wide range of issues that were at the core of production, such as land distribution and efforts to bring the cane and sugar to market. The Asian cane growers faced immense problems because these policies were either discriminatory or short-sighted. For example, colonial land policy locked the Asian farmers out of the fertile highlands in Kenya right from the start, reserving it for European settlers. In addition, European settlers were granted 999-year leases while the Asian leases were restricted to 99 years only.

These biased land policies made it difficult for the Asian cane growers to expand their farms and also to buy or sell land. As late as 1953, H.R. Bhatt was barred from selling his farm to another Asian, Mohanlal Kanji, on the grounds that selling the farm to another Asian would be prejudicial to the local Africans. Therefore, Bhatt was advised to find an African buyer, despite the fact that he had already tried and found it difficult to get an African buyer since the farm was being sold at Shs. 18,000 and most Africans could not raise the required amount.[1] Yet the same did not apply to European buyers, even when the sellers or buyers did not reside in the colony. Europeans in CN could sell to anyone as long as the land was outside the white highlands. For example, when Europeans became bankrupt or no longer wanted to engage in farming, they sold their land to willing buyers, as was demonstrated when the Asians collectively bought out farm number 1595 at Kibos that was owned by Major Carnegies.[2] Therefore, it emerges clearly that policies on land were full of discrimination and were constantly changed to suit the settlers.

Other policies, which involved restriction on procurement of farm and factory machines, also proved discouraging. When the Asian farmers wanted another mill in 1936 so as to challenge the monopoly enjoyed by the European-owned Victoria Nyanza Sugar Company, which had stopped taking their cane, the DC of K-L was not helpful. The DC sided with the European manager, despite the Asians' genuine grievances, and asserted that he would vehemently oppose the granting of permission to erect a new mill. He assured the manager that he would be supported by many administrators. Assuring the manager of the Victoria Nyanza Sugar Company, the DC wrote: "Even if the Asian farmers could establish an unimpeachable moral claim to have a mill, I should consider that the government would be failing in its duty if it allowed capital to be poured out on a project which was foredoomed, as far as one can judge, today a failure . . . in my opinion the Asian growers should be treated as peasant farmers, not as component [sic]."[3]

Such sentiments from a government official demonstrated the obstructions that the Asian farmers had to contend with. Eventually, the colonial state's officials would deny the Asian farmers permission for another mill, while during the same period, apart from being given monopoly in sugar production, the Victoria Nyanza Sugar Company was given permission to build a road to serve the transport needs of the company. The state was not very receptive to Asian cane growers while on the other hand it went out of its way to ensure that both European farmers and manufacturers got whatever they needed to spur their production.

Although the colonial government later allowed the Asians to import machinery to manufacture jaggery, even this had a restrictive regulation that was put in place to control the marketing of jaggery; the state appointed a single selling agent, which caused delays. Moreover, Asians were restricted from exporting jaggery based on the flimsy grounds that Africans could have access to it, leading to drunkenness. Such a policy restricted the sale of sugar and its by-products. By this means, the government stifled the production of jaggery, which was a major outlet for some farmers. As observed throughout the book, we find that colonial policies were characterized by contradictions, which were a major drawback to the progress of the sugarcane industry in colonial Nyanza. The colonial state discriminatively found it prudent to allow the importation of the machine for jaggery production and not for milling sugar so as to protect the European millers.

The colonial state also came up with regulations that prohibited the export of sugar as per the Government Notice No. 670 of 1937. Calls for revocation of the sugar control were made as early as 1939 as the war needs increased demand. Nonetheless, the government did not heed to the plea of the sugar millers, and the regulations came into effect in 1941 due to acute sugar shortage. Despite this, the colonial state, during the same year, imposed an excise duty on local sugar that seriously discouraged production due to the high cost.[4]

During the World War II period, however, the inadequate supply of sugar in the country caused the colonial government to change its policy and support jaggery production. Realizing that jaggery was more profitable and cheap to produce, the government shifted its stance and enacted the Defense (Control of Sugar) Regulations in 1944, which was intended to prompt a shift by cane growers and producers to jaggery production. Ironically, pressure was put on the Muhoroni factory, which was producing jaggery, to shift focus and manufacture white sugar. The shift could also have been due to the fact that most Asian sugarcane growers could easily manufacture jaggery on their farms while white sugar could only be produced at the European-owned Victoria Nyanza Sugar Company. Again, the colonial policy was contradictory as it tried to both meet demand and also protect European interests.

After World War II, the colonial government set out to solve some of the challenges facing the industry. Many obstacles were identified, such as the disposal of cane. Another problem was the manufacturers' unwillingness to accept all the cane from growers. The sugar industry also faced many problems due to contradictory government policies. For instance, the government's indirect restrictions on cultivation whereby Asians were given inadequate financial support for breaking land. As such expansion of planting was limited, and furthermore the colonial state controlled the price of sugar. In most cases, the prices offered were not commensurate with the world market; the farmers and the millers registered losses. The sugar industry lost approximately 8,000 tons of sugar during the World War II period.[5] This in essence meant that despite the fact that the sugar industry formed part of the wider colonial economy, the farmers and the millers were not major players, as they were constantly ignored by the state.

Another challenge was the problem of low-grade sugar. Although the colonial government imposed controls over the importation, distribution, and pricing of sugar, control over the quality of sugar supplied to the local market was ignored for a long time. For example, during the four months of July-October 1952, sugar offered for the local market ranged from dirty dark brown low grade to a high-grade white crystallized sugar. The dirty sugar contained impurities and bagacillo.[6] Much of the sugar sold in the month of July was found to be unfit for human consumption.[7] This was allowed to go on for a long period of time, because up to 1952, there was no clear policy on quality and sugar grading. Factories with the capacity to produce pure clean sugar were under no obligation to do so. In most cases it was the consumers in CN who were penalized by this lack of supervision. For example, sugar of 80 percent purity could be sold as that of 100 percent purity.[8]

It was also difficult for cane farmers to make production plans because of the constant change of policies. By the early 1950s, the colonial state started to pursue aggressively a policy of self-sufficiency in sugar production to offset Uganda's manipulation of the market. The state encouraged the Asian farmers to produce for the domestic market as opposed to export, and toward this end the sugar manufacturers received government subsidies to improve the quality of sugar produced.[9] What the colonial government forgot was that the quality of sugar depended also on the quality of cane, and as such it was imperative that both the farmers and the manufacturers benefit from the subsidies. This shift of emphasis by the colonial state on quality of cane was as a result of a research conducted by Douwes Dekker, a sugar expert from South Africa, who was invited to Kenya by the committee of economic coordination of the East Africa High Commission in 1959 to investigate ways of improving the sugar industry in Kenya. The results of his findings, released in 1960, emphasized the issue of quality of cane and led to the establishment of the Sugar Equalization Fund, which was financed by sugar taxes in the colony. The findings led to the restructuring of sugar prices based on the Commonwealth Sugar Agreement.[10] Hence the colonial government finally made sure that sugar prices given to Asian millers were equal to those given to Ugandan millers.

By 1961, the colonial government policy was geared toward expansion of the sugar industry. The director of agriculture commissioned a working party to come up with solutions to the problems confronting the sugar industry. One of its major recommendations was to provide loans for expansion of the industry, though only few Asian farmers were able to access the loans for farming.[11] Furthermore, the many and frequent colonial policies that were enacted to control sugar production in CN inhibited the growth of the sugar industry. Only at the very end of the colonial period was this situation altered.

ECOLOGICAL PROBLEMS AND NATURAL CALAMITIES

Agriculture was the economic backbone of Kenya during the colonial period and has remained so up to the present. And for a meaningful agricultural industry such as sugarcane to thrive, arable land is key. The Asian cane farmers were forced to farm the less fertile areas of Kenya such as Kibos-Muhoroni. The Kibos-Muhoroni area experienced ecological problems such as poor soils and unpredictable weather that made it difficult to produce sufficient sugar to compete with the two neighboring territories. The main soils found in the area were red sandy loams, black swamp soils, and gray-black clay soil, the latter being the poorest in the area, partly a result of bad soil management due to improper cultivation.[12] Secondly, the area had two rainy seasons and poor drainage, so during the long rains, crops were often damaged or destroyed due to waterlogging. For example, the heavy rains in 1946 left a lot of flooding that affected the cane and destroyed the Asian maize farms.[13]

Land, especially in Kibos, was swampy and prone to flooding during heavy rain. The Miwani area in particular had poor drainage. Thus the colonial state contacted surveyors who were working on the Kano plains between Lake Victoria and Ahero to raise the lake level to also examine the Miwani area, especially at the foot of the Nandi Escarpment, so as to put a complete drainage system in place. This was necessary since there was a bad run-off from the escarpment that caused soil erosion and flooded the land in Miwani, particularly at the factory area.[14]

One year exemplifies the difficulties that the farmers faced with the weather. The year 1947 was one of extraordinary weather with erratic rainfall. The long rains started early in January and continued until the month of June, and the rainfall exceeded the normal average of the area. This caused flooding in the low-lying area of Kibos-Muhoroni, where crops were destroyed. Additionally, the mill at Muhoroni was affected by the floods and stopped operations for two months because all the bridges were washed away. Before resuming production, new bridges had to be built.[15] However, during the second half of the year, only a few showers were experienced between July and September, and a period of drought followed from October to December that also led to crop failure.[16]

Heavy flooding was also experienced in 1952 during particularly long and heavy rains, and the Kibos-Muhoroni area was declared unsuitable for maize growing due to the waterlogging.[17] Besides this, drought was a frequent and major problem that the farmers had to contend with. As soon as the Asians started cane farming, they faced drought challenges, despite the usually steady rainfall patterns that Lake Victoria impacted the region due to effects of conventional rainfall patterns.[18] Droughts such as that in 1928 affected production because of the decrease in sucrose content.[19] The situation did not improve during the following year when the climatic conditions were also unfavorable; hence, there was a significant decrease in cane production and other export crops colony-wide.[20]

Drought was a common feature in the area and the cane crop was badly affected. Extreme drought occurred periodically such as *Kanga* and *Otonglo,* as earlier mentioned. For instance, in 1940, rainfall was not adequate at the beginning of the year, which was followed by a period of drought that was so severe that it affected the sugarcane.[21] Drought occurred at the beginning of 1940 and in September. Much of the young cane died, and the rest of the cane was in such bad shape that it was worsened by the sprouting of shoots. Those plants that were harvested had very light and hollow canes.[22]

The alternation between drought and excess rain caused havoc throughout the late 1940s and 1950s. A review of those years demonstrates the challenges. In 1945 a severe drought hit the entire colony, and harvested cane dropped from 8,187 tons to 5,760 tons.[23] This, however, was followed by adequate rainfall in the early part of 1946 that later merged into the long period of good weather that saw an increase in the harvest. While the rainfall was beneficial to the cane farmers, it destroyed the Asian maize crop due to waterlogging.[24]

In 1948 favorable weather conditions led to improvement in cane production. The good weather conditions prevailed until 1950 when again there was excess rainfall, as a result of which the acreage under cane increased, though the area had to contend with the problem of waterlogging.[25] Heavy rains were also experienced in 1952, which led to an increase in the acreage under cane production. However, it also caused more problems due to poor drainage. Then in 1953 a prolonged drought caused the crop to dry up.[26] For example, it was estimated that the area that lay to the north of the Kibos-Muhoroni-Chemelil road lost 20 percent of the crop since the farms were situated adjacent to the Nandi escarpment which, as indicated earlier, had a good climate. The farms to the south of the road, however, lost 80 percent of their crops to the drought, which was made even more severe due to the poorer ecology.[27]

Faced by these numerous ecological problems, the farmers remained resilient. Despite the fact that cane farmers increased the area under cultivation in 1952, they experienced a lot of problems with flooding and drought. Although the Asian farmers tried very hard, the ecology of the area was poorer for cane farming than was in Uganda, and thus Kenyan sugar was simply not competitive with Ugandan sugar. This fact was also arrived at by the Committee for Economic Coordination, which met in Dar es Salaam in 1955 to discuss the future of the East African sugar industry. According to the committee, Kenya produced less sugar than Uganda and Tanganyika due to its poorer climatic and soil conditions.[28] Moreover, the sugar industry remained dependent on good weather conditions. Therefore, despite the improved varieties of cane introduced to farmers, use of fertilizers, good methods of cultivation, and the vast outlay of labor, the success of the industry was still precarious because it depended on adequate rainfall, well distributed throughout the year, which was far from reliable in the Kibos-Muhoroni area.

Apart from the ecology of the area, the farmers also had to contend with natural calamities, such as the locusts that often invaded the Nyanza Province. For example, there was an onslaught of locusts in 1931 which destroyed the sugarcane plantations. The farmers around the Kibos and Miwani area suffered heavily as the locusts destroyed almost 50 percent of the cane crop.[29] After the locust invasion of 1931, the colonial state intervened and appointed G. C. Brown as the locust officer who was to ensure that the problem was contained. Farmers were also advised by the agricultural officers to destroy the locusts' eggs. However, within a few months later, in 1932, a large swarm moved from Kisumu to the Muhoroni area and by September had destroyed most of the crops. The greatest casualty was the maize crop. Sugarcane was also not spared and the most affected areas were between Kibos, Muhoroni, and Kedowa.[30] The poor harvests affected the farmers' ability to repay their loans. The drastic effect of the locust invasion contributed to the declining figures in production of the two milling factories. The sugar production was immensely affected between 1929 and 1932. Before the locust attack, both the production at Victoria Nyanza Sugar Company and Muhoroni Sugar was on the increase. However, in 1931, after the invasion, cane production was reduced almost by half. For instance, the Victoria Nyanza Sugar Company had produced over 7,000 tons of sugar by 1930 but produced only a small quantity of sugar in 1931 when it managed to produce only 3,500 tons. This trend continued and in 1932, the mill was only able to produce 2,500 tons after the locust invasion of 1931. Muhoroni also did not do better, production dropped from 1,000 tons to 250 tons in 1931.[31] In November 1932, for example, locusts were reported to have lain on

Asian farms at Kibos, and the locust officer advised the farmers to destroy the eggs.[32] The sugar industry had a slow recovery the following year as such these natural catastrophes devastated the cane farmers; yet the situation in most cases was also complicated by lack of rain.

Lack of adequate rainfall and locusts forced the farmers to incur losses, and sometimes it meant replanting in the worst-hit areas. Yet in most cases the small-scale farmers experienced financial constraints that hindered their attempts to cope with these ecological problems. Nevertheless, after the 1930s, the locust menace seemed to have been brought under control. One result of the poor ecology of the area, which occasionally was compounded by locusts' visitations, was that the Asian sugar industry produced less sugar than the two neighboring territories. The ecological situation required the cane growers to come up with strategies to succeed in the poor area, such as the intensive use of fertilizers, which added to the expense of production. Unfortunately, the colonial state did not advance the Asian cane growers the financial assistance that was so needed. The state failed to bail out farmers during crop failure through the provision of the GMR to Asian cane farmers, but the European settler farmers were given such subsidies during the difficult times. Asian cane farmers were denied such financial assistance on the flimsy grounds that cane was not an annual crop.[33] Due to the need to buy fertilizers, cane growing became expensive in the Nyanza sugar belt, yet the government's policy, instead of protecting both the farmers and consumers was solely interested in seeing that the consumer did not pay unduly higher prices for the local sugar compared with the imported sugar. As such the state seems to have neglected the sugarcane farmers.

FINANCIAL CONSTRAINTS OF THE ASIAN CANE FARMERS

Although the Asian farmers struggled with state policies and ecological issues, the Asian cane growers were also undercapitalized. This was compounded by the fact that they lacked financial assistance from the colonial government, as indicated in an earlier section. The farmers' lack of capital meant they often had insufficient funds to buy land and in most cases bought the land through loans that they then struggled to repay. Most of these loans were not from the colonial government or banking institutions but from money lenders. The farmers attempted to employ other strategies to solve the problem of cash flow, such as buying farms collectively, but even this did not reduce pressure on farmers. For example, in 1930 a number of Asians purchased farms at speculative prices hoping to gain from cane farming and repay their loans only to have their crops destroyed by the locusts in 1931. Due to this catastrophe, they needed financial assistance to get back on their feet.[34] Unfortunately, the colonial state was reluctant to assist the cane growers.

Furthermore, the small-scale farmers faced financial problems and many could not afford to buy farm implements such as tractors, especially for breaking land. The price of the tractors was way above what the small-scale farmers could afford. For example, by 1950 a high clearance tractor was priced at Shs. 29,115.[35] This was very expensive; it was only the large-scale farmers and sugar estates that could afford it. The majority of the farmers had to struggle getting help. Many relied on bullocks to cultivate the land. Even with the introduction of the machine pool system, they still relied on the bullocks since they sometimes had to have the land prepared long before their "turn" to use the machine. Cash was also needed to buy fertilizers and chemicals for weed control.[36]

Moreover, it was not easy for the Asian cane farmers to access loans from government institutions. For example, during the Great Depression, the colonial state came up with institutions to help the farmers such as the Land and Agricultural Bank established in 1931.[37] However, the help was mainly meant for the European farmers and many of Asian applications were rejected.[38] Even when the committee started to re-examine the applications after complaints by the cane farmers, the process was very slow because a majority of the members had rushed to Kakamega after the discovery of gold in that area. For instance, by the end of the year in December 1936, only one application had been dealt with.[39] It could be argued that there were some biases since the applications had to go through the local committee whose members were mainly composed of Europeans, mostly large-scale farmers such as Messrs T. Allen and J. Fullen, who could not devote much time in such matters. Hence the local committees working with the Land Bank were found to be ineffective, and discrimination made most Asians not opt for such assistance. Yet the farmers needed money to buy other things like agro-chemicals and fertilizers.[40] The farmers' situation was made worse, as already indicated in earlier chapters, when the government allowed the millers to pay farmers in 150 days rather than the previous 30 days.[41] The long wait contributed to the undercapitalization of the Asian sugar farmers, who, unlike other farmers, had continuous activities on their farm due to the ratoon crops and as such needed capital.

It was not easy for Asian farmers to get loans from the colonial government or banking institutions until late in the 1940s, when Asian-owned banks such as the Standard Bank of India and Bank of Baroda started operating in the colony. These became the main source of financing for the sugarcane farmers.[42] By the late 1940s cane growing had increased because of these institutions. However, because of the long history of undercapitalization, which affected the buying of land for expansion, mechanization, and the supply of labor, as well as the colonial state's neglect, recovery was slow.

DUMPING OF SUGAR

The problem of sugar imports from nearby countries was a significant problem for Asian farmers in Kenya. The problem was twofold: illegal importation, and legal importation that took advantage of Kenya's precarious sugar industry and led to what appears clearly as dumping of sugar in the colony. In particular, the Asian cane growers and manufacturers were constantly threatened by the incessant dumping of sugar into Kenya from Uganda, especially after South Africa withdrew from the Commonwealth due to pressure mounted on its apartheid system. Uganda took advantage as the sole supplier of sugar, and a lot of its sugar found its way into Kenya both legally and illegally. All this curtailed the efficiency of sugar production in colonial Nyanza due to interference with prices, as Ugandan sugar was cheaper, and this interfered with distribution. The acceptance of such practices by the colonial state curtailed the growth and attainment of full economic potential of the industry in CN by denying the local producers profits.

Uganda had a consistent surplus of sugar over consumption, and this contributed to the dumping of it in Kenya. During the period from 1955 to 1963, Uganda had production ranging between 10,000–50,000 tons while Kenya had a persistent shortfall over the same period.[43] This problem persisted because Uganda constantly overproduced, possibly due to the ready market in Kenya. As a result, the Miwani Sugar Mills protested to the Ministry of Commerce and Industry about the colonial government paying highly for Ugandan sugar, arguing that prices should be the same irrespective of the origin.[44] Dumping could also be attributed to lack of support by the colonial government that did not try to hinder or control the situation.

A meeting in 1955 well illustrates this problem. In the meeting held in the month of September, the Kenya government failed to stand firm and protect the cane producers in Kenya; the resolutions passed in the meeting, instead of being protective, were amended to favor the dumping of Uganda's surplus into Kenya and Tanganyika. The Uganda representative, W. Padley, categorically explained that he was negotiating on behalf of the Ugandan manufacturers and did not hesitate to lay his terms clearly that should the two territories not take up the surplus, he would tell the manufacturers to export to Sudan.[45] This was not unique as rarely did the colonial state come out openly to protect the Asian cane farmers. For example, in a previous CEC meeting on self-sufficiency, the Kenyan representative and member for agriculture and natural resources, Major F. W. Cavendish-Bentinck, did not challenge the small sugar quota allocated to Kenyan producers that limited their production and relegated the colony to be the main importer of Ugandan sugar. A prominent European settler himself, Cavendish-Bentinck was more interested in securing the wheat market for Europeans in the region than standing up for the

Asian cane farmers.[46] Such sentiments were also shared by another Kenyan representative, the civil servant K. W. S. Mackenzie, who reiterated that Kenya was anxious to import as much sugar as possible from Uganda.[47] Consequently, the European representatives failed to protect Asian cane growers at the meeting. It was no wonder that Mackenzie was so much disliked by Asian cane growers, who blamed him for the dumping of sugar. One farmer, Mr. Mohinder Singh, bitterly expressed it this way: *Mackenzi nene onego wa* (Mackenzie killed us).[48]

It seems that during the colonial period, many of the agricultural policies were tailored to favor European production and failed to protect the home industry as Uganda did. Furthermore, it seems that Kenya was hoodwinked into allowing the dumping of surplus sugar into the colony partly due to the desire to protect her wheat market and possibly due to lack of self-sufficiency, which the state was not eager to address, despite the potential. For example, in 1955–1956, Kenya became the dumping ground of most of Uganda's 9,800 ton surplus as indicated in table 6.1.

Table 6.1 shows that out of Uganda's surplus of 9,800 tons, 9,000 tons was dumped into Kenya in a span of only three months. In that same time period Tanganyika received only 800 tons.[49] Such dumping was outrageous and undermined the efforts of the Asian cane farmers in CN. During the same period, Kenya faced a deficit of approximately 5,500 tons.[50] The colony was only supposed to import the deficit but not allow excess so that Kenyan farmers would not have to compete with the cheap imported sugar, a situation that was already aggravated by the importation of sugar from Tanganyika, which started as early as 1936.[51]

The deficiency in Kenyan production was at least partly the result of the discriminatory policies against the Asian farmers on the part of the colonial state. The colonial state hindered machines and spare parts from being procured easily, which led to inadequate milling capacity. Furthermore, the colonial state was reluctant to start another sugar factory in the 1950s, even though the production of cane had outpaced the ability of the mills to process it. There was a need for another mill in the area as Miwani alone could not serve the farmers adequately. This being the

Table 6.1. Comparative Ugandan Exports to Kenya and Tanganyika, December 1955–February 1956

Year	Month	Uganda to Kenya (tons)	Uganda to Tanganyika (tons)
1955	December	2,000	400
1956	January	3,500	400
1956	February	3,500	—

Source: Extract from the Confidential Report on ECE meeting, 7th September 1955, KNA: BV/6/1704.

case, the colonial state could easily be blamed for deliberately allowing dumping of sugar in the colony to destroy the local industry that was Asian owned and financed.

LACK OF HARMONY: FARMERS AND MANUFACTURERS

If the dumping of sugar from neighboring countries was one major problem, another was the lack of cooperation between the two major players in sugar production: the Asian cane farmers and the millers (who were both European and Asian). The two sectors were at odds partly due to neglect and lack of support in supervision by the colonial state but also because of the discriminative policies that the farmers had to endure. This was because the Victoria Nyanza Sugar Company, being the only miller that took cane from farmers in CN from the mid-1920s, for some time enjoyed a virtual monopoly on milling and took advantage of the situation, supported by the colonial state. For example, around 1934 the Victoria Nyanza Sugar Company started rejecting a substantial amount of the cane that the Kibos Asian farmers had brought to them for milling. Moreover, the limited amount of cane that the mill did accept was bought at the low price of Shs. 6 per ton.[52] The constant low prices offered to cane farmers were a manifestation of capitalist exploitation where the millers' interest predominated. The mill manager, Mr. Abel, made it clear that his company was determined to grow for itself all the cane that the mill required for its production and consequently would no longer buy any cane from the farmers at the expiration of the farmers' contracts. He even suggested that the Asian cane farmers should have a license to mill their own cane, knowing very well that this could not happen due to the monopoly accorded to the company.[53] Moreover, this was contrary to the earlier position prior to 1931, when the same company had persuaded the cane planters to discontinue making jaggery and accepted all sugarcane that was offered.[54]

At that time in Nyanza the Victoria Nyanza Sugar Company was the only mill that took cane. Yet when the Asian cane farmers realized their plight and requested a license to start their own mill, the same Mr. Abel argued against it, saying that it was only a bluff to make him accept the Asian cane.[55] The problem was huge, and numerous communications took place between the DC and the PC about the issue. The problem was aggravated by the Government Notice No. 619 of 1934 that regulated the sale and distribution of sugar, since it restricted the sale of sugar and jaggery and so inhibited the disposal of cane. For example, out of approximately 45,000 tons of cane in 1934, the Victoria Nyanza mill was only prepared to take 12,000 tons.[56] The farmers were stuck with the rest, which could only be used in producing jaggery.

Another huge problem was quota allocation. In most countries the mill owners were not large-scale landowners, as was the case in colonial Nyanza; hence when a quota on a mill was imposed, it became a grower's quota. But in Kenya under the sale of sugar ordinance, the large land-owning millers would most likely absorb the whole quota from their own estates, leaving the Asian farmers with no outlet for their cane crop. Therefore the Asian farmers requested that either the growers be given a quota so that the mills could be forced to take up a portion of their cane or they be permitted to erect a mill and have their own mill quota. Before 1942 the Victoria Nyanza Sugar Company was the only mill in the region and the farmers were at the mercy of their pricing, regardless of the worldwide market prices of sugar. The quota system and the Victoria mill's monopoly was a disastrous financial combination for the farmers.

Unfortunately, sugar pricing policy has long been a problem in Kenya. This problem has continued even in independent Kenya, where government policy has remained inadequate. The low prices offered to farmers during the colonial period affected the supply of sugarcane to the factories, which in turn resulted in great underutilization of the factory capacity.[57] Moreover, pricing of sugar was critical to the industry and had to be set taking into consideration the prevailing world prices, which were based on Commonwealth Sugar Agreement Price.[58] The relationship between the internal price of sugar and the world market prices was a critical factor.[59] As such, the Asian cane farmers in the Nyanza sugar belt had genuine grievances in disposing of cane. The problem was grave and necessitated an investigation. Consequently, the DC of K-L, Hamilton-Ross, arranged for agricultural officers who visited Miwani and Muhoroni areas to assess the situation with an aim of reaching a fair solution to the Asian case.[60]

Unfortunately, there was no lasting solution, and the cane farmers kept on experiencing problems selling their cane. When Hindocha took over the Miwani mill in 1947, he installed more tramways that facilitated transportation of sugar, installed new machinery, and started to take almost all the cane from the farmers who were able to take it to the factory. This helped the farmers briefly, but even during this period, it emerged that Miwani cheated farmers of their cane by deliberately miscalculating the number of tons required to make a ton of sugar. Even the farmers were not blameless. For example, farmers belonging to the Kibos-Muhoroni Production Sub-Committee were found in 1949 to be holding up delivery of their cane to the Miwani Sugar Mills in anticipation of an increased price, though the price had not yet been increased.[61] Such activities explain the lack of trust and ineffective communication that existed between farmers and manufacturers.

By the early 1950s, Miwani started to frustrate farmers just like its predecessor, the Victoria Nyanza Sugar Company. Miwani refused to accept cane of those farmers who had no contract, something they never did before. The company also refused cane from those who complained that the company was underpaying farmers, and those who were tired of the company's arm twisting campaigned for the establishment of a second factory at Kibos.[62] The complaints against Miwani led to an investigation conducted by the AAO who ascertained that by 1953 the company had refused to accept cane from certain farmers who advocated for another factory, such as Milwant Singh and Makan Singh.[63] Yet it was only logical and proper for the farmers to request for another mill since Miwani had refused to accept cane from many farmers. Milwant Singh was forced to sell his farm out of such frustrations, though his farm was situated near the factory. The colonial state did nothing to reprimand Miwani or settle the dispute between the two parties.

Furthermore, tension emerged because the farmers felt that the manufacturers received a greater proportion of the subsidy, yet they did not always accept cane when the grower wanted to cut it.[64] On the other hand, the manufacturers had laid out eight miles of rail and added more trucks to their fleet, which involved a huge financial outlay; they claimed that the rail received as little as only three tons of cane per day from the farmers, which was not economical.[65] There was lack of cooperation between the two parties. This lack of harmony among the stakeholders was comparable to the colonial state's lack of support for the cane producers, for instances, the state's constant refusal to fix the price of cane.[66] To ease the tension between the two groups and ensure the success of the sugar industry, the colonial government could have fixed the price of cane, which would have benefited the cane growers and reduced tension between the two dissatisfied groups.

Needless to say, the insufficient number of mills impacted sugar production. Sugar mills were supposed to be within reach of the cane growers because the absence of sufficient mills prevented the potential sugar manufacturing regions from producing enough sugar for commercial transactions.[67] This affected the sugar industry in Kenya. The problems of insufficient mills and obsolete machinery persisted to the independence period in the early 1960s. It was only in 1966 that another mill, the East African Sugar Industries Limited, was opened as part of the post-independence sugar development scheme at Muhoroni.[68] Even then, the problem persisted because of the inability of the factories to synchronize cane supply from farmers with factory operations.

SUGARCANE DISEASES

Apart from the already mentioned problems, there were other major risks borne by cane growers such as diseases and fire. Sugarcane diseases were rampant in cane growing and required an urgent response to combat the infections. Most Asian cane farmers in CN had to contend with these various diseases starting as early as the late 1920s. For instance, mosaic disease was common and very serious since it attacked many varieties. Its outward characteristics included a mottling of leaves in the form of white blotches and streaks. The greatest effect it had on cane was that it reduced the yield, though the sucrose content of the juice was not much affected.

The first serious attack was noticed in 1923, which forced the colonial government to come up with regulations to control it.[69] This was done through the Disease of Plants Prevention Ordinance. The ordinance advocated for the planting of the more disease-resistant Uba variety within a prescribed period as a measure to eradicate the disease. The mosaic disease continued to draw the attention of the Department of Agriculture despite the efforts undertaken to eradicate it. For instance, the inspection report of 1927 indicated that it had spread and the area affected was larger than initially assumed.[70]

The problem of mosaic continued to bother the cane farmers in the following year, and a system of quarantine was put in place in 1928 to try to arrest its spread. Arrangements were made so that the new imported varieties meant for the experimental trials were first to be grown in quarantined plots set up under the close supervision of the Scott Agricultural Laboratories before they were released to farmers as disease-free.[71] The quarantine system made it easier for the imported cane to be inspected frequently and released to the farmers without the supervisor having to travel to the sugarcane plantations. However, it was difficult to find nurseries where suitable quarantine conditions could effectively be imposed.

The seriousness of the disease forced the mycologist, Dr. Storey of the East Africa Agricultural Research Station Amani, to visit the Asian sugarcane farms in Kibos, Miwani, Kibigori, Muhoroni, and the neighboring Koru and East Lumbwa. He carried out a campaign of thorough inspections.[72] Due to the persistent nature of the disease, farmers were advised to plough out the susceptible cane to control the disease and stop the infection of other varieties.[73] In addition, in 1929 the Department of Agriculture introduced improved varieties of cane that were more resistant to mosaic disease and had them distributed to farmers.[74] These varieties were POJ 2714, POJ 2725, POJ 2727, and POJ 2878.[75] Many farmers were encouraged to adopt the new varieties, which were immediately taken up by large-scale farmers, probably because financially they were able to embrace the changes faster. Toward the end of the year, there was a substantial reduction of the disease in the district.

Despite the efforts over the years to eliminate it from sugarcane in the district, mosaic reappeared again in 1931 as a result of action taken by certain Asians in the Kibos-Miwani area who imported from Uganda a variety known as *Kampala*, which was a non-resistant cane.[76] The eradication effort required Mr. Naismith Jones, the plant inspector, to reside in the Songhor area for most of 1931 to keep a close check on mosaic. However, the threat continued, for in 1932, mosaic attacked many farms in the Kibigori area. In response, a total ban on the importation of any varieties from the neighboring African reserves was enacted. The senior mycologist arranged for information about the risks and preventive measures to be distributed to the whole province through agricultural offices. The notices were also displayed in administrative offices, post offices, and the railway stations, and some were sent to the farmers' association.[77] The colonial government acted quickly to contain the spread of the disease by enforcing the Government Notice No. 754 of November 1932 that prohibited the planting of *Kampala* and other soft canes imported from the African reserves.[78] It also recommended that farmers remove the stool of the crop affected as a control measure, something that farmers were reluctant to do. But the colonial government believed that the benefits of destroying the Kampala cane outweighed the disadvantages.

Complete eradication of mosaic was not attained easily, and it continued to bother farmers. For example, by mid-1953, an extremely high infection rate of the mosaic virus was reported at Kisumu and Miwani farms. The infection was so severe that the much preferred variety of cane (the CO 281), which had been resistant to infections at Miwani, was also affected and transmitted the disease to other varieties.[79]

Another disease that threatened the production of sugar was the chlorotic streak disease, also known as ratoon stunting, due to its effects. The disease was characterized by the presence of a yellowish streak found in the veins on the leaf. The disease made the leaf stiff and the plants appear more erect than the healthy ones. In severe cases, chlorotic streak caused considerable stunting infection on the stool.[80] This disease was suspected to have been brought by the cane imported from Natal. By 1954, it was found also to have interfered with the trials that were taking place in some of the varieties, especially when the rainfall was not adequate. The disease continued to be a menace to farmers until the mid-1950s when a comprehensive scheme for the eradication of the disease was put into effect.[81] It was recommended that no more materials were to be imported from Natal. Due to the seriousness of the disease, the senior plant pathologist was of the opinion that a study under glasshouse conditions had to be undertaken immediately to find ways to eradicate the disease.[82] The findings established that a thermostatic bath treatment was the best solution for all planting materials. Toward such an initiative, the Department of Agriculture imported such a bath and had it installed at Kibos where supplies of seedlings were treated, while the large-scale farmers and sugar estates ordered large tanks meant for the same treatment.[83] Farmers were also advised to stop the spread of the disease by treating cuttings to be used in planting in hot water at 50 degrees Celsius for two hours.

While these measures proved effective, smut disease also proved to be problematic to farmers. This was a fungal disease whose effective treatment entailed uprooting the whole crop, which the small-scale farmers found problematic. The disease was noted by Sampura Singh to be prevalent in the B 35345 and CO 4621 varieties, and it was believed to have originated in Uganda.[84]

Several other diseases forced the destruction of sugarcane in colonial Nyanza over the years. These reduced the yield and also reduced the sucrose content of cane, which meant that a lot more cane had to be used to make sugar. The colonial state supported the sugarcane farmers to fight the diseases. These setbacks were overcome through the combined efforts of the plant pathologists and senior mycologists who worked closely with the cane farmers. This was a rare case where the colonial state stepped in and took measures to support the entire sugar industry, and not just the European farmers and mills.

MACHINERY AND MAINTENANCE

The cane farmers and the manufacturers alike also faced problems in the procurement of machinery needed in the production of cane. The inability to purchase machinery was due to both financial constraints and the colonial bureaucracy. Earlier we had seen that by the 1920s most of the small-scale cane farmers could not afford tractors and so they used ox-drawn ploughs for many years.[85] Even for those who could later afford them, it was not easy to get the machines. For example, Amar Singh, who owned 500 total acres at Kibos and Miwani, applied in 1943 for a tractor, and by the time his farm was visited by D. O. Magor in 1947, he was still waiting for one.[86]

Moreover, the tractors were expensive for individual farmers to purchase. Later they organized and started using tractors from the machinery pool collectively, but this was not as efficient as individual ownerships. Farmers had to wait patiently for their turn and yet needed the tractors before the onset of the rains since sugarcane farming in colonial Nyanza was rain-fed. It was not until 1949 that the colonial government allocated six tractors to be used by fifty-six farmers, which were far from being adequate. Furthermore, the farmers could not afford the harvesting machines and so cutting of cane was done manually, which was slow, dangerous, and difficult. The undercapitalization of the Asian cane farmers made it difficult for them to afford machines that could have brought efficiency to the industry.

Neither was it easy to get machinery for milling sugar. The colonial government refused Asian cane growers the opportunity to import machinery needed to start their own mill in the mid-1930s. To make matters worse, by 1936 there were restrictions placed against the importation of sugar machinery. This was as result of worldwide limitation of sugar.[87] These restrictions did not allow farmers to produce to the maximum capacity and were not viable since there were not enough mills to cater for the cane farmers. What was interesting was that whereas the colonial government in Kenya ignored the complaints against the limitations by the cane producers in the colony, the same regulations were challenged by the government of Uganda.

Moreover, in 1942 when the Muhoroni sugar operation was taken over by Asian miller, Premchard Raichad and Company, the new owners asked the colonial government for help in procuring two locomotives. Although the colonial government granted the request, since it recognized that agricultural machinery was badly needed to improve agriculture in the colony, the supplier based in the United States refused to release them on grounds that they could be intercepted by enemies who could remodel them for military purposes.[88] However, the machinery was supplied to the European-owned Victoria Nyanza Company. Such hurdles continued to hamper the Asian farmers and entrepreneurs. Most of the factories faced problems of old machinery that needed repairs and parts had to be brought in from overseas. For example, in 1952 the Miwani factory had to suspend its operations due to the annual overhaul of machinery, which delayed the cutting of the mature cane.[89] By early 1953 the factory had not started receiving cane from the farmers who wanted to dispose of their cane; hence farmers asked the director of trade and supplies for a permit to be allowed to manufacture jaggery instead of their cane going to waste.[90]

The lack of machinery and spare parts was a significant problem that persisted throughout the colonial period and into the independence period. Unfortunately, machinery determined the acreage put under cane. Even those who managed to get machinery faced the problem of how to get spare parts. For instance, it was found that a number of tractors in the Kibigori area were not in working condition by 1947 because they needed spare parts. Furthermore, it was found that the spare parts work done by Gailey and Roberts Company was not satisfactory.[91] Many farmers were exploited by engineering firms because of government's lack of supervision, despite the fact that the machinery and parts shortage resulted in cultivatable land lying fallow, especially among the small-scale farmers.[92]

Obsolete machinery in factories also posed a problem as it led to the proliferation of over-matured cane on farms. It took many months to repair the machinery if it was left for a long period without maintenance. For instance, the Miwani Sugar Mills suspended its operations in May 1952, and by April of the following year the factory had not yet started to receive cane from farmers.[93] In the meantime the farmers had to wait as they watched their cane over-mature in the farms. The crops not only deteriorated in milling quality, but it also prolonged the crop cycle. All these problems were partly due to the undercapitalization of the industry as the state refused to support the building of another factory. A second factory with almost the same crushing capacity was necessary, especially during the annual overhaul of machinery at Miwani, so as to continue receiving the cane and avoid wastage. Such shortcomings and lack of appreciation of the economic importance of the cane industry proved to be expensive for the cane farmers and producers as well leading to inadequate production of sugar in colonial Kenya.

IMMIGRATION PROBLEMS

Asian cane farmers also struggled to get skilled manpower to work in the farms and especially the factory. The agricultural sector formed the very backbone of Kenya's economy, and the sugarcane industry needed skilled laborers to progress. However, the immigration of skilled workers from Asia was blocked by European settlers who feared an influx of the Asians into the colony and to more agricultural settlements even outside the lowlands, something that they deeply resisted.[94] This belief was deeply entrenched within the settler society and was expressed by Lord Delamere, one of the leading settler politicians, who explained their position: "In all new countries the backbone of the country is the small man, the white colonist with small means, but there is no place for him in a country when once the Asiatic is there . . . all the vegetable growing for the towns is done by Indians, all the butchers with one or two exceptions are Indians, all small country stores are kept by Indians and most town shops, all lower grade clerks are Indians, nearly all the carpentering and building is done by Indians. They fill all occupations and trade which would give employment to the poorer white colonist."[95]

Such sentiments showed how the Europeans feared competition from the Asians. It was that fear that made the settlers strive to keep the Asian population in the protectorate at the minimum.[96] Despite such hurdles, the Asian cane producers still made efforts to get the needed skilled workers from their home countries. These included artisans, accountants, technicians, and agriculturalists, among many other professions. However, the colonial state, as early as 1919, raised obstacles to Asians coming to the colony and dubiously claimed that the restriction of Asian immigration was meant to protect African interests.

Unfortunately, this view was shared by the officials at the Colonial Office in London, except for L.S. Amery who was a parliamentary undersecretary of state after World War I and secretary of state from 1924 to 1929. He believed that the Asian agriculturalist had something to offer.[97] Amery believed that restricting Asian immigration would only hamper development of the colony. This attitude was not shared by the officials of the colonial state, though, and thus it was not easy to bring a relation or workers from India to help with the farming. Many rules and regulations were put in place that made it very difficult, and these difficulties revealed how the Asian cane farmers, right from the infancy of the industry, had to face discrimination.

Similar immigration issues kept resurfacing throughout the colonial period. The strong position assumed by European settlers at the end of World War II, for example, led to the Immigration (Control) Ordinance of 1946, which placed permanent restrictions on immigration.[98] This bill targeted Asian immigrants and did not apply to European settlers. Although the bill was non-racial in context, it was discriminatory in conception and effect.[99] The Asians protested the bill, but after two years of protracted struggle, the Immigration (Control) Ordinance was passed into law. With this law in effect, the colonial state made it very difficult for the Asian farmers to get skilled labor into the country.

The new law set up a system of entry passes for immigrants. There were three types of passes: First there was the Temporary Entry Pass (TE Pass), which was granted for only two years. This was relatively unhelpful to the farmers since the growing cycle of sugarcane ranged between five and six years.[100] Second was the Entry Permit and third the Dependents Pass for family members of needed staff. This discrimination toward Asian workers was also seen when by 1951 the Labor secretary of state, Arthur Creech Jones, and the Colonial Office supported the colonial state and stated that the exclusion of Asians from jobs was important because it was only the British who were charged with transmission of the tenets of Western civilization.[101] So the colonial state found it prudent not to restrict the European immigrants while controlling the Asian inflow into the colony on the pretext of spreading "civilization."

Therefore, it was not easy for Asian workers or their families to be granted the immigration passes or permits. There were many refusals, especially for the issuance of the TE Pass for qualified and experienced farm and factory workers, which were so needed in the industry. This was a huge problem and Miwani in particular called attention to this problem in the early 1950s. It was not only the denial that was challenged but also the unreasonable conditions set for Asian immigrant workers. For example, there was the notorious condition that a TE Pass could only be issued when the dependents of the applicant, such as his family members, were not granted a pass.[102] Miwani petitioned the colonial state in 1952, when their estate doctor, Apabhai R. Patel, was denied the privilege of having his family come with him. Similarly, when an offer of employment as Miwani's assistant plantation manager was made to an individual, someone who had trained in agriculture at the University of Bombay and had ten years of experience in the sugar industry in Uganda, the man was ordered by the colonial government to send his family away to India.[103]

The immigration policies put in place affected farming as it denied the Asian sugarcane farmers and manufacturers the much-needed workers, especially the factory skilled manpower. On the other hand, some of the

Asian immigrants were found to flout the regulations and engage in other activities not initially intended when they applied for entry, and this complicated the matter. For example, in 1953, Milwant Singh, an immigrant who had entered Kenya on a permit to run his father's farm, was found to be working for another company, Ingersoll Rand in Kisumu. Milwant also had applied for a permit for Sikandar Singh to come and manage his father's land.[104] Changing jobs like this was illegal, but only for Asian immigrants. The immigration laws were not uniform, and no one bothered with what economic activities the Europeans tried their hands at.

The immigration problem persisted throughout the colonial period and became a thorny issue among the Asians at large. The cane farmers and manufacturers continued to struggle with lack of skilled labor, especially in the factory where technicians, chemists, and engineers were needed. The inadequacy was deliberately perpetrated by the colonial state, which failed to recognize the importance of the sugar industry. It's no wonder that the state employed only one Asian sugar specialist, Sampura Singh, who had to assist more than eighty Asian cane farmers, while at the same time denied other qualified personnel entrance into the colony.[105] These immigration regulations were exclusively meant for the Asian growers and millers, as they did not apply to European farmers, and they hindered the progress of cane growing in CN.

MINOR CHALLENGES

Although the cane farmers had major problems to contend with, they experienced also minor problems like the monkey menace, theft of cane, and fire outbreaks. Monkeys and baboons destroyed cane, especially the soft varieties. This was witnessed when by mid-1950, the DC was forced to intervene and help the farmers, especially after massive destruction was reported at Miwani farms. The DC was forced to solicit the help of the local honorary game wardens who assisted in the control of the monkeys and the baboons at Miwani.[106]

Other problems such as theft and fire also existed. There were cases when cane was stolen from the farms by the local people for consumption; in other instances it was the workers who just took cane to chew, even though there were regulations on different farms that required workers not to chew cane. This occurred mostly in farms where the employers were strict and where the food ration was bad.[107] The fires were in most cases accidental, but it was feared that there were instances when they were set by dissatisfied workers, probably due to wage contestation.

These minor problems joined the other more serious ones such as discriminatory colonial policies, overregulation, ecological challenges, and the undercapitalization of the Asian farmers; and each one compounded the effect of the others. For instance, the cane farmers often did not have enough money to transport cane to the factory at Miwani (undercapitalization) and yet were barred from constructing a second mill (discriminatory policies). Despite the challenges, the sugar industry survived partly because the Asian cane farmers and manufacturers realized that the problems could not be left to the colonial state to solve alone. The Asian cane farmers knew that if they wanted to overcome these obstacles, they would have to devise means to cope with the challenges themselves.

THE ASIAN CANE FARMERS' INITIATIVES TOWARD MEETING THE CHALLENGES

The Asian cane farmers in colonial CN faced diverse problems—the non-acceptance of sugarcane by millers, the need for another milling plant, and many others. To help address these problems, the farmers organized themselves into various organizations, including the Nyanza Indian Farmers' Cooperative Society (also known as the NFCS), Agricultural Production Sub-Committees, and the Nyanza Farmers' Association (also known as the Nyanza Indian Farmers Association or NIFA). They also got support from the EAINC. These groups each tackled problems within their area of specialty. The cooperative society handled the marketing issues while the EAINC was more concerned with injustices and policy issues that affected the Asian community in Kenya. Their work complemented each other and they never worked in opposition. This cooperation by the various groups was key to finding solutions to some of the problems and facilitated the growth of the Asian sugar industry in colonial Nyanza.

The Asian farmers also organized themselves into self-help groups to solve some of the problems collectively. For example, in 1930 the Asian farmers collectively bought land belonging to Walji and sons and subdivided it among themselves, as it was easier to buy land as a group than as individuals. This collective action was effective.[108] The same method was used to purchase Major Carnegies' land, which was expensive. In these cases, pooling resources together made economic sense. Help also came from the Asian leaders. This was seen when in 1934 the Asian cane farmers experienced disposal problems; their representative in the LegCo, Shams ud Deen, led a farmers' delegation to present their grievances to Governor Sir Joseph Byrne, which was fruitful as they were granted their request.

The colonial state also offered assistance to the farmers and employed a sugar specialist, Sampura Singh, who not only gave technical advice to farmers but also carried out experimentation that solved some of the problems the farmers faced. Moreover, the colonial state, as demonstrated earlier, facilitated the experimentation on different cane varieties. This became crucial because it determined the best variety for the weather conditions of the area but also for resisting disease, as was found among the POJ varieties.[109]

The help accorded to the Asian cane farmers through the intensification of trials and experimentation contributed to the success of the sugar industry. Nevertheless, such support from the colonial state was rare. It was mostly these individual organizations formed out of necessity that helped the Asian cane farmers and manufacturers boost the sugarcane industry.

THE NYANZA FARMERS' COOPERATIVE SOCIETY (NFCS)

The Nyanza Farmers' Cooperative Society was formed in January 1936 by the Kibos Asian farmers; the society's office was based in Kisumu and the organization was composed of mainly cane farmers, as earlier stated.[110] The cooperative was formed when the Victoria Nyanza Sugar Company had started to refuse cane from the farmers and was mainly concerned with the sale of jaggery. All members sold their produce through the society, which was granted a monopoly on selling jaggery in the country and also for exportation. The society received a commission for the produce sold through it. Hence the cooperative mainly dealt with the sale of jaggery in bulk and managed to negotiate for better prices.[111] For example, before the formation of the society, the price of jaggery was low, selling at 65 cents per frasila.[112] After the formation of the society, jaggery was sold at Shs. 3 per frasila. The society even found new markets and arranged for the export of about 1,000 tons of jaggery annually to the United Kingdom, India, and Uganda.[113] The society not only sold the members' produce but also provided them with farm implements. As such, the farmers felt encouraged and increased the acreage under cane. In 1941, the cooperative made headway and sold low-grade no. 2 brown sugar to the United Kingdom where it had a market.[114] The cooperative was also concerned with the grading of the jaggery so as to improve the quality for export.

Other initiatives were seen when by 1959 many Africans had entered the sugar industry in Nyanza. The Asian growers started feeling competition from African producers. This competition was exploited by local merchants in Kisumu who formed a ring to exploit the two groups in the marketing of jaggery. However, this trend was exposed by the NFCS which recommended to the colonial administrators a joint scheme of marketing jaggery with African cane producers. Unfortunately the administrators rejected this, claiming that the Sugar Board only handled African-grown produce.[115] The suggestion of the farmers seemed viable but was not considered because the colonial state seemed in most cases not interested enough in the industry to have bothered to solve the problem.

Although the NFCS operated under a very unreceptive government; it managed to remain steadfast and championed the farmers' cause right up to the independence period.

THE AGRICULTURAL PRODUCTION SUB-COMMITTEES

The Asian cane farmers also organized themselves into various production sub-committees in their respective areas so as to help them in their farming endeavors. These sub-committees were formed in 1947 but officially registered in October 1949. They were the Kibos-Muhoroni Production Sub-committee chaired by Sham Singh; the Muhoroni-Chemelil Production Sub-committee chaired by Bhanji Valji; and the Miwani-Kibigori Production Sub-Committee chaired by M. U. Patel.[116] These organizations were active. For example, the Kibos-Muhoroni Production Sub-Committee, right from its inception in 1947, met with the Miwani Sugar Mills when Hindocha took over and tried to negotiate a new price for sugarcane on behalf of the cane growers. During the meeting, Hindocha arrived at the conclusion that since the sugar price then was Shs. 400 per ton minus excise duty, the company was ready to offer Shs. 10.40 per ton for hard cane.[117] Unfortunately, the farmers insisted on Shs. 17.50 per ton and insisted that they would continue to produce jaggery if their demands were not met.[118]

The Kibos-Muhoroni Production Sub-committee was relentless in serving its members, for immediately after meeting the Miwani ownership, they also organized and met the company management with colonial officials such as W. Lyne Watt and C. F. Atkins. They deliberated on the issue of fixing the price of sugarcane. The cane farmers' spokesman was Mr. Bresler (a European cane farmer) who insisted that the price offered for cane was low.[119] During the negotiations, the Asian cane growers made their views known. There was representation of the fifty-five Asian cane farmers who also manufactured jaggery on their farms, the Muhoroni Sugar Company, and Messrs. Bresler and Starr. In the discussion it became clear that there was fear from some of the administrators that the price of cane had to be fixed and that the solution was not the control of sale of jaggery, which if effected would still lead to a black market in the product.[120] Moreover, the Kibos-Muhoroni Sub-committee challenged the Shs. 40 per acre grant that was offered by the state for breaking land. The sub-committee argued that it was insufficient since the rate for ploughing three times and harrowing twice was approximately Shs. 72 per acre.[121]

The ceaseless efforts of the organization were witnessed again in 1950 when the Kibos-Muhoroni Production Sub-committee asked the Central

Commodity Distribution Board to give them permits to allow the farmers to export jaggery to Uganda and Tanganyika or to the Persian Gulf where there was a market. Unfortunately the chairman of the Central Commodity Distribution Board rejected the request on the grounds that if jaggery was exported then it would encourage the production of it at the expense of white sugar and that most farmers would take very little cane to the factory for sugar.[122]

Furthermore, in 1953 the sub-committee pushed for the GMR, an insurance program to protect farmers from threats to their crops, to be applied to the sugarcane crop as it was for other crops. When the GMR program was started by the colonial state, cane was not included on the grounds that it was not an annual crop like maize. However, the cane farmers argued that cane, just like other crops in the district, suffered from adversities such as locusts and drought; hence the cane farmers wanted the GMR applied due to the risks that they bore.[123] This was necessary because the Asian cane farmers had suffered considerably from losses caused by drought, locusts, and accidental fires on individual farms. These negatively affected their activities since they had no insurance like that which was accorded to maize farmers.[124]

During the same period, the Chemelil-Muhoroni Production Sub-committee also voiced opposition against the sugar legislation that controlled production of jaggery. The committee argued that, by 1953, most of its members were producing jaggery and took little cane to the factory due to the long distance to Miwani; hence farmers were anxious that the legislation could put a stop to jaggery production.[125] In the meantime, the Miwani-Kibigori Production Sub-committee also protested and boycotted a meeting called by the Miwani Sugar Mills to discuss the legislation that barred the production of jaggery.[126] It seems that the farmers did not want to put themselves in a situation where they had to take their cane to the factory with no other option because in the past they had been exploited.

Despite the efforts, the activities of the production sub-committees were marred by divisions that hindered their proper functioning. The division was based on tension and class conflict between large-scale farmers and small-scale Asian farmers who felt they were taken advantage of. For example, the Chemelil-Muhoroni Production Sub-committee had internal divisions that threatened its functions. Most of the discontent came from the small-scale farmers who rejected domination of the official positions held by large-scale farmers. In some production sub-committees, the bone of contention was the fact that some officials came from even outside the sub-committee's zone. For instance, some members of Chemelil-Muhoroni Production Sub-committee challenged the election of G. M. Vasani as their chairman because Vasani came from the Muhoroni-Songhor area. Moreover, he was a sisal farmer, representing the Muhoroni Sisal Company. Also the election of N. M. Mehta as an official in the same production body was challenged since Mehta was also a sisal farmer from Volo Sisal Limited in Kibigori.[127]

The muddle in these elections was partly due to the colonial administration who, though present, did not advise the members accordingly because of their lack of concern with the industry. However, some government officials took responsibility. For instance, R. E. Wainwright, the then DC, lamented that due to the muddle, the two successful sisal farmers were left to hold positions that could have rightly been held by cane farmers.[128] Part of the blame also lay at the feet of the cane farmers themselves who elected officials simply on the basis of their ability to speak English and the influence that they allegedly wielded. Some members believed that this would enhance their bargaining power.[129] Despite the internal disagreements, the production sub-committees were fully involved in farmers' affairs and did all they could to improve sugarcane farming in their respective areas by championing the growers' cause.

THE NYANZA INDIAN FARMERS' ASSOCIATION (NIFA)

The NIFA (sometimes known as the Nyanza Farmers' Association) also worked tirelessly for farmers. In 1934 the association sent a petition to the DC of K-L where they addressed the refusal of their cane by the Victoria Nyanza Sugar Company, though they had an agreement with the company to supply it with cane.[130] The association also objected to the long payment period of 150 days that the millers had for paying the farmers and lobbied to return to the previous payment period of 30 days. Furthermore, they demanded that the Asian cane farmers be given adequate representation of their interests in the agency that dealt with distribution and that the staff should not be confined to a single race, in this case, Europeans.

The association further presented their request to Sir Joseph Byrne for a license to import machinery and erect a mill in 1936, as discussed earlier. The farmers felt that there was a dire need for another mill in order to process their cane and also get a fair share of the sugar quota allocation.[131] The request was granted on the condition that the machinery could not exceed the capacity of crushing 5,000 tons of sugar per annum.[132] Also during the same period the association had come up with a modality that would allow all the Asian cane growers in the area to deliver cane to the proposed mill on equal terms so as to avoid the problems witnessed with the Victoria Nyanza Sugar Company.[133]

The approval of a license to important machinery was a major victory for the association. However, the victory was short-lived when the director of agriculture suddenly shortened the license from a period of three years to three months.[134] This was a devastating blow since the farmers were still in the process of raising funds. The farmers were frustrated and the association decided to act and sent a petition through the acting colonial secretary to the governor in 1937, requesting that the colonial state give the cane farmers an assurance that the license that had been awarded to Kibos farmers to erect a mill for a period of three years would not be revoked.[135] In their petition, NIFA explained clearly that the delay in getting funds soon enough to import the machinery was due to ravages caused by locust visitations and the drought that followed which had made the farmer's financial situation difficult.[136] The request was eventually granted though the plan never materialized.

The association also made a request during the same period that funds be made available for upgrading the road that served the area between Kibos-Chemelil and Miwani, where most of the Asian cane farmers were situated, so as to ease the transport problems of the cane farmers.[137] This road was eventually upgraded since funds had already been earmarked for it, but the colonial state had been dragging its feet. The association also asked the colonial state to lower the railway freight charges on jaggery that was destined for export; the request was turned down on the grounds that jaggery was not one of the essential industries.[138]

Despite the difficulties in dealing with the state, the group still stood up against what they felt was not justified and in 1946 the NIFA challenged the terms of contract agreement with the Victoria Nyanza Sugar Company. They challenged the decision of the factory to close the mill without prior information to the farmers and demanded that the farmers be given at least one week's notice of any stoppages for repairs or annual overhauls. This was important since farmers would plan not to cut cane during the said period. Moreover, they requested that the agreements be translated into Hindustani for the benefit of farmers who did not understand English.[139]

The NIFA was relentless in its pursuit to help the farmers; this was seen in 1950 when the DC for CN placed a restriction on the sale of jaggery within the district. This led to a pile-up of jaggery stock that needed an outlet. The association came up with a plan to export jaggery by use of dhows to the Arabian countries where there was an acute shortage and a ready market in addition to exporting to the United Kingdom.[140] The farmers argued that it was unrealistic for the colonial state to insist that every stick of sugarcane be converted into white sugar, especially given the lack of cost-effective transport from the Asian farms to the milling factories.[141] Furthermore, the small-scale farmers had scattered patches of cane fields that the millers found uneconomical to offer transport from, and so transportation for that cane was basically left for the farmers.

The association worked with other groups like the East African Indian National Congress (EAINC) since the farmers' predicaments affected the entire Asian community.

THE EAST AFRICA INDIAN NATIONAL CONGRESS

The Congress, formed in March 1908 at Mombasa, was modeled on the Congress of India as a response to the colonial state policy that laid emphasis on European interests. Though formed specifically to address political issues and bring together the Asian leaders of the three East African territories, it also served as a forum to express opinions and present petitions to the colonial government, the Colonial Office, and the government of India.[142] The congress was also concerned about the economic affairs of Asians in the three colonies.[143] Some of the interests included the colonial land policy. The congress was active from its inception and in 1910 passed comprehensive resolutions on Asians in the colonies and specifically attacked the segregation policy in the white highlands.

The congress also intervened to assist the Asian cane farmers in CN whenever they felt things were not right. For example, as early as 1931 the congress appealed to the colonial state to help the nascent sugar industry by reducing the railway freight charges so as to encourage the farmers. During the same year, when the cane farmers were affected by the locusts that destroyed the sugar crop, the congress asked the colonial government to help the farmers and to put in place a protective duty which by 1931 was already enjoyed by the European manufacturer, the Victoria Nyanza Sugar Company.[144] The colonial government was reluctant to commit any funding to help the farmers.

During the same period, the EAINC also asked the colonial state to investigate and fix the price of sugarcane because the prices offered were far below the cost of growing the cane.[145] But even such a request was ignored. Unfortunately, before the farmers could recover from the effects of locust visitation, there followed the prolonged drought of 1933, and the loss was overwhelming. Many farmers were frustrated and some even contemplated switching to other crops. However changing crops was not easy. They had to apply to the local agricultural credit board.

Due to these problems, the EAINC sent a petition to the colonial secretary summarizing the bad situation of the Asian farmers at Kibos-Muhoroni. The congress suggested that the colonial government bail out the farmers by providing a subsidy or loan to help the farmers to replant.[146] The colonial state adamantly refused to oblige on grounds that the loan had no security. As always, the colonial response was tainted with racial bias. For example, the director of agriculture suggested that the only remedy that the farmers had was to persuade some Asian capitalist to erect a small factory.[147] One wonders what a factory had to do with locusts' visitation. The Asian farmers were aware of this bias, and as the racial tension continued to build up, the congress battled against the anti-Asian attitudes. For instance, the Europeans had misled the Africans into believing that the Asian community wanted economic domination and should be resisted at all costs.[148]

The Asian farmers' situation was further complicated in 1947 by the independence of India and the subsequent bloody partition of India and Pakistan. The partition unleashed bitter disagreement among Hindus, Muslims, and Sikhs in various organizations in Kenya. For example, in November 1946, Shams ud Deen, a Muslim and once a staunch supporter of the Asian cane growers, not only resigned from the LegCo but also from Asian organizations due to disillusionment with the division between Muslims and Hindus.[149] The congress was threatened with division among the members and focused more on politics, and so there was not much activity in helping the cane farmers from the early 1950s.

Due to such divisions in 1952, the congress membership became limited to Kenya where most of the leaders resided. It was renamed the Kenya Indian Congress and it worked with affiliated Asian associations in Nairobi, Mombasa, Nakuru, and Kisumu.[150] Consequently by the mid-1950s, the Kenya congress was focused on politics, and information about its support for Asian cane farmers during the period is scanty. Nonetheless, it managed to forge solidarity among the Asian immigrants that was much needed for the Asian economy to survive. However, the desire to succeed in farming saw the rise of other local organizations such as the Kibos Planters' Association, which also stepped up to help the cane farmers.

KIBOS PLANTERS' ASSOCIATION

The Kibos Planters' Association, formed in the early 1930s, was a very active advocacy group. For example, in 1934 the association challenged the proposed bill Government Notice No. 619, which was intended to regulate the sale and distribution of sugar.[151] In particular the association opposed the fact that the sale of jaggery was granted to only one individual. It also challenged the colonial government to develop cane pricing procedures that would provide a specific return to both cane growers and manufacturers alike.[152] But the colonial state kept dragging its feet. It did not appreciate the economic importance of the sugar industry. The proposed bill appeared to cripple the prospects and aspirations of Asian cane farmers and protect the European manufacturers by allowing the monopolists to reap enormous profits at the expense of the Asian cane farmers.

Moreover, when the Kibos Planters' Association realized the seriousness of the cane transportation problem, the group asked the Miwani Sugar Mills to send lorries to help farmers out. Unfortunately, Miwani rejected the proposal, citing financial constraints, and the problem continued.[153] From their experience negotiating with both the manufacturers and the state, the group realized that to have a strong bargaining front they needed the support of all the cane farmers. To attain this, the planters association first sought solidarity; in a meeting in 1953, they unanimously agreed that there should be compulsory registration for all sugarcane growers so as to attain a united front necessary for bargaining, which they were able to achieve.[154]

Other concerns that the Kibos Planters' Association addressed included pricing and cane transportation. They advocated that before any prices were arrived at, it was necessary for all stakeholders to be involved in the discussion. Up till then both the farmers and the manufacturers had been largely ignored by the colonial state during the pricing process. Moreover, the two parties rarely agreed on prices of sugarcane. Therefore, to ensure that they were not taken advantage of, the members agreed to resist any efforts by manufacturers that could divide them with the purpose of exploiting them. This was important because earlier on Asian cane growers were exploited by the European and Asian manufacturers alike as the millers occasionally refused to take cane whenever the farmers wanted and also offered low prices for cane.

The Kibos Planters' Association, on behalf of growers, also challenged some of the regulations imposed by the state. For example, when the legislation to control the production of jaggery was passed, they petitioned the state to relax the regulations since altering the jaggery plants meant incurring losses. In 1953, they also asked for ample time to improve the jaggery plants that were found to be unhygienic. Due to the prolonged drought that impacted harvest, they did not have enough capital then to embark on the improvement immediately. The request was granted due to the efforts of the DAO who saw the rationale of their arguments.[155] The request shows how undercapitalization affected the cane farmers since they could not maintain their machinery as required.

Although the Asian farmers in colonial Kenya faced numerous problems, they remained resilient and continued to struggle to survive within a hostile colonial economy that required them to take the initiative into their own hands. They made sugarcane farming a success through the formation of the Kibos Farmers Association and other groups as a coping strategy. Kibos Planters Association was involved with the activities of the farmers within its locality and contributed to the success of the industry.

CONCLUSION

The ecological conditions of the Kibos-Muhoroni region presented some of the most daunting challenges of all to the Asian cane farmers. The weather alternated between droughts and excessive rainfall. Irrigation and soil conditions were difficult. And the threat of crop diseases was continual.

The farmers' ability to counter these threats was severely hampered by their undercapitalization. Discriminatory policies kept their profit margins low, and the government and European banks would not extend loans to help them rebound from catastrophes. In addition, there were continued obstacles to purchasing additional land.

The farmers were able to solve some problems. For example, they were able to market jaggery successfully through the NFCS and even negotiated for better jaggery prices. And they increasingly responded by forming advocacy groups and associations to represent their interests to the colonial government. And in some cases the government was helpful. For instance, the problem of diseases was contained due to the cooperation of the cane growers and the quick response by the colonial state. Though most of the diseases were not completely eradicated, they were brought under control. In other cases, their advocacy efforts failed, as was the case with the dumping of cheap Ugandan sugar on the Kenyan market.

The farmers also formed economic cooperatives that would allow them to collectively purchase land when available or rent equipment. One constant problem was the disposal of harvested cane when it was rejected by the mills. However, the farmers were able to deal with the problem through collectively withdrawing from the delivery of cane to the mills and instead producing jaggery on their own. This not only saved losses but also forced the manufacturers to accept cane. Other problems, such as the difficulty in buying tractors to break land, were overcome through pooling resources and using them on rotational basis. The undercapitalization was also ameliorated when several Asian banks started operating in the region by the mid-1940s.

The collectives and increased capitalization did not solve all their problems, as the costs of production continued to rise and in most cases the prices offered for cane did not make up for the increased cost. But despite the challenges faced by the cane farmers, they had a clear view of what to do under the circumstances. They did not give up but resolved to help themselves wherever possible. Through their organizations and collective efforts, they were able to counter many of the obstacles that presented themselves. Their determination and resilience saw them overcome many challenges, and their efforts allowed the sugar industry to remain vibrant during the colonial period.

NOTES

1. KNA: DC/KSM/1/19/270, H.R. Bhatt to Chairman, Miwani-Kibigori Production Board, June 17, 1953.
2. KNA: NZA/3/2/34, NIFA to DC K-L, October 11, 1934.

3. KNA: PC/NZA/3/2/34, DC K-L to PC Nyanza, February 27, 1936.
4. KNA: BV/6/612, Chief Secretary to Director of Agriculture, October 9, 1941.
5. *East African Standard*, May 9, 1947.
6. KNA: BV/6/1704, Confidential Memorandum on the Sugar Industry from Ag. Administrator of East Africa High Commission to Member for Agriculture and Natural Resources, March 3, 1954.
7. KNA: BV/6/1704, Confidential Memorandum.
8. KNA: BV/6/1704, Confidential Memorandum.
9. KNA: AE/3/925, Report for the Committee of Economic Coordination for East African Sugar Industry, December 17, 1952.
10. KNA: AE/3/916, Report for the East Africa Sugar Manufacturers Association on Quality of Sugar in East Africa, March 5, 1960.
11. KNA: BV/125/10, Minutes of the first meeting of the working party set to investigate the development of the Sugar Industry, August 14, 1961.
12. KNA; AK/3/15, Soil Chemist, Scott Agricultural Laboratories to Director of Agriculture, January 3, 1951.
13. *DAAR 1946*, 9.
14. KNA: DC/CN/1/3/1, K-LDAR 1947.
15. KNA: AK/11/173, Muhoroni Sugar Company to SAO Nyanza, June 10, 1947.
16. KNA: DC/CN/1/3/1, CK and K-LDARs, 1947.
17. *DAAR 1952*, 34.
18. Paul Tiyambe P. Zeleza, *A Modern Economic History of Africa: The 19th Century*, vol. I (Dakar: Codesria, 1993), 25.
19. *DAAR 1928*, 17.
20. *DAAR 1929*, 25.
21. KNA: AK/2/17, CK Agriculture AR 1940.
22. KNA: AK/2/17.
23. *DAAR 1945*, 14.
24. *DAAR 1946*, 9.
25. *DAAR 1950*, 6.
26. KNA: AK/11/51, Amar Singh, to Member for Agriculture and Natural Resources, February 23, 1953.
27. KNA: AK/11/51, DAO CN to Director of Agriculture, February 27, 1953.
28. KNA: BV/6/1704, Draft paper of the CEC on the future of the East African Sugar Industry, 1955.
29. KNA: DC/CN/1/4/3, K-LDAR 1931.
30. KNA: PC/NZA/3/1/465, K-LD Intelligence Report, September 1932.
31. KNA: DC/CN/1/4/3, K-LDAR 1932.
32. KNA: PC/NZA/3/1/465, K-LD Intelligence Report, November 1932.
33. KNA: AK/11/51, Board of Agriculture to DC CN, February 6, 1953.
34. KNA: DC/CN/1/4/3, K-LDAR 1931. See also KNA: DC/KSM/1/3/123 for the Report on Land and Agricultural Bank recipients, 1932–1935.
35. KNA: AK/3/15, PAO Nyanza to Director of Agriculture, March 30, 1950.
36. KNA: AK/3/15, C. C. Webster to A. Lacey, August 12, 1952.
37. Kanogo, "Kenya and the Depression," 117–118.
38. KNA: DC/CN/1/4/5, K-LDAR 1936.
39. KNA: DC/CN/1/4/5, K-LDAR 1936.
40. Mohinder Singh (O.I.), June 22, 2009.
41. KNA: PC/NZA/3/2/34, DC K-L to Ag. PC Nyanza, October 13, 1934.
42. KNA: DC/KSM/1/3/122, Agricultural Production on Asian Farms, 1947.
43. Frank, *The Sugar Industry*, 2.
44. KNA: AE/3/916, Miwani Sugar Mills to PS Commerce and Industry, July 28, 1961.
45. KNA: BV/6/1704, Minutes of meeting to discuss arrangements regarding sugar and economic cooperation held in Dar es Salaam, September 7, 1955.
46. KNA: BV/6/1704, Minutes of meeting.

47. KNA: BV/6/1704, Minutes of meeting.
48. Mohinder Singh (O.I.), July 26, 2009.
49. Mohinder Singh (O.I).
50. KNA: BV/6/1704, Sugar supplies; East Africa High Commission Memorandum, September 2, 1955.
51. KNA: DC/CN/1/4/5, K-LDAR 1936.
52. KNA: PC/NZA/3/2/34, PC Nyanza to Colonial Secretary, November 27, 1934.
53. KNA: PC/NZA/3/2/34, Asian Cane Growers 1931.
54. KNA: PC/NZA/3/2/34, Victoria Nyanza Sugar Company to PC Nyanza, May 13, 1931.
55. KNA: PC/NZA/3/2/34, DC K-L to Pilling, February 3, 1936.
56. KNA: PC/NZA/3/2/34, DC K-L to Ag. PC Nyanza, October 13, 1934.
57. Heyer, "The Marketing System," in Heyer et al., eds., *Agricultural Development in Kenya*, 336.
58. Frank, *Sugar Industry*, 2.
59. Frank, *Sugar Industry*, 1.
60. KNA: PC/NZA/3/2/34, DAO CN to Pilling, confidential, January 27, 1936.
61. KNA: PC/NZA/3/22/16, R .O. Hennings to DC CN, August 12, 1949.
62. KNA: AK/11/51, DAO CN to Member for Agriculture and Natural Resources, April 30, 1953.
63. KNA: AK/11/51, DAO CN to member for Agriculture.
64. KNA: AK/11/30, DAO CN to PAO Nyanza, May 14, 1952.
65. KNA: AK/11/30, DAO CN to PAO.
66. Muyumbu, "A Historical Study of State Policies," 51.
67. William A. Ezra, "Sugar and Coffee: A History of Settler Agriculture in 19th Century Liberia," Ph.D. diss., Florida International University, 2002, 92.
68. Wanyande, "Management Politics in Kenya's Sugar Industry," 126.
69. *DAAR 1926*, 14.
70. *DAAR 1927, 13.*
71. *DAAR 1928*, 95.
72. *DAAR 1928*, 187.
73. *DAAR 1928*, 195.
74. *DAAR 1929*, 26.
75. *DAAR 1932*, 132.
76. KNA: PC/NZA/3/2/34, AO Nyanza to Mycologist Nairobi, August 12, 1932.
77. KNA: PC/NZA/3/2/34, Senior Mycologist to AO Nyanza, December 21, 1932.
78. *DAAR 1933*, 130.
79. KNA: AK/11/51, East African Agriculture and Forestry Research Organization to Director of Agriculture, June 16, 1953.
80. *DAAR* 1933, 30.
81. KNA: AK/2/5, Report by M. E. H. Vickers on Export Crops 1957.
82. KNA: AK/2/5, Report.
83. KNA: AK/2/4, Assistant Director of Agriculture, Nyanza to Director of Agriculture, December 16, 1957.
84. KNA: AK/11/52, Report by Wiehe on a visit to Potential Sugarcane areas in Kenya, November 6, 1956.
85. KNA: AK/3/15, Notes on Sugar Industry 1948.
86. KNA: PC/NZA/3/22/1/8, Report by Magor on a visit to Asian farms in the area of Kibos-Miwani-Kibigori-Chemelil, October 6–10, 1947.
87. KNA: PC/NZA/3/2/34, Secretariat Nairobi to S. H. Fazan, January 20, 1936.
88. KNA: BV/6/605, East African Sugar Industry, Muhoroni to Director of Agriculture, October 7, 1942.
89. KNA: AK/11/30, DAO CN to PAO Nyanza, May 14, 1952.
90. KNA: AK/11/51, A .G. Dalgleish to A. Pritam, April 18, 1953.
91. KNA: PC/NZA/3/22/1/8, Report by Magor, October 6–10, 1947.
92. Odhiambo, "The Sugarcane Procurement Problems in Kenya," 34.

93. KNA: AK/11/51, A. G. Dalgleish to A. Pritam, April 18, 1953.
94. R. D. Wolff, *The Economics of Colonialism, Britain* and *Kenya 1870–1930* (London: Yale University Press, 1974), 95.
95. Wolff, *The Economics of Colonialism*, 95.
96. Wolff, *The Economics of Colonialism*, 95.
97. Maxon, *Struggle for Kenya*, 162.
98. Seidenberg, *Uhuru and the Kenya Asians*, 34.
99. Seidenberg, *Uhuru and the Kenyan Asians*, 34.
100. KNA: AK/11/30, Miwani Sugar Mills to PAO Nyanza, August 10, 1952.
101. Gregory, *South Asians*, 189.
102. Gregory, *South Asians*, 189.
103. Gregory, *South Asians*, 189.
104. KNA: DC/KSM/1/19/270, R .E Wainwright to Milwant Singh, January 19, 1953.
105. *DAAR 1948*, 32. For the number of Asian farmers that Singh was to work with, see KNA: AK/11/50, Report by Lyne Watt on Sugar Investigational work, August 1947.
106. KNA: AK/11/30, DC CN to Game Department, July 20, 1950.
107. Mohinder Singh (O.I.), June 22, 2009.
108. Fearn, *An African Economy*, 105.
109. KNA: AK/3/15, Director of Agriculture to PAO Nyanza, July 1, 1950.
110. KNA: PC/NZA/3/2/34, Department of Agriculture Nyanza to H. G. Pilling, February 3, 1936.
111. KNA: DC/CN/1/4/5, K-LDAR 1936.
112. KNA: PC/NZA/1/3/122, Minutes from the meeting of NFCS held on April 6, 1940.
113. Osamba, "Colonial African Labour," 81.
114. KNA: BV/6/613, Dalgety and Co. to Director of Agriculture, January 8, 1941.
115. KNA: BV/14/401, Hughes-Rice to PS Ministry of Agriculture, May 12, 1959.
116. KNA: DC/KSM/1/19/270, N. G. Hardy to Production Sub-committees, November 2, 1949.
117. KNA: AK/11/50, Minutes from the Special Meeting of Kibos–Muhoroni Production Sub-Committee held on July 18, 1947.
118. KNA: AK/11/50, Special Meeting.
119. KNA: AK/11/50, Special Meeting.
120. KNA: AK/11/50, Deputy Price Controller Nyanza to Kenya Commodity Distribution Board, April 10, 1947.
121. KNA: AK/11/50, Minutes of the Kibos-Muhoroni Production Sub-Committee, September 6, 1949.
122. KNA: DC/KSM/1/19/270, N. G. Hardy to Kibos-Muhoroni Production Sub-Committee, May 10, 1950.
123. KNA: AK/11/51, DAO CN to Secretary, Board of Agriculture, January 5, 1953.
124. KNA: AK/11/51, DAO CN to Director of Agriculture, March 6, 1953.
125. KNA: AK/11/51, G. M. Vasani, Chairman, Chemelil-Muhoroni Production Subcommittee to DAO CN, June 12, 1953.
126. KNA: AK/11/51, Laljibhai Hindocha to Miwani-Kibigori Production Sub-committee, May 14, 1953.
127. KNA: PC/NZA/3/22/1/8, Report by Magor, October 6–10, 1947.
128. KNA: AK/11/13/13, DC CN to Chief Executive Officer, Board of Agriculture, April 21, 1952.
129. KNA: PC/NZA/3/22/1/8, Report by Magor, October 6–10, 1947.
130. KNA: PC/NZA/3/2/34, NIFA to DC K-L, October 11, 1934.
131. KNA: PC/NZA/3/2/34, R. G. L. Rumus to Sir Joseph Byrne, February 18, 1936.
132. KNA: PC/NZA/3/2/34, Ag. Colonial Secretary to NIFA, December 3, 1937.
133. KNA: PC/NZA/3/2/34, Ag. Colonial Secretary to NIFA.
134. KNA: PC/NZA/3/2/34, Ag. Colonial Secretary to NIFA.
135. KNA: PC/NZA/3/2/34, Ag. Colonial Secretary to Director of Agriculture, October 6, 1937.

136. KNA: PC/3/2/34, NIFA to Sir Robert Brooke-Popham, September 28, 1937.
137. KNA: PC/NZA/3/2/34. DC K-L to PC Nyanza, October 1, 1937.
138. KNA: PC/NZA/3/2/34, SAO Nyanza to Colonial Secretary, October 6, 1937.
139. KNA: AK/11/50, Minutes of Meeting of NIFA with Manager of Victoria Nyanza Sugar Company, June 22, 1946.
140. KNA: DC/KSM/1/19/270, NIFA to Chairman, Central Commodity Distribution Board, March 14, 1950.
141. KNA: DC/KSM/1/19/270, M.C. Kashap to Chairman, Central Commodity Distribution Board, March 4, 1950.
142. Gregory, *India and East Africa,* 94.
143. Gregory, *India and East Africa,* 94.
144. Gregory, *India and East Africa,* 94.
145. KNA: PC/NZA/3/2/38, Resolutions Passed by the EAINC, 1931.
146. KNA: DC/CN/1/4/5, K-LDAR 1936.
147. KNA: DC/CN/1/4/5, K-LDAR 1936.
148. Seidenberg, *Uhuru and the Kenya Asians,* 33.
149. Siedenberg, *Uhuru and the Kenyan Asians,* 58.
150. Siedenberg, *Uhuru and the Kenyan Asians,* 15.
151. KNA: PC/NZA/3/2/34, DC K-L to PC Nyanza, October 13, 1934.
152. KNA: PC/NZA/3/2/34, NIFA to DC K-L, October 11, 1934.
153. KNA: AK/11/30, PAO Nyanza to Member for Agriculture and Natural Resources, May 2, 1952.
154. KNA: AK/11/51, Kibos Planters Association to DAO CN, April 16, 1953.
155. KNA: AK/11/51, Kibos Planters.

SEVEN

Social and Economic Impact of Asian Cane Farming in Colonial Nyanza

As we have seen, the Kenyan cane industry was a relentless battle for the Asian farmers who forged a living from sugarcane farming. This chapter turns from the Asian farmers to evaluate the impact of the sugar industry among the local people, specifically at Kisumu and its environs. Asian sugarcane farming in colonial Nyanza (CN) had a tremendous effect on the region, effecting both economic and social transformations in the region. Commercial cane farming was adopted in earnest by the local African population and this in turn led to emerging social differentiation, especially among the pioneer group of African cane farmers, who were able to make great strides despite a slow take-off. Commercial cane farming subsequently spread to other parts of western Kenya and continued into the independence period. The sugar industry contributed to the introduction of wage labor in CN, and Asian cane farmers influenced the African traders who became involved in the distribution of sugar in the African reserves.

The chapter also examines the social contribution of Asians in Kisumu and its environs, specifically the Asians' role in education and health facilities and the provision of other social amenities for the local people as a way of giving back to the community. The chapter concludes with a consideration of both its positive and negative contributions to African progress, including the devastating toll that the use of the Asians' jaggery in the production of alcohol took on the African reserves.

ASIAN IMPACT ON COMMERCIAL SUGARCANE FARMING: AFRICAN CANE FARMERS, 1930S–1963

The settlement of Asians at Kibos led to the subsequent introduction of commercial cane growing that had a lasting effect on the area and its environs. This commercial cane growing impacted the area both positively and negatively in many ways.

The Nyanza sugar industry owes its origin to the Asian community. The Asians were the first to grow sugarcane commercially in western Kenya. Initially many farmers supplied cane to the Victoria Nyanza Sugar Company, which was founded by G. R. Mayers, the Australian pioneer who set up the company as the first large-scale sugar operator in Nyanza province in 1923. Three years later the Madhvani Group International of India, led by Muljibhai Madhvani, established a second sugar factory, the Associated Sugar Company at Ramisi in the coast province. These were privately owned firms that relied on their large-scale plantations as well as sugarcane produced by Asian farmers to feed the factories. This Asian sugar production was one of the first pioneering industries in colonial Kenya that was established by external linkage. These two companies remained in operation in Kenya from 1920s to mid-1960s, though the Victoria Nyanza Company was purchased by an Asian capitalist in 1947. Other sugar factories were also established at Muhoroni, Sukari, and Manoni. However, the production of these factories was not adequate to meet internal demand and so 65 percent of the sugar consumed in Kenya was imported from outside.[1]

Asians contributed to the growth of commercial agriculture in CN through their pioneering work in cane growing and through the introduction of export crops such as sugarcane. Asians also aided the economy through experimentation of various crops.[2] In the African areas, plans for sugar development were contained in the Swynnerton Plan of 1954, which provided for the development of a jaggery industry with a capacity of producing up to 2,600 tons.[3] This Asian component in Kenya's sugar industry continued after independence, and in 1966 the East African Sugar Industries (Muhoroni) was established by the Kenyan government as part of the Million Acre Scheme through the help of Agency for International Development (AID). The management of the firm was entrusted to the Mehta Group, an Asian multinational company that was also involved in the construction of the firm. Mehta Group International also managed another sugar company in South Nyanza that was built in 1979: the Sony Sugar Company.

Foreign players have continued to play a crucial role in cane production in independent Kenya. This was seen when the Chemelil Sugar Factory was also established with outside funding from Britain; consequently its management was placed for quite a while in the British hands through Booker Agricultural Technical Services of Britain. This was followed by Mumias Sugar Company in 1973. Eager to attain self-sufficiency in sugar production, the government of Kenya also set up the Nzoia Sugar Company in 1978, in the then-Western Province. The Nzoia Sugar Company was started with foreign aid from France, and the company placed under the management of a French company, French Technisure. Given the viability of cane production in Kenya, it was expected that the full utilization of the factory capacity in these sugar companies would meet the local demand for sugar and save the country monies spent on importation, which could then be used for pressing issues such as public health. Unfortunately, this has not been the case and Kenya still imports sugar. However, the Asian sugar industry left a long-lasting legacy in the Nyanza and Western region of Kenya.

AFRICAN AGENCY: ADOPTION OF SUGARCANE FARMING

The impact of cane growing in Kenya cannot be underestimated. Sugarcane was grown by the Africans, especially Nubians in Kibigori, who were influenced by Asians as early as the 1930s. For example, the Nubian communities in Lumbwa and Kibigori grew cane using the *Kampala* variety. However, the variety was found to be prone to mosaic diseases. This problem was compounded by the fact that most of the Nubians were poor and could not afford the newer, more disease-resistant varieties. They also were initially ignored and could not get access to the new varieties that were being used by the Asian farmers.[4] Other African communities in the neighboring regions, such as the Luos of CN, also took to cultivation of sugarcane. However, the take-off for the Africans in CN was a slow venture since most people regarded it as an "Asian crop."[5]

Initially, the colonial government was unsupportive of African cane farming. However, by the 1950s, the colonial state changed its policy and supported experimentation into various cane varieties. It also began to encourage African farmers to pick up cane farming so as to boost the sugarcane industry. The peasant farmers were given free seeds and a subsidy. For example, in 1951 alone, the colonial government provided the African cane farmers with 150 tons of seeds. Unfortunately, the rains came late that year, resulting in crop yields that were not very good.[6] However, the first ratoon was better, and the African farmers in 1953 got approximately Shs. 8,993 from cane sold to the Miwani Sugar Mills.[7]

As more Africans became involved in commercial cane farming, the farmers around Chemelil formed a cooperative society in 1954 to assist in transporting their cane to the factory. During the same period, the African cane growers managed to sell 130 tons of cane worth Shs. 5,163 to the Miwani Sugar Mills.[8] Thanks to the government's provision of free cane seedlings and subsidy on transportation, more Africans started growing cane in Kibos-Muhoroni area and in the rest of Nyanza generally. By 1957 the number of African cane farmers in CN who were registered in the cooperative society had increased to forty-seven. They collectively cultivated 51 acres of land.[9] By 1960 the area under cane grown by the African farmers in the area rose to 300 acres. There continued to be significant growth since by 1963 farmers in East Kano location alone cultivated 700 acres of cane.[10]

Therefore, this trend shows that African farmers were taking to sugarcane as a commercial crop. Some farmers prospered and even bought more land to cultivate sugarcane and employed workers, in most cases casual laborers, to help them.[11] The entrance of Africans into cane cultivation brought prosperity to some people. For example, farmers such as Eli Onyuro prospered. From the proceeds he bought more land and a tractor to intensify his cane farming, which proved rewarding since he not only educated his children but also built himself a good brick house. This was by no means a small achievement in the early 1960s.[12] With this example in mind, we find that with commercial cane farming, social differentiation was enhanced among the Africans in CN. The beginning of cane growing by Africans also led to competition, especially with the Asian small-scale cane farmers who now had to compete with the African cane farmers for labor and in the production and marketing of jaggery. The competition caused the NFCS to ask the colonial administrators in the district to allow the joint marketing of jaggery with African cane producers. The request was turned down on the grounds that there was a special board that handled only African produce.[13]

Despite the fact that Africans picked up commercial cane farming, the number of cane farmers remained small. There were factors that contributed to this. First, sugarcane took longer to mature than grains like maize that matured within a short while. Maize was also very marketable. Secondly, transportation was a huge problem; potential African farmers shied away from cane production as it was difficult to find a cheap, reliable means of transport to ferry their produce to Miwani, especially for cane growers who were situated far from the factory.

Despite all this, we find that the success of the Asian cane farming in colonial Nyanza continued to attract many peasant producers. After independence many Africans took to cane farming in CN. For example, the number of African small-scale farmers increased, and by 1964 such cane growers were organized into four zones: Mitete and Chemasis (both in Nandi) and two more in Luoland. The Luoland area was further divided into the God-Abuoro and Songhor zones.[14] Some of the small-scale farmers had as little as one acre of cane but were very keen to start commercial farming.[15]

Moreover, the Kenya government, in its endeavor to attain self-sufficiency in sugar production, inaugurated the Muhoroni scheme where commercial cane farming began in earnest. The government went further and established another factory, the East African Sugar Industries (Muhoroni), in 1966, as stated earlier. Another factory, the Chemelil Sugar Company, was established in 1968.[16] The establishment of these two factories enhanced commercial cane farming in CN as they served the areas that were hard hit by transportation problems during the colonial times due to their distance from the Miwani Sugar Mills.

Although the independent state had plans to develop the sugar industry in CN by giving land to poor people, like its predecessor, the colonial state, its policies were contradictory. This was evident when many rich people who came from far also got land in the scheme. These straddlers operated as absentee landlords, leading to low productivity of the land.[17]

Be that as it may, the independence years witnessed the expansion of commercial cane growing to the rest of Nyanza as Africans in areas such as Konyango, Sare, and Oyani in South Nyanza started cane farming. Meanwhile, in the then North Nyanza commercial cane farming also started in Kakamega, Mumias-Wanga, and Kabras. In these areas, the climate was more conducive to cane growing, and African cane farming progressed so quickly that by 1957 the farmers were demanding their own factories and the colonial state made efforts to attract outside capital for such ventures.[18]

Therefore, we see that as a result of the humble beginning of commercial farming by Asians in Kibos-Muhoroni; the Nyanza region in particular was transformed into a thriving sugar-belt. This was in large part due to the resilience of Asian cane farmers in the face of the colonial government's lack of support and neglect of the industry's potential.

AFRICAN WAGE LABOR

Asian cane farming also had a significant impact upon labor practices among Africans. The Asian cane plantations employed many African people as discussed in the preceding chapter. Most laborers in the sugarcane plantations came from the neighboring reserves. As early as 1923, Africans were found to be working in the sugarcane farms and also at the Victoria Nyanza Sugar factory, where approximately 1,500 Africans were employed on a monthly contract.[19] Therefore, the introduction of the Asian commercial sugarcane farming introduced a labor market and set in place the process of turning some Africans into wage laborers.

As the sugar industry expanded, more opportunities emerged for Africans to join wage labor. For example, in 1927, additional Africans joined the labor force at work on the new Nottidge and Allen sugar plantation and factory, which employed 400 laborers. By 1928 the total number employed at the two large estates was approximately 2,000.[20] Over the years, other challenges made the Africans to seek wage labor in the sugar industry. For instance, it was found that during the Great Depression many people ventured into wage labor as the prices for their home-grown commodities fell. The effect of the Depression and famine caused by locusts in early 1930s also had great influence on workers to enter the labor force. The effects were so severe that people came out on their own. The sugar factories no longer had to rely on the professional recruiting agents.[21] The transition to wage labor was also abetted by the colonial government, who raised taxes on the Africans both to increase their own revenue during the Depression and also to force them into labor for the farms and factories.[22]

Furthermore, as maize was the crop mostly grown by the peasants, the collapse in maize prices during the Depression affected many households. The price of maize fell by 50 percent in 1931, which led to a reduction in household incomes.[23] Yet the Africans still had to meet their tax obligations. Therefore, by 1934 the number of Africans seeking employment was still on the rise. For example, Muhoroni increased the number of its laborers from 400 to 500, while the small-scale sugarcane growers were found to have engaged between 600 and 700 laborers.[24] This translated into approximately 2,500 Africans employed in the sugar industry annually. Those who secured employment in Asian and European farms during the Depression of the 1930s benefited. Such engagement offered the Africans the opportunity to diversify the family's source of income during decreases in commodity prices and during famine periods. The low market prices for maize during the period made wage labor to be more viable in supplementing the household income during the difficult times. Indeed, wage labor kept on increasing as the sugar industry expanded. By 1947, Miwani alone had a labor force of about 4,200 workers, while by 1960 Miwani had employed approximately 4,700 laborers as it moved into the new phase of producing refined sugar.[25]

As the colonial era wore on, many Africans sought wage labor as the pressure on land increased. For example, at the prime of the sugar industry by 1960, it was estimated that the large-scale and small-scale farms employed between 5,000 and 8,000 Africans annually.[26] There were other reasons that led to the increase in wage labor. First was the fact that weather was erratic and so employment was seen as security because crops failed and people suffered. Secondly, wage labor was seen as a reliable source of cash accumulation, which was important to meet the family obligations such as dowry payments. Although many Africans saw agriculture as a means of subsistence, others entered wage labor as a source to accumulate wealth.[27]

The increase in wage labor significantly enhanced the CN economy because monies sent home to relatives from wages earned allowed families involved in peasant farming to purchase farm implements. The contribution of wage labor to household economies also led to the expansion of trade in CN as peasant farmers within the region felt motivated by the opportunity of a ready market. Many farmers produced more and sold the produce to the large estates for rations that were needed for the laborers. As the sugar industry increased due to the volume of local trade, there was more money in circulation, and the purchasing power of Africans increased too. The introduction of wage labor acted as stimuli for both trade and peasant production. The remittances were indeed significant as these not only supplemented peasant household income but also provided capital for expansion of trade as well as a means of wealth acquisition and accumulation.[28]

AFRICAN SUGAR TRADERS

The Asians not only farmed sugarcane in Kenya but they also introduced the distribution and sale of sugar. When sugarcane crops were destroyed due to effects of locusts and drought, especially in the early 1930s, many small-scale Asian farmers resorted to being petty traders on their farms. They sold sugar and other goods to the Africans in the reserves. African traders picked up the practice from the Asian hawkers and became involved themselves in the sale and distribution of sugar in the African reserves.[29] These hawkers were issued licenses by the CN African District Council, and their activities threatened other African shopkeepers in the market centers.[30] African sugar traders were also scattered in different market centers in various locations. Eventually, the traders stopped hawking and erected stalls to sell sugar in various market centers.

Over the years, the numbers of such African traders increased and many were members of the Nyanza African Chamber of Commerce, which was concerned with the interests of the African traders. For example, by 1957, the chamber sent a petition to the DC highlighting the plight of the African sugar traders in Chemelil and East Kano who were experiencing difficulties in getting their sugar allocation to supply the African consumers in the area.[31] Nevertheless, many Africans were involved in the sale of sugar in various markets in different divisions and locations. For example, there were many African traders in Alego location alone by 1954. Alego location had twenty-eight sugar traders among whom was one woman, Mrs. Rhoda Obougo. Majority of these traders sold one bag per month, but there were a few that sold between two to four bags per month. It is clear that the distribution was good since there were sugar traders in almost all the markets and trading centers in the location. Going by the numbers of Alego sugar traders alone, one can deduce that indeed many Africans were involved in the sale and distribution of sugar in the African reserves.

Needless to say, the sale and distribution of sugar was hampered by the limited working capital available to buy sugar. Not all traders could buy large amounts. Many African customers suffered from a lack of sufficient sugar supply.[32] This necessitated the location councils to intervene when shortages were acute. For example, in 1957 the Seme Location

Council was forced to replace the former sugar distributor with Mr. Josephat Amolo, who had sufficient funds to facilitate distribution.[33] During the same period in the same location, shortages were so acute that the DC was obliged to increase sugar allocation with an additional fifty-two bags.[34] Despite these difficulties, it is clear that African traders were involved in the distribution of sugar, having been influenced by the Asians who used to hawk it in the African reserves.

By late 1950s, there were over 200 African sugar traders in CN. However, there were some retailers found in the African reserves that were allocated sugar but failed to take up full allocations, as indicated in table 7.1.

Table 7.1 shows how some African traders rarely took all the allocation. The collectors with the biggest drops were Eliakim Mbalo, Josphat Abwao, and Naftali Otiende while those who performed well were Jairo Mbume, Tolo Achola Jason Okeyo, and Kusa Trading. Between November 1957 and January 1958, 606 bags were not collected.[35] These uncollected allocations impacted distribution in the sense that the consumers

Table 7.1. African Sugar Distributors, Uncollected Sugar, January 1958

Traders	Permit No.	Location	Amount Uncollected for 3 Months		
			Nov	Dec	Jan
Eliakim Mbalo	45	East Kano	57	27	30
Josephat Abwao	44	Seme	56	56	105
Tolo Achola	170	West Kano	3	2	7
Christopher Owiro	48	Kisumu	—	8	8
Jason Okeyo	160	Asembo	—	—	3
Ojwang Warinde	57	South Ugenya	—	4	13
Elisha Wagade	49	Kisumu	4	—	9
Benjamin Alouch	53	Kajulu	22	16	9
Fresto Ogindo	168	West Kano	23	13	10
Naftali Otiende	47	Kisumu	—	2	19
Amwata Onono	55	Nyakach	11	—	21
Nicholas Okoth	43	Alego	6	—	21
James Owek	46	Kisumu-lela	—	20	—
Jairo Mbume	2	Samia	1	1	1
Kusa Trading	59	Asembo	6	6	6

Source: Extract from the letter from the DC CN, to District Officers, Ukwala, Maseno, Bondo, Winam and Nyando, 1958, KNA: DC/KSM/1/3/58.

did not get the sugar and the holding wholesale stores had to deal with the problem of storage.

The reasons for failure in collecting sugar allocations by the African distributors were varied. The majority cited limited capital as the main reason for reducing their stock. This was because the majority relied on whatever capital they had at hand to make the purchases. Secondly, some of the African distributors were frustrated by the difficulties in transporting sugar from the stores to their shops due to bad roads, especially during the rainy season. The problem was compounded by the fact that most did not own their own vehicles. When traders constantly failed to collect their allocation, the DOs could reduce the allocation of an area or reallocate sugar to traders within those divisions who had the capacity and were more efficient in distribution.

African involvement in the sale and distribution of sugar thus was not without problems, but it was significant and it provided an important economic support to people in CN. It was an alternative source of income to the households. Those involved in the trade could use the resources to meet family obligation such as paying taxes and school fees and even expand agriculture through acquisition of better tools from the monies obtained from trade. Furthermore, the Asian sugar industry, through its influence on the sale and distribution of sugar in the African reserves, also facilitated the adoption of new diets among the people. Many people started drinking tea instead of porridge or used both.

THE SOCIAL IMPACTS OF SUGARCANE GROWING IN NYANZA

Asians also contributed immensely to the social life in Kisumu. With the monies they received from cane sale, they joined hands with their brothers running shops in Kisumu and put up social amenities such as schools, hospitals, social halls, club facilities, and temples. They also introduced new architectural designs of building houses. However, one of the most valuable services that the Asians offered to the people of Kenya was the construction of numerous libraries and reading rooms in major towns, including Kisumu. Some of these libraries started modestly in small rooms and later became big reading rooms. Some were turned into institutions that hosted functions. It is ironic that the same Asians who were denied land by the Europeans contributed land to the community later. For example, A. M. Jeevanjee gave out large tracks of land to the city of Nairobi for the purpose of building a market and a municipal park.[36] This park still exists to date.

CONTRIBUTION TOWARD EDUCATION

In the early years of the colonial rule, the colonial state did not pay much attention to the educational needs of the growing Asian population. Therefore, the Asians, specifically the Hindus, stepped in and made plans to educate their children.[37] The Asian schools adopted not only a religious but also a linguistic pattern.[38] This was not evidence of Asian disunity but rather portrayed their religious inclinations. When schools were built, religious education linked with vernacular languages became a major consideration. By 1920, a number of Asian schools had been opened in Kisumu, and owing to the deep-seated religious differences of the various sects, many schools initially began and were run on a sectarian basis.

At Miwani there was a small Asian school that operated at a building constructed by the Sikh farmers of the area. At Muhoroni the school was built by public contributions, and the colonial state initially paid half of the teachers' salaries.[39] However, in CN most of these schools were concentrated at Kisumu town. Even the Asian farmers at Kibos-Muhoroni preferred to send their older children to school in Kisumu. Some of the schools built in the district included the Arya Samaj Girls School that by 1936 had a staff of one principal and three teachers. The school had an average of sixty-one Asian girls, and provided education from sub-standard A (pre-school) to standard 6. Instruction was conducted in both Hindi and English. By 1938 another school, the Siri Guru Singh Sabha Girls, was operational and had approximately thirty-five girls and two teachers with classes going up to standard 5. The medium of instruction in this school was Gurmukhi.[40] The school in 1938 was able to get a grant in aid amounting to £59 from the colonial government.[41] Other grants and donations were received from rich people within the Asian community.

Other schools got external support. For example, schools built by the Ismaili Khoja sect got grants from His Highness the Aga Khan.[42] These grants were instrumental in the building and maintenance of schools in Kisumu town such as the Aga Khan nursery school and the Aga Khan Girls School, which had their own premises. The latter school was by far the best in Kisumu; it had five teachers and approximately sixty-five students. The school performed well in the junior Cambridge examinations. The medium of instruction was Gujarat up to standard 5, and the upper classes used English. Also there was the Ismaili Girls School, which was also managed by the H. H. Aga Khan Education Board and provided education to approximately sixty-five pupils belonging to the Khoja sect; it had four teachers and was housed in a former hostel.

Apart from the Asian-owned schools there was also the Government Asian School in Kisumu that was the largest of all. The school was housed at the Sikh Library and the Sanatan Hindu Union Building. This was a boy's school that provided education for Asian boys of all denominations. Due to lack of schools for the Christian Goan children, fifteen Goan boys and two girls were admitted also in the school in 1936.[43] At the time, the school included classes from sub-standard (nursery) to elementary and also secondary. This school had approximately 211 pupils whereas the primary section had 79 pupils and Secondary 10 in 1938. Here the medium of instruction was English from standard 1 upward.[44] This school introduced hockey, cricket, and scouting among its activities.

Meanwhile, Asian Muslims also had their own school in Kisumu that was open to Arabs and African Muslims. It was housed in a building owned by the Moslem Association during the 1930s, and the furniture was donated by the Muslim community. It had a total of ninety-one boys in 1936, among them seventeen Arabs and three Somalis with the rest Asians and a few Muslim Nubians.[45] The boys here were taught the Koran, Arabic, and arithmetic.[46] These separate Asian schools benefited from a healthy and friendly rivalry that led to improvement of both the facilities and the academic performance.[47]

Needless to say, in colonial Nyanza Asians left a mark in the various schools that they built. Although these schools were initially meant for Asian children, as independence approached they were opened to a few African students. The Asians contributed to the building of schools not only by collecting monies and building materials but also by offering scholarships. Among them contribution toward social welfare was considered as a duty owed to their gods and their communities.[48] The Asians had a belief that it was honorable for a person to sacrifice his own comfort and donate money and time so that health and education services might be spread.[49] Through such contributions, including the buying of books and stationery and payment of other fees that were levied in schools, the Asian cane growers and manufacturers in the Kibos-Muhoroni area by extension could be said to have contributed to the building and maintenance of schools as most of their children attended schools in Kisumu, especially at the upper primary and the secondary level. With the political transformations at independence, these different Asian schools eventually were opened to all races.[50] Today, the schools that are still operational, especially those built in Kisumu, are among the best performers in the area.

CONTRIBUTIONS TO HEALTH FACILITIES

The Asian initiative of offering social services in medicine began during the Depression period due to the need created when some Asians living in major urban centers like Nairobi and Mombasa lost their jobs and businesses. Such people and their families found themselves in precarious situations when medical issues arose. To address such problems the

Asian community came together, especially in the major towns of Nairobi, Mombasa, and Kisumu, and formed a voluntary organization known as the Social Service League in early 1930s.[51] The league provided free medical help from trained Asian doctors and donated medicine. This social service soon became permanent as Asians were contributing toward it, and soon afterward the league opened dispensaries and hospitals in various parts of the country, including Kisumu.[52]

Asians used their money to build clinics and hospitals that served even the local communities before independence. The community continued to expand the medical facilities; by the late 1950s, an Aga Khan Hospital opened in Kisumu. Other branches would later be opened in Nairobi and Mombasa. Initially serving Asians, these hospitals were later opened to the public. Many Asian groups contributed to building such amenities; specifically the Ismailia community set an excellent example in providing medical facilities.[53] Another hospital was also built by the Asians in Kisumu. This was the Jalaram Nursing Home, which was built partly by funds donated by two brothers, Sheth Haridas Chhagnal and Morzaria Chhagnal of Messrs Haridas Chhagnal Ltd of Kericho, who donated Shs. 5,101 toward the building cost in memory of their father.[54] The opening of the nursing home was performed by Devjibhai Karmshi Hindocha, the manager of Miwani Sugar Mills, on April 4, 1948. Later, in 1957, the Miwani Sugar Mills and the Hindocha family trust donated Shs. 25,000 and 10,000, respectively, to the nursing home. Meanwhile, the Guru Nanak Dispensary was built in Kisumu by the Siri Guru Singh Sabha fraternity.[55]

These developments in social amenities were made possible by Asian capital. Moreover, it was the capital accrued from the sugar industry that led to the construction and maintenance of some hospitals in Kisumu and its environs as demonstrated by the donations to the Jalaram Nursing Home by D. K. Hindocha, and the trust fund he established. These health facilities would later benefit the Africans and other races in CN as well as people living in Nyanza Province.

SOCIAL HALLS AND ARCHITECTURE

The Asians also put up buildings in Kisumu that introduced new architectural styles in the area. The communal buildings were erected by the different sects/castes. It is in these buildings where some of the early schools were housed such as Siri Guru Singh Sabha and Sanatan Hindu Union; the schools rented the buildings annually at a rent of £252.[56] The Asians living in Kisumu and its environs also built sports clubs such as the Sikh Club, Simba Club, and Goan Institute where various sporting events took place. The Sikhs, who were mostly the farmers at Kibos, invested their hard-earned money in putting up social amenities where they could spend leisure time after hard work on the farms.[57] These venues have continued to serve all races after independence as social centers.

During the year 1938, a significant Asian activity included the erection of the Kisumu clock tower by the family of the late Kasim Lakha. The tower was presented to the Kisumu council in his memory. This clock tower was unveiled by the Duke and Duchess of Gloucester.[58]

There were also various social halls built to host functions such as parties and wedding ceremonies, for instance the Simba Club.[59] Most of these halls at independence were opened to all races and are now used as recreation facilities. Other halls were built by the Asian community as donations to institutions in the area. For example, when the Siriba teachers college was built, Asians, courtesy of the Hindocha Trust, built the expansive and beautiful Hindocha Hall as a way of giving back to the community. This building is now part of Maseno University.[60]

The assessment of the Asian contribution in Kisumu and its environs would not be complete without mentioning the type of architecture that emerged in Kisumu. The cane farmers and producers sometimes invested in the building of residential houses around Kisumu and Kibos. Many who had children at school in Kisumu built such houses and settled part of their family there to avoid traveling long distances daily to school.[61] The designs were oriental in style in which many rooms were built within an enclosed wall with a main entrance. For example, Vajv Sundra, a resident of Kisumu in 2009, still lived in the family house in Kisumu where he was born in 1945. His grandfather was the first to arrive in Kenya with the coolies who built the Kenya-Uganda railway while his father came to Kisumu in 1924. They both worked for the Victoria Nyanza Sugar Company in 1930s as skilled workers; later they moved to Kisumu and started body work for vehicles. Together the two were able to build the family residence in 1934.[62] Now the house is occupied by Mr. Vajv, his wife, and his mother; his children are adults and live in Britain and Canada. The Asian housing style also influenced Africans who later copied the designs in building of rental houses in Kisumu and the neighboring areas such as Ahero.

SOCIAL INTERACTION

Socially Asians did not intermarry with Africans nor were there any known liaisons such as those found among the Asians who settled in Zanzibar and Dar es Salaam that gave rise to the *jotawa,* a derogatory term for Afro-Asians in East Africa.[63] However, in the interior of Kenya, Bharati points out that the Sikhs, who came earlier from India with no wives and had lived and traded with the Maasai, were known to have established relationships with Maasai women that produced children.[64]

But generally no interracial marriages took place, especially in the Kibos-Muhoroni area and its environs. However, this should not be misinterpreted as racial discrimination since Asians did not even intermarry among themselves due to their strict religious and caste system.[65] For example, when Tom Mboya (Kenyan Trade Unionist) once asked the Asian leader J. Nehru how he expected the Asians to be accepted in Kenya if they could not marry Africans, Nehru replied, "we do not even marry each other, why should we marry non-Asians."[66]

Such sentiments showed how the Asians were a closed society within which interracial marriage was unimaginable. Moreover, the Asians believed that Africans were spendthrifts who could not handle money and goods properly; hence they would be incompatible with Asian families. Asians often looked down upon the Africans, referring to them as *kala kala,* meaning "black black."[67] At the same time, Africans also mistrusted Asians, especially when it came to trade; they perceived them as people who exploited Africans. They also looked down upon the Asians and referred to their women as *banyani* and Sikhs generally as *kalasingah* (derived from Kala Singh).[68] Despite this, the two communities lived in harmony in Kisumu and its environs.

NEGATIVE ASPECTS OF ASIAN SETTLEMENT

Poor Farming

Despite the prosperity brought by the Asian sugar industry, it also impacted the area negatively in some ways, both social and economic. For example, the establishment of the Asian sugar industry led to more people being involved in wage labor at the expense of agriculture in the reserves, which suffered in turn. The sometimes semi-forced participation of African males in wage labor meant that they were not at home to clear lands for family cultivation. Also the women of the family were often overburdened, taking over, for example, the responsibility of looking after livestock, which was traditionally a male task. Moreover, some of the laborers were engaged on three- or six-month contracts, which duration was bound to affect agriculture especially during the farming season. This could also have led to instances where a piece of land was overused, leading to further problems of soil erosion.

It is generally recognized that CN was, during the colonial period, characterized by poor agricultural performance. The poor performance continued during the postwar period. For example, the production of food commodities, like maize, sorghum and eleusine, declined in CN between 1946 and 1951: In 1946 CN produced 4,690 tons of maize and 272 tons of sorghum while in 1951 it produced only 3,385 tons of maize and 107 tons of sorghum.[69]

Various reasons have been advanced by scholars for this situation. For example, Professor Anyang' Nyong'o argued that the absence of the labor force had devastating consequences on the African economy in CN.[70] However, the decrease in peasant production in CN during the postwar period could also be attributed to other factors. First, there was the decline in demand for the peasant produce as the war came to an end. Second, there was increased competition from Gusii peasant producers. The Gusii highlands, unlike CN, have good ecology and by 1946, the region produced large volumes of maize that was exported to other parts of the colony.[71] This increase was due to the pressure from the colonial state that was eager to avert the ravages of food shortages. The farmers were also encouraged by the good market prices offered for maize, which had shot up from Shs. 10 per bag by the end of the war to Shs. 21 per bag in 1949.[72] Thus the higher prices of maize, supported by good weather, acted as an incentive to the Gusii peasant farmers who produced more for market. For example, in 1949 the Gusii had a record harvest of maize that was sold outside the district, a trend that continued over the years, transforming the Gusii highlands into a major exporter of maize.[73] This in turn generated competition to the peasant farmers in CN, which partly explains the decline in production during the period.

Loss of Land

The stiff competition from the Gusii and the erratic climate of drought-prone Kibos-Muhoroni area might have influenced many peasant farmers to seek alternative sources of income. Many became migrant workers in other areas, a process which led to some land being underdeveloped. With more workers laboring away from home and cultivating less acreage in family plots, land became a commodity for sale. Consequently the value of land increased in the area as more Asian farmers sought land for expansion. For example, the price of an acre of land in Kibos increased to £12–15 by 1956.[74]

Another effect of the Asian sugar industry in CN was the stress brought about by population pressures beginning around the 1940s. The Luo people needed land for expansion, and the land around Kibos-Miwani was much contested because of its suitability for cane farming. This was tricky for the colonial administrators since they had settled the Asians there so as to lock them out of the fertile Rift Valley. So many white administrators felt that alienation of land to Asians was more beneficial to the area than allowing the Africans to use the land for subsistence farming.[75] However, it didn't have to be Asians alone getting the land for cane; neither did it mean that Africans could not have taken up to cane farming if given an opportunity with state support, as was done for cotton farming in the district.

Furthermore, the occupation of the land by Asians, especially on the fertile lands at the foothills of the Nandi Escarpment, denied the Luos of Kajulu the opportunity to expand to the only fertile lands near Kajulu. This became problematic as the population increased and their reserve was restricted to the much drier area. The settlement of the Asian sugar-cane farmers affected the Kajulu area, which became densely populated as a result of the alienation.[76] By 1948 Kajulu registered the highest population density in the whole of CN district as indicated in table 7.2.

Table 7.2 shows how in 1948 Kajulu had the smallest area in CN but also the highest population density per square mile, followed by Kisumu. It seems that the land alienated to the Asian cane farmers in Kibos denied the Kajulu people room for expansion as their population increased and led to a land squeeze over time. Furthermore, being a patrilineal society, subdivisions of land took place among all male members of Luo households; hence land was scarce and household production was affected. And since part of the alienated land at the foot of the Nandi Escarpment was the most fertile land in the area, the Kajulu people were denied an opportunity to have the best land for farming. Consequently, land shortages resulting from Asian settlement affected peasant production and

Table 7.2. Distribution and Density of African Population CN District, 1948

Location	African Population	Area in Square Miles	Density per Square Mile
Alego	55, 859	211	265
Asembo	19,153	58	330
Gem	54,961	168	327
Kadimu	7,145	100	71
Kajulu	7,970	14	569
East Kano	36,601	155	236
West Kano	36,783	112	328
Nyakach	36,793	143	257
Kisumu	34,860	84	415
Sakwa	20,831	162	128
Samia	42,521	206	206
Seme	33,829	108	313
North Ugenya	33,930	103	329
South Ugenya	8,368	23	364
Uyoma	17,822	99	180

Source: Fearn, *An African Economy*, 237.

forced many people in Kajulu area to enter wage labor in Kisumu town and at the nearby Asian farms.[77]

Drunkenness

Another negative phenomenon that emerged was that of drunkenness, which was prevalent in the African reserves. Asian farmers produced jaggery and used it to make their alcoholic drinks in small quantities. Meanwhile the farmers at Kibos sold jaggery to Africans who not only consumed it as sugar but also used it to manufacture Nubian gin.[78] Africans had also learned to manufacture jaggery from the Asians, and they also used it to make *tembo* (African local beer).[79] The colonial administration did not like the impact of jaggery production by Africans and associated the sale of molasses and jaggery from the Miwani and Muhoroni factories to the drunkenness that led to lawlessness in the area as people engaged in brewing these illicit drinks.[80] The problem was made worse by the fact that the youth, both male and female, started drinking too.

By the 1940s the sale of the Nubian gin and other local brews was a major problem in the African reserves. There was excessive drinking, especially at the Nubian settlement at Kibos, which was so popular that even Asians were found to be frequenting the drinking dens. This led the police to carry out raids in the area.[81] Unfortunately, it was difficult for the police to make arrests and obtain evidence of the sale of native liquor because the Africans found in the drinking houses denied sale or purchase of the liquor since in most cases money was not exchanged. Payment was made later and many drank on credit. Such cunning methods frustrated the police efforts.[82]

The DCs were also alarmed at the increase of the consumption of the liquor by the Africans in the reserves. For example, in 1938 the DC of K-L district reported that young men had taken to more drinking than was the case ten years earlier because the trade in liquor was flourishing.[83] Later, it was found that large quantities of molasses and jaggery purchased by the Africans from the Asian sugarcane farms were used in making traditional liquor.[84] To curb the drinking problem the administration placed restrictions on the sale of jaggery to Africans, but even such a move was not effective since gin could also be made from ordinary sugar. Hence sugar started to be rationed in the reserves so that there would be no excess sugar that could be used to make the Nubian gin.[85] As the situation worsened, the native tribunals heard many cases of drunken behavior, and even headmen stepped up and tried to prohibit the supply of liquor to unmarried males and females in an effort to control bad behavior, such as lack of respect by the young people some of whom started engaging in theft of livestock, that resulted from excessive drinking.[86]

The brewing of illicit drinks continued to be widespread in the African reserves all over CN after the war, and this led to many problems. For example, in 1955 the DO of Bondo division lamented that there were a considerable number of assaults due to increased drinking of the Nubian gin manufactured in the division. According to the DO, the effect started to be felt after the increase in sugar allocation in the division. Through the DO's effort, the sugar allocation was reduced from 134 bags to 60 bags per month in the division so that there would be no excess for use in brewing.[87]

Moreover, the sale of Nubian gin not only offered competition but also interfered with the sale of licensed local beer. Many of the licensed traders who had beer shops that sold native beer complained that the manufacture of Nubian gin affected their business since most people went to the Nubian villages to drink even when the manufacture of it was prohibited by the colonial government.[88] Nonetheless, the problem of excessive drinking in the reserves continued for a long period of time. The availability of jaggery and sugar for residents of CN obviously played a part in producing this problem.

CONCLUSION

The arrival of a large community of foreigners and the establishment of a new industry cannot help but affect the world into which they arrive. The introduction of Asian sugarcane farming to colonial Nyanza had enormous impacts in the region and on the indigenous population, both positive and negative.

Economically, the introduction of commercial cane farming in the area spread to Africans, who as a result experienced economic growth and social differentiation. The sugar industry also led to increased earnings from wages, as those who could not cultivate the crop were involved as wage laborers on the sugarcane farms while some were distributors of the produce. The cultivation of sugarcane, wages, and sale of sugar raised people's purchasing power and led to other benefits as the accumulated monies were used to increase peasant agricultural production through the buying of farm implements. Others purchased land while some met family obligations such as payment of taxes and dowry or education of children.

The success of the Asian farming community led to social contributions as well. Successful Asian farmers and associations built schools, hospitals, health centers, and social halls. Though many of these were originally intended for the Asian community alone, after independence they were opened to all Kenyans and benefited Africans and others in the region.

There were negative consequences to the emergence of the sugar industry as well. These included the new taxes intended to force Africans into wage labor and the land squeeze in the Kibos-Muhoroni area that halted the African expansion specifically to the fertile area near the Nandi Escarpment. Land was taken from Africans to provide farms for the Asians and Europeans, and Africans were often shunted into reserves, where traditional land management techniques were disrupted and overpopulation became a new threat. In addition, the abundance of jaggery from Asian farms led to the production of an alcoholic beverage, whose widespread consumption created social problems such as drunkenness, assaults, and stock theft.

However, that said, the positive effects that resulted from the introduction of the Asian commercial sugarcane farming in CN was the creation of wealth for the African people in the area, the economic development of the Western sugar-belt, and the possibilities for further growth were a great benefit to the local community. And the industry continues to support the Kenyan economy today.

NOTES

1. KNA: PC/NZA/3/2/34, Nyanza Indian Farmers Association to Sir Robert Brooke-Popham, September 28, 1939. The Association sent a petition giving the history of sugarcane farming in Nyanza. Also KNA: BV/6/613, Sugar Export Quota, 1938.
2. Fearn, *An African Economy*, 106.
3. KNA: BV/14/401, Sugar growing in Nyanza.
4. KNA: DC/KSM/1/19/270, AO Nyanza to Mycologist, Nairobi, August 12, 1933.
5. Thomas Okeyo (O.I.), Kibos small scale farmer, July 3, 2009.
6. KNA: DC/KSM/1/3/22, Sugar Production Report 1954.
7. KNA: DC/KSM/1/3/22, Sugar Production.
8. KNA: DC/KSM/1/3/22, Sugar Production.
9. Osamba, "Colonial African Labour," 166.
10. KNA: DC/KSM/1/3/32, African Cane Production, 1963.
11. Osamba, "Colonial African Labour," 167.
12. Osamba, "Colonial African Labour," 167.
13. KNA: BV/14/401, Hughes-Rice to PS Ministry of Agriculture, May 12, 1959.
14. Grace Nandi (O.I.), July 4, 2009.
15. Grace Nandi (O.I).
16. Shadrack Nasongo, "Multilateral Imperialism in Kenya's Sugar Industry: The Political Economy of Transnational Corporations," *Nigerian Journal of International Affairs*, 26 (2000): 57–77.
17. Grace Nandi (O.I.).
18. KNA: AK/11/52, L. H. Brown to Registrar of Co-operative Societies, December 23, 1957.
19. *NADAR 1923*, 28.
20. *NADAR 1928*, 99.
21. *NADAR 1933*, 129.
22. *NADAR 1933*, 122.
23. Kanogo, "Kenya and the Depression," 116.
24. *NADAR 1934*, 174.
25. *East African Standard*, July 25, 1960.

26. Chanan Singh (O.I.), July 26, 2009.
27. Margaret Jean Hay, "Economic Change in Luoland: Kowe, 1890–1945," Ph.D. diss., University of Wisconsin, 1972, 226–230.
28. Fearn, *An African Economy*, 55.
29. KNA: DC/KSM/1/3/57, DC CN to Senior Health Inspector, November 2, 1956. Hawking is referred to here as the selling of goods by moving from place to place.
30. KNA: DC/KSM/1/3/58, Nyanza African Chamber of Commerce to DC CN, January 23, 1958.
31. KNA: DC/KSM/1/3/57, Nyanza African Chamber of Commerce to DC CN, February 4, 1957.
32. KNA: DC/KSM/1/3/57, D O Maseno to D C CN, February 26, 1958.
33. KNA: DC/KSM/1/3/57, D O Maseno to D C CN, December 13, 1957.
34. KNA: DC/KSM/1/3/57, A. G. Pearsall to D. K. Hindocha, December 19, 1957.
35. KNA: DC/KSM/1/3/58, DC CN to DOs: Ukwala, Bondo, Maseno, Winam, and Nyando, February 20, 1958.
36. McCormack, *Asians in Kenya*, 23.
37. Hollingsworth, *The Asians of East Africa*, 151.
38. KNA: PUB/1/1/6, EAINC Paper "New Plans to blot out the very existence of Asians," 53.
39. KNA: PC/NZA/3/22/1/8, Report on a visit to Asian farms, 1948.
40. KNA: DC/CN/1/5/4, K-LDAR 1938.
41. KNA: DC/NC/1/5/4, K-LDAR 1938.
42. Hollingsworth, *The Asians of East Africa*, 151.
43. KNA: DC/CN/1/4/5, K-LDAR, 1936.
44. KNA: DC/CN/1/5/4, K-LDAR, 1938.
45. KNA: DC/CN/1/4/5, K-LDAR, 1936.
46. KNA: DC/CN/1/4/5, K-LDAR, 1936.
47. KNA: PUB.1/1/1/6, EAINC Paper "New Plan to blot out the very existence of Asians," 1923, 54.
48. KNA: PUB.1/1/1/6, EAINC, 52.
49. KNA: PUB.1/1/1/6, EAINC, 52.
50. KNA: PUB.1/1/1/6, EAINC, 54.
51. KNA: PUB.1/1/1/6, EAINC, 54.
52. KNA: PUB.1/1/1/6, EAINC, 54.
53. Delf, *Asians in East Africa*, 47.
54. Vajv Sundra (O.I.), July 5, 2009.
55. Vajy Sundra (O.I).
56. KNA: DC/CN/1/4/5, K-LDAR 1936.
57. Mohinder Singh (O.I.), July 26, 2009.
58. KNA: DC/1/5/4, CKAR 1938.
59. Vajv Sundra, (O.I.), July 5, 2009.
60. Mohinder Singh (O.I.), July 26, 2009.
61. Vajv Sundra (O.I.), July 5, 2009.
62. Vajy Sundra. Vajv acknowledges that he is a Kenyan Asian and proud of the fact. He says he only visits India the way he visits his children in America. Having been born, raised, educated, worked, and now ageing in Kenya, to him Kenya is home.
63. Bharati, *Asians in East Africa: Jayhind and Uhuru*, 160.
64. Bharati, *Asians in East Africa: Jayhind and Uhuru*, 163.
65. Bharati, *Asians in East Africa: Jayhind and Uhuru*, 160.
66. Bharati, *Asians in East Africa: Jayhind and Uhuru*, 160.
67. Bharati, *Asians in East Africa: Jayhind and Uhuru*, 159.
68. Bharati, *Asians in East Africa: Jayhind and Uhuru*, 169. The term *Banyani* has been used in some scholarship to refer to the Asian money lenders, especially those that were found at the coast.
69. Gavin Kitching, *Class and Economic Change in Kenya*, 132.
70. Anyang' Nyong'o, "Middle Peasantry," 108–109.

71. Maxon, *Going Their Separate Ways*, 229.
72. Maxon, *Going Their Separate Ways*, 230.
73. Maxon, *Going Their Separate Ways*, 230.
74. KNA: AGR/6/4576, Sugar Production in Nyanza, 1956.
75. Fearn, *An African Economy*, 105.
76. Fearn, *An African Economy*, 105.
77. KNA: DC/CN/1/4/5, K-LDAR 1936.
78. KNA: DC/KSM/1/3/57, DC CN to PC Nyanza, August 23, 1954.
79. Gregory, *South Asians*, 279.
80. KNA: PC/NZA/3/15/142, PC Nyanza to Chief Secretary, March 26, 1949.
81. KNA: PC/NZA3/15/142, PC Nyanza to Superintendent of Police Nyanza, confidential, June 18, 1946.
82. KNA: PC/NZA/3/15/142, Superintendent of Police Nyanza to PC Nyanza, January 7, 1938.
83. KNA: PC/NZA/3/15/142, DC K-L to PC Nyanza, November 29, 1938.
84. KNA: PC/NZA/3/15/142, PC Nyanza to Chief Secretary, January 20, 1949.
85. KNA: PC/NZA/3/15/142, Nubian Gin and Application of Sugar Ordinance to Central Kavirondo.
86. KNA: PC/NZA/3/15/142, Chief John Paul Olola to DC CK, June 21, 1948.
87. KNA: DC/KSM/1/3/58, DO Bondo to DC CN, April 4, 1955.
88. KNA: PC/NZA/3/15/42, Chief John Paul Olola to DC CK, June 21, 1948.

EIGHT

Conclusion

The sugar industry in Western Kenya and Central Nyanza in particular has always had the potential to play a significant role in the country's economic development. However, in the pre-independence period, the colonial government largely ignored the potential of the sugar industry. The colonial administration put in place policies that were contradictory, sometimes helping the industry, sometimes hindering it, but almost always guided by considerations other than the health of the sugar industry itself.

Asian sugarcane farming began in earnest in Kenya in 1903. Around this time the first wave of Asians began migrating to Kenya's coast, eventually settling inland once the railroad was established. With the completion of the Kenya-Uganda railway, Asians began to establish sugarcane farms in the area of Kibos-Muhoroni. The Asian farmers had several choices of crops including maize and sisal, but focused their efforts on sugarcane.

From the earliest settlements right up to independence, Asian cane farmers in Kenya faced substantial obstacles to their success. In this conclusion, we will review the most serious obstacles to their success, discuss their coping mechanisms, and review the effects, both positive and negative, of Asian cane farming on the local and regional economy.

OBSTACLES

One of the most difficult obstacles to the success of Asian cane farmers was racial discrimination on the part of the colonial state, which put in place policies that favored the European settlers. Asian farmers were barred from acquiring land in the fertile white highlands. Instead, they were forced to farm in the less fertile area around Kibos. While land policy was the most obvious area of discrimination, such preferential treatment for European farmers was a constant in all areas of colonial policy, a fact that put the Asian community at a long-term disadvantage. For example, the colonial government built rail lines and reduced rail rates for European farmers but not Asian farmers. High transportation costs and the unreliability of transport were problems that dogged the Asian farmers for decades.

A second serious obstacle to success was the severe undercapitalization of the Asian farming community. Discriminatory practices such as land policy meant that the Asian farmers had less profit to reinvest in their industry. More seriously, however, the colonial government failed to extend financial and infrastructure support to the Asian farmers, as they did to the European farmers. As a result, Asian farmers faced constant problems in areas like transportation, since some farms were often far from the mills and there were no rail freight subsidies and insufficient trucking to take their cane to the mill.

Undercapitalization also stemmed from the discriminatory banking practices in the colony. Banks would not offer loans to the Asian farmers as they did to the Europeans, and it was not until the establishment of Asian-owned banks in the late 1940s that Asian farmers had freer access to capital to improve their operations. For example, Asian farmers could rarely afford tractors and so were limited to the use of the ox-plough to bring land under cultivation. This was compounded by the colonial government's onerous restrictions on their purchase of machinery and spare parts.

A third serious challenge to their success was the ecological conditions of the area to which their farming was restricted. The area was prone to excessive rainfall, poorly drained soil, and droughts and these conditions were unpredictable. Other challenges such as locust infestation could decimate a year's crop with little warning. The farmers' production was therefore erratic, and there was no support from the colonial administration to help them weather the ups and downs of climatic conditions. While the European farmers participated in an insurance program called the GMR to help them ride out bad years, no such insurance program was available to Asian farmers.

A fourth obstacle to success was the European monopoly on milling factories. For most of the early period of cane farming, there was only one mill in the area, the European-owned Victoria Nyanza Sugar Company. The existence of only one mill meant that farmers had to accept whatever terms the mill offered them for their product. The mill often offered prices for the cane that were well below world market prices. At times the mill would summarily refuse to accept the cane at all, and the farmers were left with crops that could not be sold. Even later, when an Asian company bought the mill, the existence of only one mill in the area meant that these problems, rooted in the mill's monopoly power, continued.

Time and again the colonial government refused permission for the farmers to erect a second mill. When the mills refused to accept the cane, the farmers were reduced to producing jaggery themselves, at a much lower profit.

A fifth obstacle was the insufficient labor market. The farms required many workers to plant and harvest, but the demand for workers could be erratic. In times of good weather and during certain points in the crop cycle, huge numbers of workers were needed; in times of difficulty, the demand for labor dropped precipitously. The erratic demand, poor pay, and poor living conditions for workers made it difficult to retain workers, who often preferred to return to their home farms or plots. In times of extreme labor shortage, the colonial government put in place taxes and other measures intended to virtually force the African population into wage labor for the benefit of the farmers.

A sixth obstacle was the divided loyalty of the colonial government. The government did occasionally enact policies to help the sugarcane farmers. One example is its intervention to force African workers into wage labor. Another is the testing of different varieties of cane to determine which were most resistant to disease, flooding, and drought. The government's efforts to quarantine plants with the mosaic disease, for example, and to distribute information about the best varieties for the region were positive initiatives for the sugar industry.

However, these policies were counteracted by others that were in direct opposition to the interests of the sugar industry. One of the most serious was the Kenyan colonial officials' opening of the Kenyan market to cheap sugar from Uganda and Tanganyika. The dumping of Ugandan sugar, in particular, lowered prices, and the colonial government refused to erect tariffs on the incoming sugar. This free access to the Kenyan market was provided so that Uganda would in turn provide free market access to Kenyan wheat, a crop that was dominated by European farmers. Again, rather than seeing the potential for Kenyan sugar industry as a whole, the colonial state narrowly focused on European interests. And since the apparatuses of the colonial state; its committees, departments, and policy makers, were dominated by Europeans, the Asian farmers had little chance of challenging the policies.

COPING MECHANISMS

Despite these serious obstacles to their success, the Asian cane farmers persisted not only to found an entirely new industry in Kenya but to become successful in their own right.

Some coping mechanisms included the migration of family members to Kenya to help with operations (despite the restrictions that the colonial government put in place for family migration), staggering of the planting seasons, and the pooling of resources. In addition, the eventual establishment of an Asian bank and Asian-owned mill helped mitigate the effects of discriminatory practices in lending and buying.

The poor working conditions and low wages on farms of all types led to the emergence of labor organization and trade unionism as a collective bargaining method by the cane workers and factory workers. However, by the early 1960s, labor supply had stabilized as the industry expanded and started producing refined sugar, fully financed by Asian capital. The fact that laborers were sufficient toward the end of colonial rule, and even came from the neighboring countries, indicated that the sugar industry was indeed well established as part of the wider colonial economy, despite the neglect by the colonial government.

Despite the locust infestations of the early 1930s, the interwar period was one of steady growth for Asian farmers. By 1940 the industry was well established, and the increased demand for sugar during World War II led to a period of expansion for the Asian farmers. In contrast to the previous period's lack of interest, during the war the colonial state supported intensification of sugarcane farming due to pressure from the United Kingdom to fulfill the war needs.

The farmers used the increased demand during the war to solidify and expand their production, and by the mid-1940s the industry was thriving. The most significant manifestation of this was the taking over of the Victoria Nyanza Sugar Company in 1947 by an Asian company led by Hindocha, which renamed the company the Miwani Sugar Mills. Although the colonial state had long opposed the establishment of an Asian-owned mill, the farmers did not give up. The entrance of the Asian capital into the manufacturing sector revolutionized cane farming and infused life into the sugar industry in CN as more Asian cane farmers were contracted to supply cane to the factory than when the Miwani mill was under European ownership. This expansion reached its peak by the mid-1950s as Asian cane farmers overcame some obstacles to establish a vibrant industry in Central Nyanza.

The Asian farming communities' efforts were aided by the colonial government's change of agricultural policy in the 1950s. By this time the colonial state came to appreciate the exploitative nature of the Ugandan manufacturers and realized that European settler farming alone could not sustain the economy. In response, the colonial government began to provide technical assistance to farmers and started to conduct trials on the different varieties of cane to determine which strains were most suited to regional conditions.

Although the farmers were aided by the changes in colonial policy, they had to rely on their own resourcefulness to address other obstacles. For example, they continued to be locked out of loans through which they might have bought machinery and improved their operations. In response, they pooled their resources to both rent machinery from European farmers and, when possible, buy machinery that they would share among themselves. This did not solve all their problems, as individual farmers had to wait their turn for the machinery, and sometimes the machinery was unavailable when they needed it. But it was a huge improvement over previous arrangements. In a similar way, the Asian farmers and millers cooperated to establish better transportation options, from hiring trucks to constructing a trolley system to take their cane from the farms to the mill. The type of cooperative initiative spread out into many areas of concern. For example, the Asian farmers began to stagger growing seasons so that the cane would not all mature at the same time.

As the Asian farmers began to appreciate the importance of collective efforts, they established political and economic associations such as the Nyanza Indian Farmers' Cooperative Society (also known as the NFCS), Agricultural Production Sub-Committees, and the Nyanza Farmers' Association (also known as the Nyanza Indian Farmers Association or NIFA) to support their interests. These associations would work alongside Indian associations such as EAINC to petition the colonial state on specific policy issues. This was particularly important since the colonial state had complete control over the pricing, sale, and distribution of cane and sugar. For example, when the colonial state allowed mills to pay farmers for the cane at 150 days rather than the usual 30 days, the Asian farmers petitioned for a return to the shorter timeline. They likewise petitioned for tariffs on Ugandan sugar and for price points that reflected world market prices.

THE SUGAR INDUSTRY'S EFFECT ON SOCIAL AND ECONOMIC LIFE

The establishment and growth of a new industry in a country such as Kenya was bound to transform the economy and social organization of the region in numerous ways. From the Asian sugar industry's earliest years in the 1900s through its expansion at midcentury, it affected the community in ways large and small, negative and positive.

In the most general sense, the Asian sugar industry contributed to the economic growth of CN through the agrarian transformations that took place. Although sugarcane was initially viewed as solely an "Asian crop," the Luo community later adopted cane farming. Some peasant farmers excelled, leading to the emergence of social differentiation. Those who could not farm took to the trading of sugar, and this contributed to the household economies that led to the further development of CN. Although the colonial government was slow to see its potential, in later years it recognized that the sugar industry could be a major positive economic force in Kenya.

The sugar industry also transformed labor practices in the region. At the time of the Asian farmers' settlement at Kibos, the indigenous economy was diverse but generally focused on subsistence needs. African communities had an abundance of land that was managed through its tenure system and the kinship mode of production. The indigenous economy would be transformed over the following decades by the establishment of colonial rule and the sugar industry. The rise of wage labor was both a blessing and a curse, as it provided additional income and cash flow to African workers, allowing them to supplement their subsistence farming, but also led to oppressive practices such as the levying of taxes intended to force Africans into wage labor when the sugar industry's labor needs were high.

In addition, the rise of the sugar industry brought land shortages in some parts of CN, such as the Kajulu area, that were densely populated. Previously, land had been collectively managed by the African communities in ways that were sustainable for them and governed by clear rules. The selling off of land for sugar development and the colonial government's manipulation of land distribution led to population pressure in many of these communities. The African communities were also affected by the new abundance of sugar to make alcoholic drinks that ultimately led to serious social problems.

Despite these negatives, the Asian sugar industry brought many benefits to the region, especially as the farmers and millers began to prosper from the midcentury on. They built hospitals, schools, and social centers that enhanced the towns in which they operated. Though often used just for the Asian community, with time these facilities served all Kenyans.

CURRENT PROSPECTS FOR THE SUGAR INDUSTRY IN KENYA

The sugar industry in Kenya has come a long way from its establishment in the early 1900s. Unfortunately, many of the problems that beset the industry in its early days persist today. These include lack of adequate transport, the dumping of sugar, undercapitalization, and obsolete machinery. There are also environmental factors such as drought, pests, and diseases. The most important issue is still the lack of the political will in the government to support the sugar industry. Small-scale farmers especially face problems such as the high cost of renting farm machinery, among others.[1]

The Kenyan government should learn from the past and guard the interests of the cane farmers and millers so that Kenya can attain self-sufficiency in sugar production or at least reduce the amount of importation. By addressing challenges such as mismanagement, bad governance, poor infrastructure, and lack of political will to support the industry, the state can give much-needed support to an industry that has the potential to play an important role in the national economy.

For the industry to fulfill its potential in western Kenya's sugar-belt, several steps must be taken. First, the state should support irrigation of cane farms from the Nyando River. Additionally, the Kenyan government needs to review its policies to ensure that it does not favor interest groups that are behind the persistent dumping of cheap sugar into the country. Instead state policies should be geared toward promoting sugarcane production at the farm level through provision of subsidies and incentives to encourage intensification of sugarcane farming, as well as imposing tariffs on cheap foreign sugar. Also, the state should encourage block farming, a more viable form of production than individual farming by small-scale farmers; this would reduce production costs through collective mechanized farming and make Kenyan sugar competitive.[2] Finally, it should examine quota allocations to make sure that they are optimized for the success of the industry.

Such measures would begin to address many of the contradictory policies and legacy of neglect from the colonial era. With the proper attention and nurture, the sugar industry should be able to fulfill its potential as a major economic engine in the Kenyan economy.

NOTES

1. Odhiambo, "The Sugarcane Procurement Problems in Kenya," 35.
2. Anyang' Nyong'o, "Modernize Sugar Production."

Bibliography

KENYA NATIONAL ARCHIVES (KNA), NAIROBI

AGRICULTURAL OFFICE, KISUMU (AK)

AK/2/4. Nyanza Province: Agricultural Quarterly Reports (1957).
AK/2/5. Nyanza Province Annual Reports (Dec. 1956–1957).
AK/2/6. Nyanza Province Annual Report (1956).
AK/2/7. Nyanza Province Annual Report (1955).
AK/2/17. Central Kavirondo Annual Reports (1934–1935).
AK/2/19. Central Kavirondo Monthly Reports.
AK/3/15. Investigation Work General Crops-Sugar.
AK/3/16. Investigation Work General Crops-Sugar.
AK/11/30. Sugar General (1949–1952).
AK/11/50. Agricultural Office Kisumu, Sugar General (1949).
AK/11/50. Sugar General (1935–1949).
AK/11/51.Sugar General (1953).
AK/11/52. Sugar General (1956–1958).
AK/11/73. Jaggery General (1940–1953).
AK/11/74. Jaggery General (1954–1955).
AK/13/13. Farming Indian Areas (1950–1953).

MINISTRY OF AGRICULTURE

AG/2/2. Sugar Ordinance Cap 134.
AG/2/19. Sugar Ordinance: Development of Sugar Industry in Kenya.
AG/16/186. The Sale of Jaggery: Sugar General.
AG/25/70. The Victoria Nyanza Sugar Company's Tramway Ordinance (1923).
BV/6/178. Crops and Production Sugar Jaggery Marketing Control.
BV/1/249. Grants Muhoroni Sugar, Ltd.
BV/6/613. Sugar International Restriction Scheme (1939–1942, Vol. II).
BV/6/615. Colonial Sugar Committee.
BV/6/619. Ordinance Sugar Control Ordinance.
BV/6/1704. International Sugar Agreement Regulations and Control of Sugar.
BV/6/1713. Sugar.
BV/14/401. Ranger Farm Management, Jaggery Including Disposal.
BV/111/119. Report of Kenya Sugar Industry.
BV/125/27. Sugar Research Station Kibos.

MINISTRY OF COMMERCE

AE/3/916. Secret Sugar Policy.
AE/3/952. African Chamber of Commerce (1944–1950).

DISTRICT COMMISSIONER, KISUMU/KISUMU-LONDIANI

DC/CN/1/3/1. Central Kavirondo Annual Report (1947).
DC/CN/1/3/2. Central Kavirondo Annual Report (1946).
DC/CN/1/4/1. Kisumu-Londiani Annual Report (1921).
DC/CN/1/4/3. Kisumu-Londiani Annual Reports (1931–1932).
DC/CN/1/4/4. Kisumu-Londiani Annual Report (1933).
DC/CN/1/4/5. Kisumu-Londiani Annual Report (1936).
DC/CN/1/5/1. Kisumu District Annual and Quarterly Reports (1909–1918).
DC/KSM/1/3/22. Sugar Production Report 1954.
DC/KSM/1/3/32. African Cane Production 1963.
DC/KSM/1/3/57. Crop Production.
DC/KSM/1/3/58. Sugar Distribution.
DC/KSM/1/3/128. Production and Settlement.
DC/KSM/1/19/270. Sugar Ordinance (1949–1950).

LABOUR DEPARTMENT

Lab 9/256. Labour for Sugar Industry.
Lab 9/1895. Labour Protests at Miwani.
Lab 9/ 2132. The Miwani Sugar Mills 1921–1926.

PROVINCIAL COMMISSIONER—COAST

PC/COAST/1/1/3. Railway Survey (1892–1895).
PC/COAST/1/17/44. Return of Railway Employees Deaths and Accidents (1890–1901).
PC/COAST/1/18/44. Uganda Railway (1912).
PC/COAST/1/1/21B. Outward File: Uganda Railway.
PC/COAST/1/14/101. Labour for Uganda Railway.

PROVINCIAL COMMISSIONER—NYANZA

PC/NZA/1/1/2. Nyanza Province Annual Reports (1906–1907).
PC/NZA/1/3/43/1. Uganda Railway General Matters (1925–1927).
PC/NZA/2/2/35. Crown Lands Rent Premier (1932–1943).
PC/NZA/2/12/108. Crop Production (1943–1951).
PC/NZA/2/19/35. Hut and Poll Tax Receipts from Muhoroni.
PC/NZA/3/1/465. Intelligence Reports Kisumu-Londiani Districts (1931–1935).
PC/NZA/3/2/34. Indian Cane Growers (1931–1958).
PC/NZA/3/2/182. Commodity Distribution (1947–1950).
PC/NZA/3/3/22. Sugar Industry (1940–1950).
PC/NZA/3/7/1. Lime.
PC/NZA/3/10/2/1. Proposed Asiatic Schools in Kisumu (1924–1926).
PC/NZA/3/13/8. Native Affairs (1940–1955).
PC/NZA/3/13/35. District Labour Committee (1942–1943).
PC/NZA/3/15/142. Native Liquour (1931–1958).
PC/NZA/3/22/1/2. Kibos Indian Settlement (1910–1916).
PC/NZA/3/22/1/5. Kibos Indian Settlement (1910).
PC/NZA/3/22/1/9. Kibos Indian Settlement (1910–1912).
PC/NZA/3/22/4/1. Sugar Factories (1925–1928).
PC/NZA/3/41/3/1. Buildings Kisumu Districts (1940–1955).
PC/NZA/4/4/88. Kisumu-Londiani Annual Report (1938).

PUB.1/1/1/6. EAINC Paper (1923).

COLONY AND PROTECTORATE OF KENYA

Department of Agriculture Annual Report, 1919–1920. Nairobi: Government Printer, 1921.
Department of Agriculture Annual Report, 1921. Nairobi: Government Printer, 1922.
Department of Agriculture Annual Report, 1922. Nairobi: Government Printer, 1924.
Department of Agriculture Annual Report, 1923. Nairobi: Government Printer, 1924.
Department of Agriculture Annual Report, 1924. Nairobi: Government Printer, 1925.
Department of Agriculture Annual Report, 1925. Nairobi: Government Printer, 1926.
Department of Agriculture Annual Report, 1926. Nairobi: Government Printer, 1927.
Department of Agriculture Annual Report, 1927. Nairobi: Government Printer, 1928.
Department of Agriculture Annual Report, 1928. Nairobi: Government Printer, 1929.
Department of Agriculture Annual Report, 1929. Nairobi: Government Printer, 1930.
Department of Agriculture Annual Report, 1931. Nairobi: Government Printer, 1932.
Department of Agriculture Annual Report, 1932. Nairobi: Government Printer, 1933.
Department of Agriculture Annual Report, 1933. Nairobi: Government Printer, 1934.
Department of Agriculture Annual Report, 1934, Vol. II, Nairobi: Government Printer, 1936.
Department of Agriculture Annual Report, 1935, Vol. I. Nairobi: Government Printer, 1936.
Department of Agriculture Annual Report, 1936, Vol. I. Nairobi: Government Printer, 1937.
Department of Agriculture Annual Report, 1937, Vol. I. Nairobi: Government Printer, 1939.
Department of Agriculture Annual Report, 1938, Vol. I and II. Nairobi: Government Printer, 1939.
Department of Agriculture Annual Report, 1945. Nairobi: Government Printer, 1946.
Department of Agriculture Annual Report, 1946, Parts I and II. Nairobi: Government Printer, 1948.
Department of Agriculture Annual Report, 1947, Parts I and II. Nairobi: Government Printer, 1949.
Department of Agriculture Annual Report, 1948, Parts I and II. Nairobi: Government Printer, 1950.
Department of Agriculture Annual Report, 1949, Parts I and II. Nairobi: Government Printer, 1951.
Department of Agriculture Annual Report, 1950, Vol. II. Nairobi: Government Printer, 1952.
Department of Agriculture Annual Report, 1952, Vol. I. Nairobi: Government Printer, 1953.
Department of Agriculture Annual Report, 1956, Vol. I. Nairobi: Government Printer, 1957.
Native Affairs Department Annual Report, 1923.
Native Affairs Department Annual Report, 1926–1928.
Native Affairs Department Annual Report, 1931–1938.
Native Affairs Department Annual Report, 1939–1945.
Native Affairs Department Annual Report, 1950.
Kenya Registrar General Annual Report, 1960.
Report of the Kenya Land Commission. London: His Majesty's Stationary Office, 1934.

PAPERS

Osamba, Joshia Otieno. "A Forgotten Minority: The Case of Asian Sugar Plantation Owners in Nyanza Province, Kenya." Paper presented at the Historical Association of Kenya Annual Conference, Moi University Eldoret, Kenya, July 1998.

DISSERTATIONS/THESES

Bennett, Charles. "Persistence Amid Adversity: The Growth and Spatial Distribution of the Asian Population in Kenya 1902–1963," Ph.D. diss., Syracuse University, 1976.

Bhatt Purnima Mehta. "A History of Asians in Kenya, 1900–1970," Ph.D. diss., Howard University, 1976.

Breen, Rita. "The Politics of Land: Kenya Land Commission (1932–1933) and Its Effects on Land Policy in Kenya," Ph.D. diss., Michigan State University, 1976.

Ezra, Allen William. "Sugar and Coffee: A History of Settler Agriculture in 19th Century Liberia," Ph.D. diss., Florida International University, 2002.

Haribhai, Patel H., "Race, Class, and Citizenship in Uganda: Power, Politics and the Indian Minority, 1900–1972," Ph.D. diss., University of California, 1977.

Makana, Nicholas. "Changing Patterns of Indigenous Economic Systems: Agrarian Change and Rural Transformation in Bugoma District 1930–1960," Ph.D. diss., West Virginia University, 2005.

Muyumbu, Gilbert H, "A Historical Study of State Policies in Kenya Sugar Industry 1940–2006," M.A. thesis, Egerton University, 2006.

Nangulu, Anne K. "Food Security and Coping Mechanisms in Kenya's Marginal Areas: The Case of West Pokot," Ph.D. diss., West Virginia University, 2001.

Ndege, Peter O. "Struggle for the Market: The Political Economy of Commodity Production and Trade in Western Kenya 1929–1939," Ph.D. diss., West Virginia University, 1993.

Okumu, John J. "Kenya 1900–1930: A Study in Conflict," Ph.D. diss., University of California, 1966.

Osamba, Joshia. " Colonial African Labour in Asian Owned Sugar Plantations in Kisumu District 1901– 1963," M.Phil. thesis, Moi University, 1996.

Seidenberg, Dana April. "The Asians and Uhuru: The Role of a Minority Community in Kenya Politics, 1939–1963," Ph.D. diss., Syracuse University, 1979.

Shanguiya, Martin. "The State Ecology and Society in Western Kenya: Soil Erosion and Conservation in Vihiga, 1930–1950," Ph.D. diss., West Virginia University, 2007.

Talbott, D. Irvin. "Agricultural Innovation and Policy Change in Kenya in 1930s," Ph.D. diss., West Virginia University, 1976.

Zarwan, John I. "Indian Businessmen in Kenya during the Twentieth Century: A Case Study," Ph.D. diss., Yale University, 1977.

BOOKS

Aloo, Ojuka, and William Ochieng, eds. *Politics and Leadership in Africa.* Nairobi: East African Literature Bureau, 1974.

Andrew, C. F. *The Indian Question in East Africa.* London: Swift Press, 1921.

Atieno-Odhiambo, E. S. *Siasa: Politics and Nationalism in East* Africa. Nairobi: Kenya Literature Bureau, 1981.

Bank for Reconstruction and Development. *Economic Development of Kenya.* Baltimore: Johns Hopkins University Press, 1965.

Bates, Robert H. *Markets and States in Tropical Africa: The Political Basis of Agricultural Policies.* Berkeley: University of California Press, 1984.

———. *Beyond the Miracle of Market: The Political Economy of Agrarian Development in Kenya.* Cambridge: Cambridge University Press, 1989.

———. *Beyond the Miracle of Market: The Political Economy of Agrarian Development in Kenya.* 2nd ed. Cambridge: Cambridge University Press, 2005.

Bates, R. H., and M. F. Lofchie, eds. *Agricultural Development in Africa: Issues of Public Policy.* New York, Praeger, 1980.

Bartlett, H. Moyse. *The King's African Rifles.* Aldershot: Gale and Polden Press, 1956.

Beinant, W., and J. McGregor, eds. *Social History and African Environment.* Oxford: James Currey, 2003.

Bennet, George. *Kenya: A Political History.* London: Oxford University Press, 1963.

Berman, Bruce. *Control and Crisis in Colonial Kenya: The Dialectic of Domination.* London: James Currey, 1990.

Berman, B., and J. Londsdale. *Unhappy Valley: Conflict in Kenya and Africa, Book One: State and Class.* London: James Currey, 1992.

Bharati, Agehananda. *Asians in East Africa: Jayhind and Uhuru.* Chicago: Nelson-Hall Company Publishers, 1972.

Bunker, Stephen. *Peasants Against the State: The Politics of Market Control in Bugisu Uganda.* Chicago: University of Chicago Press, 1991.

Chattopadhyaya, Haraprasad. *Indians in Africa: A Socio-Economic Study.* Calcutta: Bookland Private, Ltd., 1970.

Chazan, Naomi, and T. M. Shaw, eds. *Coping with Africa's Food Crisis.* Boulder, CO: Lynne Reinner, 1988.

Chirot, Daniel, and Anthony Reid, eds. *Essential Outsiders: Chinese and Jews in the Modern Transformation of Southeast Asian and Central Europe.* Seattle: University of Washington Press, 1997.

Clayton, Anthony, and Donald Savage. *Government and Labor in Kenya 1895–1963.* London: Frank Cass and Company, Ltd., 1974.

Collier, Paul, and Deepak Lal. *Labour and Poverty in Kenya 1900–1980.* New York: Oxford University Press, 1986.

Debroy, B., and A. U. Khan, eds. *Integrating the Rural Poor in Markets.* New Delhi: Academic Foundation Press, 2004.

Delf, George. *Asians in East Africa.* London: Oxford University Press, 1963.

Dilley, Majorie. *British Policy in Kenya Colony.* New York: Thomas Nelson and Sons, 1937.

Ellis, Frank. *Peasant Economics: Farm Household and Agrarian Development.* Cambridge: Cambridge University Press, 1988.

Eliot, Charles. *The East African Protectorate.* London: Edward Arnold Publishers, 1905.

Enloe, H. Cynthia. *Ethnic Conflict and Political Development.* Boston: Little, Brown and Company, 1972.

Fearn, Hugh. *An African Economy: A Study of the Economic Development of the Nyanza Province of Kenya, 1903–1953.* London: Oxford University Press, 1961.

Frank, Charles S. *The Sugar Industry in East Africa.* Nairobi: East African Publishing House, 1965.

Fumagalli, Carl T. "An Evaluation of Development Project Among the East African Pastoralists." In *Social Sciences Involvement in African Development Planning,* edited by Philips Stevens. Waltham: Crossroads Press, 1978.

Ghai, Dharam P., ed. *Portrait of a Minority: Asians in East Africa,* rev. ed. London: Oxford University Press, 1970.

Gregory, Robert. *India and East Africa: A History of Race Relations within the British Empire, 1890–1939.* Oxford: Clarendon Press, 1977.

———. *Quest for Equality: Asian Politics in East Africa 1900–1967.* New Delhi: Orient Longman, Ltd., 1993.

———. *South Asians in East Africa: An Economic Social History 1890-1980.* Boulder, CO: Westview Press, 1993.

Hancock, William K. *Survey of British Commonwealth Africans: Problems of Nationality*. Latham: Oxford University Press, 1942.

Harlow, V., and E. M. Chilver, ed. *History of East Africa*. Oxford: The Clarendon Press, 1965.

Heyer, Judith, J. K. Maitha, and W. M. Senga, eds. *Agricultural Development in Kenya: An Economic Assessment*. Nairobi: Oxford University Press, 1976.

Higgins, B. *Economic Development: Principles, Problems, and Policies*. New York: W.W. Norton and Company, 1968.

Hill, M. F. *Permanent Ways: Story of Kenya-Uganda Railway and Harbours*. Nairobi: East Africa Railways and Harbours, 1949.

Hill, Polly. *The Migrant Cocoa Farmers of Southern Ghana: A Study in Rural Capitalism*. Oxford: International African Institute, 1997.

Himmelstrand, U., Kabiru Kinyanjui, and Edward Mburugu, eds. *African Perspective on Development: Controversies Dilemmas and Openings*. Nairobi: East African Educational Publishers, 1994.

Hoffenberg, A. Alice. *International Firms and Labour in Kenya 1945–1970*. London: Frank Cass and Company, 1971.

Hollingsworth, L.W. *The Asians of East Africa*. London: Macmillan and Company, Ltd., 1960.

Ikiara, Gerishon. *Development and Society: The Dynamics of Social Change*. New York: St. Martins Press, 1964.

Kanogo, Tabitha. *Squatters and the Roots of Mau Mau, 1905–1963*. London: James Currey, 1987.

Kitching, Gavin. *Class and Economic Change in Kenya: The Making of an African Petit Bourgeoisie 1905–1970*. New Haven, CT: Yale University Press, 1980.

Leys, Colin. *The Rise and Fall of Development Theory*. London: James Currey, 1996.

Leys, Norman. *Kenya*. London: Frank Cass Company, 1973.

Lord Hailey. *An Africa Survey: A Study of Problems Arising in Africa South of Sahara*. London: Oxford University Press, 1938.

Makoha, Joseph. *The District Focus: Conceptual and Management Problems*. Nairobi: American Press Research Bureau, 1985.

Mangat, J. S. *A History of the Asians in East Africa 1886 to 1945*. Oxford: Clarendon Press, 1969.

Maxon, Robert. *East Africa: An Introductory History*. 2nd ed. Morgantown: West Virginia University Press, 1994.

———. *Going Their Separate Ways: Agrarian Transformation in Kenya 1930–1950*. London: Associated University Presses, 2003.

———. *John Ainsworth and the Making of Kenya*. Lanham, MD: University Press of America, 1980.

———. *Struggle for Kenya: The Loss and Reassertion of Imperial Initiative 1912–1923*. London: Associated University Presses, 1993.

Maxon, R., and T. Ofcansky. *Historical Dictionary of Kenya*. 2nd ed. London: Scarecrow Press, 2000.

McCormack, Richard T. *Asians in Kenya: Conflict and Politics*. Brooklyn: Theo. Gaus' Sons, Inc., 1971.

McGregor, W. Ross. *Kenya from Within: A Short Political History*. London: Cass Publishers, 1968.

Morris, H. S. *The Indians in Uganda*. London: Weidenfeld and Nicolson, 1968.

Mosley, Paul. *The Settler Economies: Studies in the History of Kenya and Southern Rhodesia*. Cambridge: Cambridge University Press, 1983.

Norman, Humphery. *Liguru and the Land: Sociological Aspects of Some Agricultural Problems of North Kavirondo*. Nairobi: Government Printer, 1947.

Novak, D. E., and R. Lekachman, eds. *Development and Society: The Dynamics of Social Change*. New York: St. Martin's Press, 1964.

Ochieng W. R., ed. *A Modern History of Kenya 1895–1980*. Nairobi: Evans Brothers (Kenya) Ltd, 1989.

Ochieng, W. R. *An Outline History of Nyanza Up to 1914*. Nairobi: East African Literature Bureau, 1974.
Ochieng, W. R., and R. M. Maxon. *An Economic History of Kenya.* Nairobi: East African Educational Publisher, 1992.
Orvis, Stephen. *The Agrarian Question in Kenya*. Gainesville: University Press of Florida, 1997.
Pandit, Shanti. *Asians in East and Central Africa*. Nairobi: Panco Publishers, 1963.
Pearson, Scott, et al., eds. *Agricultural Policy in Kenya: Application of the Policy Analysis Matrix*. Ithaca, NY: Cornell University Press, 1995.
Ravallion, Martin. *Markets and Famine*. Oxford: Clarendon Press, 1987.
Richards, P. *Indigenous Agricultural Revolution: Ecology and Food Production in West Africa*. London: Hutchinson and Company, 1985.
Rodney, Walter. *How Europe Underdeveloped Africa*. Dar es Salaam: Tanzania Publishing House, 1992.
Rothchild, Donald. *Racial Bargaining in Independent Kenya*. New York: Oxford University Press, 1973.
Seidenberg, Dana April. *Uhuru and the Kenya Indians: The Role of a Minority Community in Kenya Politics 1939–1963*. New Delhi: Vikas Publishing House, 1983.
Shillaro, Priscilla, M. *A Failed Eldorado: Colonial Capitalism, Rural Industrialization, African Land Rights in Kenya, and the Kakamega Gold Rush 1930–1952*. Lanham, MD: University Press of America, 2008.
Sorrenson, M. P. K. *Origins of European Settlement in Kenya*. Nairobi: Oxford University Press, 1968.
Stichter, Sharon. *Migrant Labour in Kenya: Capitalism and African Response 1895–1975*. Harlow: Longman Publishers, 1982.
Swainson, Nicola. *The Development of Corporate Capitalism in Kenya 1918–1977*. Berkeley: University of California Press, 1980.
Timmer, C. Peter. *Getting Prices Right: The Scope and Limits of Agricultural Price Policy*. Ithaca, NY: Cornell University Press, 1986.
Van Zwanenberg, R. M. *Colonial Capitalism and Labour in Kenya*. Nairobi: East African Literature Bureau, 1975.
Van Zwanenberg, R. M., and Anne King. *An Economic History of Kenya and Uganda*. Atlantic Highlands, NJ: Humanities Press, 1975.
Wagner, Gunter. *The Bantu of Western Kenya with Special Reference to Vugusu and Logoli: Economic Life*. London: Oxford University Press, 1956.
Wolff, R. D. *The Economics of Colonialism, Britain and Kenya 1870– 1930*. London: Yale University Press, 1974.
Worthington, S., and E. B. Worthington. *Inland Waters of Africa*. London: Macmillan and Co Ltd, 1933.
Zeleza, Tiyambe. *A Modern Economic History of Africa: The Nineteenth Century*. Dakar: CODESRIA, 1993.

JOURNALS

Anderson, M. David. "Master and Servant in Colonial Kenya 1895–1939." *Journal of African History* 41, no. 3 (2000): 459–485.
Anyang' Nyong'o Peter. "The Development of Middle Peasantry in Nyanza." *Review of African Political Economy* 8, no. 20 (1981): 108–120.
Berman, Bruce, and John Lonsdale. "Coping with Contradictions: The Development of the Colonial State in Kenya 1895–1914." *Journal of African History* 20, no.4 (1979): 487–505.
Meredith, David. "The British Government and the Colonial Economic Policy 1919–1939." *The Economic History Review* 28, no. 3 (1975): 484–499.

Nasongo, Shadrack. "Multilateral Imperialism in Kenya's Sugar Industry: The Political Economy of Transnational Corporations." *Nigerian Journal of International Affairs* 26, no. 1 (2000): 56–77.

Peterson, William. "A General Typology of Migration." *American Sociological Review* 23, (1958): 256–266.

Plessis, S., and Sophia du Plessis. " Explanation for Zambia's Economic Decline." *Development Southern Africa* 23, no. 3 (2006): 327–369.

Odhiambo, M. O. "The Sugar-Cane Procurement Problems in Kenya: The Case of the Nyanza Sugar Belt." *East African Agricultural and Forestry Journal* 53, no. 1 (1987): 29–41.

Olielo, Tom. "Food Security Problems in Various Income Groups of Kenya." *African Journal of Food, Agriculture, Nutrition and Development* 13, no. 4 (2013): 1–13.

Wanyande, Peter. "Management Politics in Kenya's Sugar Industry: Towards an Effective Framework." *African Journal of Political Science* 6, no. 1 (2001): 123–140.

NEWSPAPERS

Anyang', Nyong' P. "Modernize Sugar Methods to Compete with Fellow Cane Growers." *East African Standard* (Nairobi, Kenya), September 13, 2015.

EAINC. "New Plan to Blot Out the Very Existence of Asians," *Indian Review*, 1920, 54.

Indian Overseas Association. *East Africa Pamphlet* 6, no. 45 (1921).

INTERVIEWS

Joseph Juma, Kibos, 06/02/2009.
Chanan Singh, Kibos, 06/27/2009.
Mohinder Singh, Kibos, 06/22/2009, 07/03/2009, 07/26/09.
Harjit Singh, Kibos, 06/22/2009.
Ranuhan Singh, Kibos, 06/27/2009.
Martin Okeyo, Miwani, 06/24/2009.
Martin Otieno, Chemelil, 06/20/2009.
Joseph Otieno, Chemelil, 06/26/2009.
J. S. Cheema, Kibos, 06/23/2009.
B. Rama Mohan, Kibos, 06/22/2009.
Vitalis Ogola, Kibos, 06/22/2009.
Hezborn Atudo, Kibos, 06/06/2009, 06/23/2009.
C.B.Ombogo, Muhoroni, 06/25/2009.
Absalom Apondi, Muhoroni, 07/25/2009.
Peter Abwogo, Muhoroni, 06/21/2009.
Joseph Okongo, Muhoroni, 06/21/2009.
Hezborn Muynze, Kibigori, 07/01/2009.
J. Kiprono, Kibigori, 07/01/2009.
Grace Nandi, Chemelil, 07/04/2009.
Manase Nandi, Chemelil, 06/20/2009.
Thomas Okeyo, Kibos, 07/03/2009.
Tom Ogada, Chemelil, 07/06/2009.
Michael Juma, Chemelil, 07/06/2009.
Zakaria Olang, Miwani, 07/26/2009.
Vajv Sudra, Kisumu town, 07/05/2009.
Amira Sudra, Kisumu town, 07/05/2009.
Rajiv Khan, Kisumu town, 07/05/2009.

Index

About the Author

Dr. Godriver Wanga-Odhiambo is an assistant professor of African history at Le Moyne College, Syracuse, New York. She is a citizen of Kenya where she attained her undergraduate and master's degree at Moi University Kenya. Dr. Wanga-Odhiambo attained her Ph.D., at West Virginia University. She specialized in the use of archival sources to write history. Her research focuses on economic history of colonial Kenya as well as the state of refugee women in conflict zones in Africa.

www.ingramcontent.com/pod-product-compliance
Ingram Content Group UK Ltd.
Pitfield, Milton Keynes, MK11 3LW, UK
UKHW041126071026
15587UKWH00009B/259